HVAC SYSTEMS

OPERATION,
MAINTENANCE,
& OPTIMIZATION

SAMUEL C. MONGER

PRENTICE HALL

Prentice-Hall International (UK) Limited, *London*
Prentice-Hall of Australia Pty. Limited, *Sydney*
Prentice-Hall Canada, Inc., *Toronto*
Prentice-Hall Hispanoamericana, S.A., *Mexico*
Prentice-Hall of India Private Limited, *New Delhi*
Prentice-Hall of Japan, Inc., *Tokyo*
Simon & Schuster Asia Pte. Ltd., *Singapore*
Editora Prentice-Hall do Brasil, Ltda., *Rio de Janeiro*

© 1992 by
PRENTICE-HALL, INC.
Englewood Cliffs, NJ 07632

10   9   8   7   6   5   4   3   2
Printed in the United States of America

Library of Congress Cataloging-in-Publication Data
Monger, Sam, [date]
    HVAC optimization handbook / Samuel C. Monger
        p.   cm.
    Includes index.
    ISBN 0–13–446493–1
    1. Heating—Equipment and supplies—Maintenance and repair.
    2. Ventilation—Equipment and supplies—Maintenance and repair.
    3. Air conditioning—Equipment and supplies—Maintenance and
repair.   4. Mechanical efficiency.   I. Title.
    TH7015.M65   1992
    697—dc20   92–10683
                                                        CIP

ISBN 0-13-446493-1

PRENTICE-HALL
BUSINESS & PROFESSIONAL DIVISION
A SIMON & SCHUSTER COMPANY

About the Author

Samuel C. Monger is vice president of Monger Consulting Services. Mr. Monger is also an adjunct instructor at San Diego City College where he teaches heating, ventilating, air-conditioning and refrigeration design and fluid flow dynamics. Over the last ten years Mr. Monger has conducted more than a hundred HVAC seminars and in-hours training programs across the country. During this time he has taught over 3,500 engineers, technicians and contractors how to optimize HVAC systems for energy efficiency and occupancy comfort. Mr. Monger's wealth of technical knowledge is based on 21 years of hands-on field testing and evaluation of HVAC systems.

About the Author

Samuel C. Monger is vice president of Monger Consulting Services. Mr. Monger is also an adjunct instructor at San Diego City College where he teaches heating, ventilating, air-conditioning and refrigeration design and fluid flow dynamics. Over the last ten years Mr. Monger has conducted more than a hundred HVAC seminars and in-hours training programs across the country. During this time he has taught over 3,500 engineers, technicians and contractors how to optimize HVAC systems for energy efficiency and occupancy comfort. Mr. Monger's wealth of technical knowledge is based on 21 years of hands-on field testing and evaluation of HVAC systems.

What This Book Will Do for You

The heating, ventilating and air-conditioning (HVAC) systems designed and installed from the 1950s through the 1980s were energy intensive. There were many reasons for this. First of all, being energy intensive was not a bad thing—utility costs were very low. Second, the energy bill was based on consumption, but not in the same way as it is today. Until the oil embargo, the more energy you used, the lower the cost per unit of electricity or natural gas. Because the energy costs were low and, then as today, more attention was given to initial costs than to ongoing utility costs, architects and owners wanted less space dedicated for the mechanical systems. The architects were concerned about aesthetics and the owners were concerned about usable or salable space. This meant that the mechanical systems were squeezed into smaller and smaller spaces. Therefore, ductwork was smaller. Smaller ducts increased duct resistances in the system, which meant that motors and fans had to be bigger to overcome the increased resistances.

During this same time, HVAC and mechanical engineers were developing innovative mechanical systems with more emphasis placed on air than water as a means of transporting energy. Then, to add to the inefficiencies of the systems, buildings were being constructed with less insulation. Thermal losses increased greatly. At the same time that insulation factors were being ignored, internal design temperatures were changing. Inside heating design for winter conditions went from approximately 70°F to 75°F. Summer cooling conditions went from 80°F down to 72°F or less. Thus began the tradition of cooling lower than the heating temperature. Engineers were deliberately designing for simultaneous heating and cooling to increase occupancy comfort. The results of all these changes were that systems were overdesigned and overpowered.

What is the condition of HVAC systems today? The general status of HVAC systems is that equipment is obsolete and in disrepair. Operating personnel are undertrained, and maintenance crews are understaffed. If this is the case—and it is—what actions are needed to optimize and make the most effective use of our systems? To answer this question we should understand the three basic reasons why HVAC systems need optimization and take steps to do the actual optimization.

HVAC systems need optimization because:

- They simply wear out. This is certainly commonplace, but in theory unnecessary. If a system is well maintained it should never wear out. The purpose of preventive and predictive maintenance is to replace parts so that the system remains operational.

- There are newer technologies and better ways to achieve results.
- The system does not perform (or never did perform) properly—energy-wise or comfort-wise. This may be because of deficiencies in the initial design or installation, or because of inadequate or improper maintenance. Both problems are the fault of management.

To optimize our HVAC systems we need to:

- Understand the design intent of our systems.
- Understand the operating characteristics of the system's fans, pumps, compressors and other energy-using components. For instance, the fans used in a typical commercial establishment consume about 30 to 35% of the total energy. Depending on the type of system, the mechanical refrigeration compressors use approximately 15% of the total energy. Water pumps (boiler, chiller and distribution pumps) use about 10%. The building's lighting and process equipment (computers, etc.) account for most of the rest of the energy consumed.
- Do a verification of system performance and energy usage.
- Make our systems safer, more energy efficient and more oriented on occupancy comfort.
- Train HVAC mechanics and operators to higher levels.
- Develop an attitude that optimization is a team effort. The team will consist of owners, engineers, architects, consultants, technicians, mechanics, operators, utility companies and energy supplies—"TEAM" (Together Everyone Accomplishes More).

HVAC Systems is intended for HVAC engineers, designers, facilities engineers, plant engineers, chief engineers, utility engineers, energy managers, energy management technicians, energy auditors, HVAC technicians, HVAC mechanics and operating personnel, refrigeration mechanics and air and water balancing technicians.

The book takes you on a guided tour of the various HVAC systems and their components. It uses numerous drawings and examples to show you how the components and systems should operate, how to test the systems for actual operating conditions and how to improve operation and performance. This is an exceptional source book packed with useful checklists, equations, tables, charts, curves, forms and definitions.

This handbook will help you to:

- Better understand your HVAC systems and components.
- Do a verification of system performance and energy usage.
- Understand what you can do to optimize your systems for safety, energy efficiency and occupancy comfort.
- Set HVAC goals.
- Train your technicians, mechanics and operators.

In this handbook you will learn:

- How to verify operating capacity and performance of the central air system (Chap. 1).

- How to set up an agenda to determine system performance, including what documents and forms to use (Chap. 1).

- How to conduct a verification of system performance, including verifying motor operation, where and how to measure fan pressures, determining the amount of outside air and verifying coil performance from air and water temperatures (Chap. 1).

- The characteristics of air distribution ductwork, dampers, air valves, diverters, terminal boxes and outlets (Chap. 4).

- How to calculate fan motor horsepower, fan efficiency and fan tip speed (Chap. 2).

- How to use the fan laws to determine system performance (Chap. 2).

- How to use fan performance curves and fan multirating tables to predict fan performance (Chap. 2).

- How to plot a system curve and determine the operating point of fans in series and parallel (Chap. 2).

- How to take velocity pressure readings for a Pitot tube traverse in the duct (Chap. 3).

- How to measure air flow at the outlets (Chap. 3).

- How to correct instrument readings for changes in density (Chap. 3).

- How to determine the operating capacity and performance of the water side of the heating, ventilating and air-conditioning system (Chap. 5).

- How to measure water pressures at the pump and in the piping to determine pump and system performance (Chap. 5).

- Pumps and pump characteristics, including multiple pump arrangements (Chap. 6).

- How to calculate pump power and pump efficiency (Chap. 6).

- How to use the pump laws to determine system performance (Chap. 6).

- How to use pump curves to predict pump performance (Chap. 6).

- How to plot a system curve and determine the operating point of pumps in series and parallel (Chap. 6).

- The characteristics of piping systems, water filtration, water flow control, flow meters, pressure control components, air control components and heat exchangers (Chap. 7).

- How to verify the performance of electrical systems, heating systems, AC refrigeration systems and HVAC control systems (Chap. 8).

- About heating, AC refrigeration and HVAC control components (Chap. 8–13).

- A plan for setting and accomplishing HVAC optimization goals (Chap. 14).

- Optimization and retrofit opportunities for increasing system performance for occupancy comfort and energy usage reduction (Chap. 15–16).

- How to reduce or eliminate simultaneous heating and cooling, duct leakage, overpowered systems and thermal losses (Chap. 15–16).

- About retrofitting to a VAV system (Chap. 15–16).

- About optimizing single zone, multizone, reheat and dual duct systems (Chap. 15–16).

- How to test, adjust and balance constant volume and variable air volume systems (Chap. 17–22).

- How to test, adjust and balance water systems (Chap. 23).

- About pneumatic, electric and electronic-direct digital controls and control systems (Chap. 12–13).

- The characteristics and applications of various types of fans, problems encountered with the various fan types, how fans are rated, how fans are selected, how to establish a system curve and what it means, the problems associated with fan system effect, fan inlet and outlet conditions, what to do to correct system effect, fan laws, speed vs. pressures (fan and duct) and horsepower (Chap. 2).

- The various types of ductwork (rectangular, round, flat-oval, sizing, aspect ratios, galvanized, wrapped duct, lined, fiberglass and flex duct), duct design, duct components and fittings and their effect in the system (transitions, elbows, turning vanes, dampers, diverting devices), the various types of air distribution devices (diffusers and grilles, supply, return and exhaust), characteristics, surface effect, pattern control, effective areas. Terminal boxes (single duct, dual duct, pressure independent, pressure dependent, fan powered, system powered, constant air volume and variable air volume) (Chap. 4).

- About the various types of VAV systems (bypass, system powered, fan powered, pressure independent, pressure dependent, single duct and dual duct) (Chap. 4).

- About the types of pumps (single suction, double suction, inlet and discharge sizes), pump characteristics, pump curves, establishing impeller size and performance curve, establishing system curves, operating characteristics, pumps in series, parallel pumps, pump laws (speed or impeller size vs. TDH and horsepower) (Chap. 6).

- The characteristics of pipe systems (direct and reverse return systems, primary and secondary systems), the advantages and disadvantages of each, components (expansion/compression tanks, air vents, air separators and proper location in the system), valves (type and application), manual valves (gate, butterfly, ball, globe and balancing), automatic valves (three-way, two-way, balanced, mixing, diverting, normally open, normally closed, quick opening, linear and equal percentage), automatic valve application (bypass or mixing), terminals or coils (types, applications, piping, heat transfer, parallel flow and counter flow) (Chap. 7).

- How to use the pump as a flow meter, pump pressures, how to take readings and determine gpm, how to use and deal with problems with installed valves or coils as flow meters, instrumentation and flow charts (Chap. 5).

- Design and testing of laboratory fume hoods (Chap. 25).

Contents

CHAPTER 6 **Central Water System Components—Pumps,** *104*

CHAPTER 7 **Water Distribution Components,** *120*

CHAPTER 21   **Testing, Adjusting and Balancing Constant Air Volume—Multizone, Dual Duct, and Induction Systems,** *324*

CHAPTER 22   **Testing, Adjusting and Balancing Pressure Independent and Pressure Dependent Variable Air Volume Systems,** *330*

CHAPTER 1 Verification of System Performance Air Side—Central System

In this chapter you will learn how to verify operating capacity and performance of the central air system, which includes the fan, the motor and the coil, and how to set up an agenda to determine system performance including what documents and forms to use. The verification of system performance will include verifying motor operation, where and how to measure fan pressures, determining amount of outside air, and verifying coil performance from air and water temperatures.

SETTING UP AN AGENDA TO DETERMINE CENTRAL AIR SYSTEM PERFORMANCE

Not all the documentation listed below will be available for every system, but the more information you get, the better your understanding of the system, will be. With greater "system understanding" you will be able to make better recommendations for system optimization. Attempt to obtain from the mechanical contractor:

Engineering drawings

Shop drawings

"As-built" drawings

Schematics

Previous air balance reports

From the mechanical contractor or the component manufacturer, get:

Equipment catalogs

Fan description and capacities

Fan performance curves

Recommendations for testing equipment

Operation and maintenance instructions

Terminal box description and capacities

Before beginning the actual field measurements, study the mechanical plans, equipment specifications and catalogs to become familiar with the air handling system and its design intent. For clarity, be sure that all equipment and air distribution is designated correctly on the drawings and reports. While reviewing the documents make note of any piece of equipment, system component or any other condition that should specifically be investigated during the field verification of performance.

Report Forms Needed

As applicable, prepare the following report forms for each air system:

Motor Data and Test Sheet

Drive Data and Test Sheet

Air Handling Equipment Data and Test Sheet

These forms are provided and described later in this chapter.

VERIFYING THE PERFORMANCE OF THE CENTRAL SYSTEM

To determine the operating capacity and performance of the central air system:

Verify the operation of the motor.

Take drive information.

Take speed and pressure measurements of the fan.

Measure amount of outside air.

As applicable, take air temperatures.

Verifying Motor Operation

To determine the operating performance of the motor take voltage, current and power factor readings. Voltage and current readings can be made using a portable clamp-on volt-ammeter. Generally, all electrical readings are taken at the motor control center or the disconnect box. The measured voltage should be plus or minus 10% of the motor nameplate voltage. If it's not, note it on the report. The measured current should not be over nameplate amperage. If it is, reduce the fan speed until the current reading is down to maximum rated nameplate amperage. Power factor measurements are done with a power factor meter, an ammeter and a wattmeter, or an ammeter and the electrical utilities' watt-hour meter.

Record nameplate motor speed. The motor operating speed is generally not measured. The nameplate rpm is recorded on the report sheet as constant operating speed, unless the motor has a variable frequency drive (*VFD*). If the motor is VFD it is noted on the report.

Check the installed motor thermal overload protection devices if the motor or fan is new or changes have been made recently. Check the rotation of motors to ensure that fans are rotating in the correct direction. Certain centrifugal fans will produce measurable pressures and some fluid flow, sometimes as much as 50% of design, even when the rotation is incorrect. In axial fans, if the motor rotation is incorrect, the air flow will reverse direction. To check rotation, momentarily start and stop the fan motor to "bump" the fan enough to determine the direction of rotation. There's usually an arrow on the fan housing showing proper rotation. However, if there's no arrow, view double inlet centrifugal fans from the drive side and single inlet fans from the side opposite the inlet. This will let you determine proper rotation and whether the wheel is turning clockwise or counterclockwise. If the rotation is incorrect, reverse the rotation on a three-phase motor by changing any two of the three power leads at the motor control center or disconnect. On single-phase motors change rotation, as applicable, by switching the internal motor leads within the motor terminal box.

Checklist for Verifying Motor Operation

- Take voltage, current and power factor readings.
- Record nameplate motor speed.
- Check the installed motor thermal overload protection devices.
- Check the rotation of motors.
- Record the following motor and starter nameplate information on the motor data and test sheet (Fig. 1.1):

> Manufacturer
>
> Frame size
>
> Horsepower
>
> Phase
>
> Hertz
>
> Motor speed, rpm
>
> Service factor
>
> Voltage
>
> Amperage
>
> Power factor
>
> Efficiency
>
> Starter size
>
> Overload protection

FIGURE 1.1.   MOTOR DATA AND TEST SHEET

Project:		Engineer/Contact:		
	Specified	Actual	Specified	Actual
Fan or Pump No.				
Motor Information				
Manufacturer				
Frame				
Horsepower				
Phase				
Hertz				
Motor Speed rpm				
Service Factor				
Voltage				
Amperage				
Power Factor				
Efficiency				
Brake HorsePower				
Starter Size				
Thermal OLP				
Notes:				
Fan or Pump No.				
Motor Information				
Manufacturer				
Frame				
Horsepower				
Phase				
Hertz				
Motor Speed rpm				
Service Factor				
Voltage				
Amperage				
Power Factor				
Efficiency				
Brake HorsePower				
Starter Size				
Thermal OLP				
Notes:				

Taking Drive Information

Stop the fan and remove the belt guard. Read the information off the motor, fan sheaves and belts. Also, measure the shaft sizes and the distance between the center of the fan and motor shafts. This is a good time to also measure and record the slide adjustment on the motor frame or motor adjustment on the all-thread rods. The motor slide or the all-thread is for adjusting belt tension. For instance, if a sheave needs changing and there is space available on the motor frame or all-thread, you may be able to move the motor forwards or backwards, or up or down, so that the old belt will fit. If an adjustment is not possible, a change in sheave size will create the need to install a different size belt.

Before we go further, let's define some drive component terms:

- The *fan sheave* is the driven pulley on the fan shaft.

- The *motor sheave* is the driver pulley on the motor shaft. The motor sheave may be either a fixed or adjustable groove sheave.

- *Adjustable groove sheaves,* or simply, *adjustable sheaves* are also known as variable speed or variable pitch sheaves. An adjustable sheave means that the belt grooves on the sheave are movable.

- A *fixed sheave* means that the belt grooves are non-movable. Fixed sheaves are normally used on the fan. Generally, after fans have been air balanced for the proper air flow, adjustable motor sheaves are replaced with fixed motor sheaves. The reason is that, size for size, fixed sheaves are less expensive than adjustable sheaves and there's less wear on the belts.

Here are some other terms referring to belts that need defining:

- *V-belts*—There are two types of V-belts generally used on HVAC equipment. Light-duty, fractional horsepower (*FHP*) belts (sizes 2L through 5L), and heavier-duty industrial belts (sizes "A" through "E"). Fractional horsepower belts are generally used on smaller diameter sheaves because they're more flexible than industrial belts for the same equivalent cross-sectional size. For example, a 5L belt and a "B" belt have the same cross-sectional dimension, but because of its greater flexibility, the 5L belt would generally be used on light-duty fans that have smaller sheaves. The general practice in HVAC design is to use belts of smaller cross-sectional size with smaller sheaves instead of large belts and large sheaves for the drive components. Multiple belt sheaves are used to avoid excessive belt stress.

- *Pitch diameter* is a measurement that refers to where the middle of the V-belt rides in the sheave groove.

- A *matched set* of belts is a set of belts whose exact lengths and tensions are measured and matched by the belt supplier in order for each belt to carry its proportionate share of the drive load. I have rarely seen a true set of matched belts.

Now that we have an understanding of some of the important drive component terms, let's go back and continue to get information from the sheaves and belts. After you have the belt guard off, write down the quantity of belts, the name of the belt manufacturer and the belt size or belt number imprinted or stamped on the belt. Check the outside of the sheave for a

FIGURE 1.2

Variable Speed Sheaves				
Diameters			Part Number	
Pitch Range		Outside	Sheave	Bushing
"A" Belts 6.9–8.0	"B" Belts 7.0–8.4	8.68	3MVP70B84P	P2

Companion Sheaves				
Diameters			Part Number	
Pitch	Pitch	Outside	Sheave	Bushing
"A" Belts 18.0	"B" Belts 18.4	18.75	3MVB184Q	Q1

Bushing Bores	
Bushing No.	Bore Range
P2	¾"–1¾"
Q1	¾"–2¹¹⁄₁₆"

stamped part number which shows the sheave size. For example, on the fan sheave you might find the word Browning and the numbers and letters 3MVB184Q (Fig. 1.2). On the motor sheave, you might find, 3MVP70B84P (Fig. 1.2). Looking in the manufacturer's catalog (Browning, in this example), you'd find that the numbers and letters mean that the fan sheave has 3 fixed grooves and can have either a "B" belt at a pitch diameter of 18.4 inches or an "A" belt at a pitch diameter of 18.0 inches. The outside diameter is 18.75 inches and the bushing size range is Q1. The letters and numbers on the motor sheave show that it's a 3 groove adjustable sheave with a pitch diameter range of 7.0 to 8.4 inches for a "B" belt or a range of 6.9 to 8.0 inches for an "A" belt. The outside diameter is 8.68 inches and the bushing size range is P2. The bushing sizes, Q1 and P2 show that they'll fit shaft sizes from ¾ to 2 ¹¹⁄₁₆ inches (Q1) and ¾ to 1¾ inches (P2).

If there's no part number on the sheave, measure the outside diameter and then refer to the manufacturer's catalog to find the corresponding pitch diameter. Most manufacturers list both pitch diameter and outside diameter in their catalogs. If you can't get the pitch diameter from the catalog use a tape measure or rule to measure, as close as possible, the approximate pitch diameter. Or, use the drive equation to calculate pitch diameter if the motor and fan speeds are known and the diameter of one of the sheaves is known. The drive equation ($Pd_M \times RPM_M = Pd_F \times RPM_F$) states that the motor sheave pitch diameter times the motor speed is equal to the fan sheave pitch diameter times the fan speed.

Checklist for Taking Drive Information

• For safety when taking drive information, stop the fan and lock the motor disconnect switch so only you have control over starting the fan.

• Remove the belt guard.

• Record drive information on the Drive Data and Test Sheet (Fig. 1.3).

• Record sheave information. Look on the outside of the sheave for a stamped part number

FIGURE 1.3.   DRIVE DATA AND TEST SHEET

Project:		Engineer/Contact:		
	Specified	Actual	Specified	Actual
Fan Designation				
Drive Information				
Fan Shaft Size				
Motor Shaft Size				
Shaft Center Distance				
Fan Sheave Size				
Fixed or Adjustable				
Motor Sheave Size				
Fixed or Adjustable				
Belt Manufacturer				
Number of Belts				
Belt Size				
Amt. of Motor Adjust.				
Alignment				
Tension				
Notes:				
Fan Designation				
Drive Information				
Fan Shaft Size				
Motor Shaft Size				
Shaft Center Distance				
Fan Sheave Size				
Fixed or Adjustable				
Motor Sheave Size				
Fixed or Adjustable				
Belt Manufacturer				
Number of Belts				
Belt Size				
Amt. of Motor Adjust.				
Alignment				
Tension				
Notes:				

which indicates the sheave pitch diameter. If there's no part number on the sheave, measure the outside diameter and then refer to the manufacturer's catalog to find the corresponding pitch diameter. Calculate unknown pitch diameter from fan and motor speeds and one known sheave pitch diameter.

- Measure and record shaft sizes, and distance between the shafts.
- Record the number of belts, manufacturer and size. Note belt tension, and alignment.
- Record adjustment available on the belt tension adjustment rod (all-thread) or slide on the motor frame. If a different sheave size is required to optimize the system you may be able to move the motor forwards or backwards, up or down, to fit the change in sheave size without having to buy a new belt.
- Replace the belt guard.

Measuring Fan Operation

From prints or specifications, total the air quantities for all supply outlets and compare with the design capacity for the supply fan. Do the same for return fans and exhaust fans. Reconcile any differences. Use a direct contact (chronometric or digital) or noncontact (photo or strobe) tachometer to take fan speed.

Take total pressure and/or static pressure readings in the fan cabinet at the fan inlet, or in the fan inlet duct and outlet duct. Take static pressure readings across filters and coils.

To take fan pressure(s), first drill test holes in the fan cabinet, inlet and outlet duct, or anywhere else as needed. A $\frac{3}{8}$ inch hole is normally drilled to accommodate the standard Pitot tube which is $\frac{5}{16}$ inch in diameter. In some instances, the test hole may need to be larger, depending on the thickness of the material of the air handling equipment and it's associated ductwork. Use tubes such as the standard Pitot tube, the "pocket"Pitot tube or a static pressure tip to sense the air pressure. Connect the sensing tube via a hose(s) to an air pressure gauge. Commonly used air pressure gauges are liquid filled, electronic or dry type manometers. Air pressure in HVAC systems is measured in inches of water gauge (in. wg) or inches of water column (in. wc). Record all pressure readings on the Air Handling Equipment Data and Test Sheet (Fig. 1.4).

Checklist for Verifying Fan Operation

Record the following nameplate and actual information on the Air Handling Equipment Data and Test Sheet (Fig. 1.4):

Manufacturer

Serial number

Model number

Fan speed, rpm

Rotation

Capacity, cfm

FIGURE 1.4.   AIR HANDLING EQUIPMENT DATA AND TEST SHEET

Project:		Engineer/Contact:		
Fan Designation				
Location				
Area Served				
	Specified	Actual	Specified	Actual
Fan Information				
Manufacturer				
Rotation				
Capacity cfm				
Efficiency				
Type				
Wheel Size				
Tip Speed				

Fan Pressure: Fan SP, Fan TP, Total SP or External SP

	Specified	Actual	Specified	Actual
Inlet TP				
Outlet SP				
Fan SP				
Inlet TP				
Outlet TP				
Fan TP				
Inlet SP				
Outlet SP				
Total SP				
External Inlet SP				
External Outlet SP				
External SP				

Pressure Differential: Filter, Heating Coil, Cooling Coil

	Specified	Actual	Specified	Actual
Filter SP Entering				
Filter SP Leaving				
Filter SP Diff.				
Htg. Coil SP Entering				
Htg. Coil SP Leaving				

FIGURE 1.4.   (*Continued*)

Project:		*Engineer/Contact:*		
Htg. Coil SP Diff.				
Clg. Coil Wet or Dry				
Clg. Coil SP Entering				
Clg. Coil SP Leaving				
Clg. Coil SP Diff.				

Fan Designation

Air Volume Cubic Feet Per Minute

	Specified	*Actual*	*Specified*	*Actual*
Fan Total				
Outlet Total				
Outside Air Total				
Return Air Total				

Air Temperatures: Dry Bulb, Wet Bulb; Relative Humidity (%RH)

	Specified	*Actual*	*Specified*	*Actual*
Supply Air DB				
WB Supply Air				
SA Relative Humidity				
Return Air DB				
WB Return Air				
RA Relative Humidity				
Outside Air DB				
WB Outside Air				
OA Relative Humidity				
Mixed Air DB				
WB Mixed Air				
MA Relative Humidity				
Entering Clg. Coil DB				
Leaving Clg. Coil DB				
WB Entering Clg. Coil				
WB Leaving Clg. Coil				
Cond. Space % RH				

FIGURE 1.4. (*Continued*)

Project:		Engineer/Contact:		
System Condition				
Fan				
Duct				
Notes:				

How to Take Total Pressure Readings at the Fan

Total pressure (TP) is the sum of the static pressure (SP) and the velocity pressure (VP) taken at a given point of measurement ($TP = SP + VP$). Total pressure may be greater than or less than atmospheric pressure and can carry either a positive (+) or negative (−) sign. To take total pressure on the inlet or negative side of the fan connect the Pitot tube total pressure port to the low or negative (−) side of the instrument (Fig. 1.5). To take total pressure on the discharge or positive side of the fan connect the Pitot tube total pressure port to the high or positive (+) side of the instrument (Fig. 1.6). Total pressure readings may be taken at the inlet of the fan to determine fan total pressure (Fig. 1.7) or fan static pressure (Fig. 1.8).

How to Take Static Pressure Readings at the Fan

Static pressure (SP) is the pressure or force within or outside the air handling unit or duct that exerts pressure against all the walls of the unit or duct. Static pressure may be greater than or less than atmospheric pressure and may carry either a positive (+) or negative (−) sign. To take static pressure on the inlet or negative side of the fan connect the Pitot tube static pressure port to the low or negative (−) side of the instrument. To take static pressure on the outlet or positive side of the fan connect the Pitot tube static pressure port to the high or positive (+) side of the instrument.

FIGURE 1.5.   TAKING TOTAL PRESSURE AT THE FAN INLET

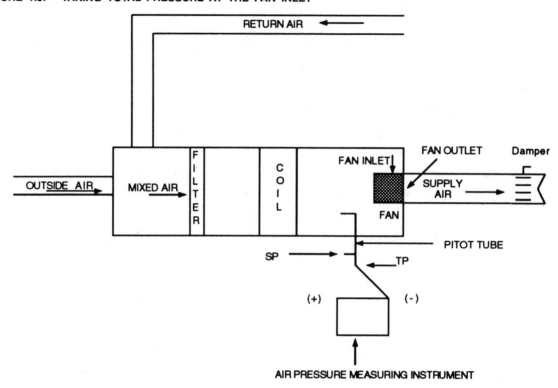

FIGURE 1.6.   TAKING TOTAL PRESSURE AT THE FAN OUTLET

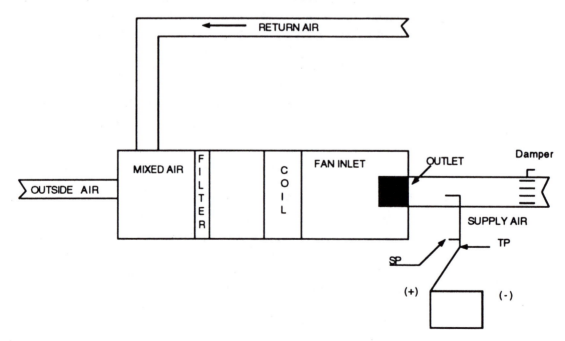

FIGURE 1.7.   TAKING FAN TOTAL PRESSURE

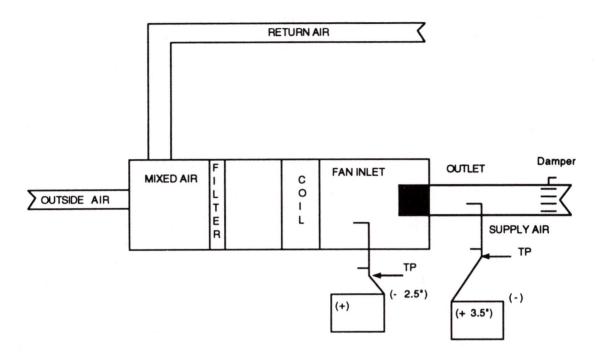

FIGURE 1.8.   TAKING FAN STATIC PRESSURE

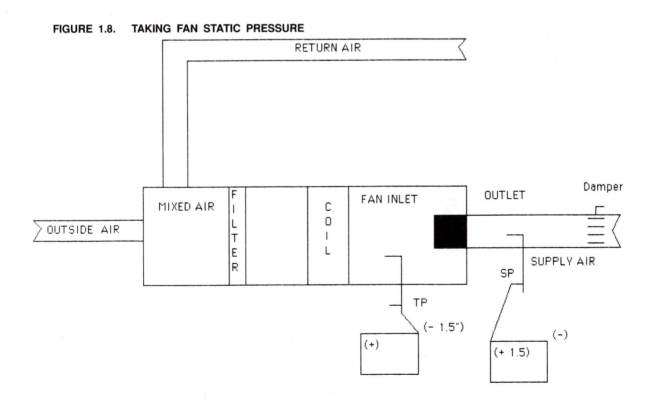

Measuring Fan Total Pressure and Fan Static Pressure

Fan total pressure (TP_F) is the measure of total mechanical energy added to the air by the fan. It is the rise in total pressure from the fan inlet to the fan outlet as expressed in Eq. 1.1.

Equation 1.1: Fan total pressure

$$TP_F = TP_O - TP_I$$

TP_F = fan total pressure
TP_O = total pressure at the fan outlet
TP_I = total pressure at the fan inlet

To determine fan total pressure simply *add* the inlet total pressure (a negative number) to the outlet total pressure. In Fig. 1.7, the inlet total pressure is −2.5 in. wg and the outlet total pressure is +3.5 in. wg. Therefore, the fan total pressure is 6.0 in. wg [+3.5 − (−2.0)].

Fan static pressure is defined as the fan total pressure minus the fan velocity pressure ($TP_F - VP_F$). Fan velocity pressure (VP_F) is equal to the velocity pressure at the outlet of the fan (outlet velocity pressure). Outlet velocity pressure (VP_O) is defined as the pressure that corresponds to the outlet velocity of the air as it leaves the fan. Outlet velocity (OV) is the theoretical value of the uniform velocity that would exist at the fan outlet. Outlet velocity is found using Eq. 1.2. Fan velocity pressure is found from Eq. 1.3.

Equation 1.2: Fan outlet velocity

$$OV = \frac{\text{cfm}}{\text{area}}$$

OV = outlet velocity at the fan outlet in feet per minute (fpm)
area = fan outlet area. The gross inside area of the fan outlet expressed in square feet.
cfm = cubic feet per minute air flow volume

Equation 1.3: Fan velocity pressure

$$VP_F = \left[\frac{OV}{4,005} \right]^2$$

VP_F = fan velocity pressure
4,005 = constant for standard air

Equation 1.4: Fan static pressure

$$SP_F = SP_O - TP_I$$

derived from:
$$SP_F = TP_F - VP_F$$
$$TP_F = TP_O - TP_I$$
$$VP_F = VP_O$$
$$SP_F = TP_O - TP_I - VP_O$$
$$TP_O = SP_O + VP_O$$
$$SP_F = SP_O + VP_O - TP_I - VP_O$$

SP_F = fan static pressure
SP_O = static pressure at the fan outlet
VP_O = velocity pressure at the fan outlet

From Eq. 1.3, fan static pressure can be restated as the difference between the static pressure at the fan outlet and the total pressure at the fan inlet ($SP_F = SP_O - TP_I$). However, for field measurements you will take inlet total pressure and add it to outlet static pressure. In Fig. 1.8, the inlet total pressure is −1.5 in. wg and the outlet static pressure is + 1.5 in. wg. Therefore, the fan static pressure is 3.0 in. wg [+1.5 − (−1.5)].

Fan total pressure and fan static pressure may be used with the fan curve to evaluate the fan's performance. This will be demonstrated in Chap. 2.

Measuring Fan Total Static Pressure and Fan External Static Pressure

Fan total static pressure (*TSP*) is defined as the rise across the fan calculated from adding the static pressure measurements at the fan inlet and the fan outlet. Fan total static pressure is expressed in Eq. 1.5.

Equation 1.5: Fan total static pressure

$$TSP = SP_0 - SP_1$$
$$TSP = \text{fan total static pressure}$$
$$SP_O = \text{static pressure at the fan outlet}$$
$$SP_I = \text{static pressure at the fan inlet}$$

In Fig. 1.9, the inlet static pressure is − 1.0 in. wg and the outlet static pressure is + 2.3 in. wg. Therefore, the total static pressure is 3.3 in. wg [+2.3 − (−1.0)].

External static pressure (*ESP*) is the total static pressure measured in the duct external of the air handling unit, as seen in Eq. 1.6.

Equation 1.6: External static pressure

$$ESP = SP_O - SP_I$$
$$ESP = \text{static pressure external of the unit}$$
$$SP_O = \text{static pressure in the outlet duct just external to the unit}$$
$$SP_I = \text{static pressure in the inlet duct just external to the unit}$$

FIGURE 1.9.   TAKING FAN TOTAL STATIC PRESSURE

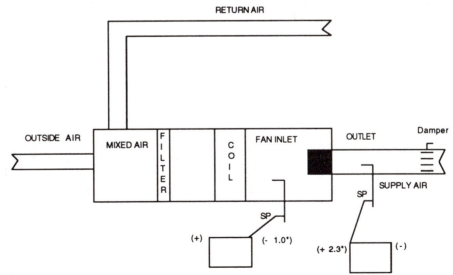

FIGURE 1.10.   TAKING EXTERNAL STATIC PRESSURE

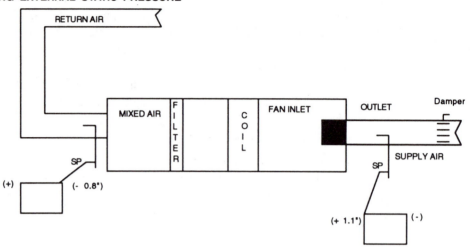

In Fig. 1.10, external static pressure equals 1.9 in. wg [+1.1 − (−0.8)]. Measured external static pressure is used by the design professional to indicate the system curve. System curves are discussed in Chap. 2.

How to Take Static Pressure Drop Across Filters and Coils

Static pressure readings should also be taken at the inlet (entering air side) and outlet (leaving air side) of the coil(s) for static pressure drop across the coil (Fig. 1.11) and the inlet (entering air side) and outlet (leaving air side) of the filter for static pressure drop across the filter (Fig. 1.12). Static pressures may also be measured at various locations in the system such as:

FIGURE 1.11.   TAKING STATIC PRESSURE DROP ACROSS THE COIL

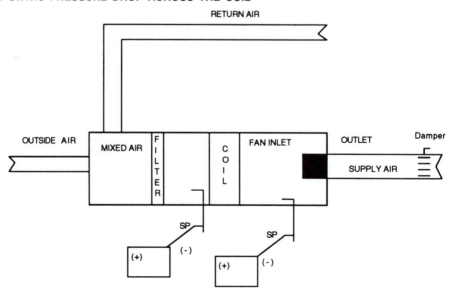

FIGURE 1.12.   TAKING STATIC PRESSURE DROP ACROSS THE FILTERS

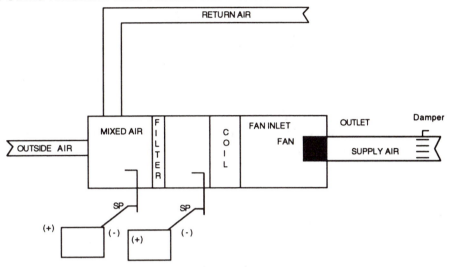

- At Pitot tube traverse points in the duct.

- Across any system component such as volume dampers, smoke and fire dampers, elbows, etc.

- Anywhere a restriction is suspected.

- At all points necessary to establish a static pressure profile of the air handling unit or duct system.

Quantifying the Amount of Outside Air

To determine the operating performance of the HVAC system we need to quantify the total supply air, the total return air, and the total outside air.

Determining the Amount of Outside Air from Direct Readings

The preferred method of determining the amount of outside air is through direct air readings using a Pitot tube traverse in the outside air duct. However, in many cases this is not practical. The next best way to determine the amount of outside air is to measure the air volume in the return duct with a Pitot tube traverse. Also measure the air volume in the supply duct with a Pitot tube traverse. Then subtract the total air in the return from the total air in the supply duct. This calculation will result in the amount of outside air (outside air = supply air − return air). Another method is to use a capture hood or anemometer at the outside air opening or louver. However, these readings generally will be difficult to take and are usually inaccurate. Another direct reading method is to read all the supply air outlets and subtract the total of all the return air inlets. Once again, this is difficult and time consuming, and not always accurate.

Determining the Amount of Outside Air from Temperature Readings

When direct air readings are not practical, the amount of outside air can be determined by measuring air temperatures in the mixed air plenum or supply air duct, and the outside air and return air ducts. Take temperatures when the temperature difference between the outside air and the return air is the greatest (e.g., summer or winter). The temperature in the return duct, supply duct and outside duct normally will be a one point reading because the air is one temperature. This is usually not the case in the mixed air plenum. Here you will need to take multiple temperatures to get an average mixed air temperature. One method is to take readings at the center of each filter and average the readings. Use a digital or multi-sensor thermometer. Also, by taking a number of readings across the mixed air plenum, you will be able to tell if there is air stratification. That is, a different air temperature on one side of the filters (from either the outside or return air streams) than on another part of the filters. Air stratification can cause poor heat transfer across the cooling and/or heating coil and in cold climates, may cause the freezestat to stop the fan or even worse, may cause a water coil to freeze. Use Eq. 1.7 or 1.8 to determine the percentage of outside air.

Equation 1.7: Percentage of outside air

$$\%OA = \frac{RAT - MAT}{RAT - OAT} \times 100$$

Equation 1.8: Percentage of outside air

$$\%OA = \frac{RAT - [SAT - 0.5(TSP)]}{RAT - OAT} \times 100$$

$\%OA$ = outside air (cfm) percent
RAT = return air temperature
MAT = mixed air temperature
0.5 = ½° per inch of static pressure (use 0.5 when the motor is out of the air stream, when the motor is in the air stream, use 0.6)
TSP = total static pressure across the fan
OAT = outside air temperature
SAT = supply air temperature

Example 1.1: To determine the percent of outside air entering the fan unit, use a digital thermometer to take a temperature traverse across the filter section. The average mixed air temperature is 67°F. The return air temperature is 77°F and the outside air is 35°F. The amount of outside air is calculated to be 23.8%.

$$\%OA = \frac{RAT - MAT}{RAT - OAT} \times 100$$

$$\%OA = \frac{77 - 67}{77 - 35} \times 100$$

$$\%OA = \frac{10}{42} \times 100$$

$$\%OA = 23.8$$

Example 1.2: Supply air temperature is taken to determine the percent of outside air entering the unit. The supply air temperature is 72°F. The return air temperature is 77°F and the outside air is 95°F. The pressure rise across the fan is 2 in. TSP. The motor is out of the air stream. The calculated amount of outside air is 33.3%.

$$\%OA = \frac{RAT - [SAT - 0.5\,(TSP)]}{RAT - OAT} \times 100$$

$$\%OA = \frac{77 - [72 - 0.5(2)]}{77 - 95} \times 100$$

$$\%OA = \frac{6}{18} \times 100$$

$$\%OA = 33.3$$

Verifying System Performance by Temperature

Generally, using temperatures to verify system performance is, at best, only an approximation. If temperature measurements are used for this purpose be sure that the instruments are recently calibrated and used only according to manufacturer's instructions. In addition, all metal surfaces to be measured must be clean. Finally, patience is a must. Allow enough time for the instruments to register properly.

Where To Take Temperatures

In some cases the same test holes used for the air pressure readings may be used to take temperatures. Temperatures are generally taken in the following locations:

- In the return air duct
- In the outside air duct
- In the mixed air plenum
- Before and after heat exchangers such as water and refrigerant coils, electrical heating apparatus, etc.
- Before and after the fan
- In the supply air duct
- In the supply air outlets
- In the return air inlets

How to Determine the Performance of Heat Exchangers Using Air and Water Temperatures

For heat exchangers where only dry bulb temperature is measured such as across electric heaters, heating coils, dry cooling coils, air-cooled condenser coils, etc., use Eq. 1.9.

Equation 1.9: Air, sensible heat.

$$Btuh_s = cfm \times 1.08 \times \Delta T$$

$$
\begin{aligned}
Btuh_s &= \text{Btu per hour sensible heat} \\
cfm &= \text{cubic feet per minute} \\
1.08 &= \text{constant} \\
\Delta T &= \text{dry bulb temperature difference between the air enter-}\\
&\quad\text{ing and leaving the coil.}
\end{aligned}
$$

For heat exchangers where only wet bulb temperature is measured, such as across wet cooling coils, use Eq. 1.10.

Equation 1.10: Air, total heat.

$$\text{Btuh}_T = \text{cfm} \times 4.5 \times \Delta h$$

Btuh_T = Btu per hour total heat
cfm = cubic feet per minute
4.5 = constant
Δh = change in total heat content of the supply air, Btu per pound (from wet bulb temperatures and psychrometric chart or table of properties of mixtures of air and saturated water vapor)

For water heat exchangers use Eq. 1.11.

Equation 1.11: Water, total heat

$$\text{Btuh} = \text{gpm} \times 500 \times \Delta T_w$$

Btuh = Btu per hour, water
gpm = water volume in gallons per minute
500 = constant
ΔT_w = Temperature difference between the water entering the heat exchanger and the water leaving the heat exchanger.

Example 1.3: Verify coil performance from air and water temperatures. The entering water temperature on a wet cooling coil is 45°F. The leaving water temperature is measured at 58°F. The water flow through the coil has been directly measured at 34 gpm. The air flow

FIGURE 1.13.   LOCATIONS FOR TEMPERATURE AND PRESSURE READINGS IN THE DUCT AND AIR HANDLING UNIT

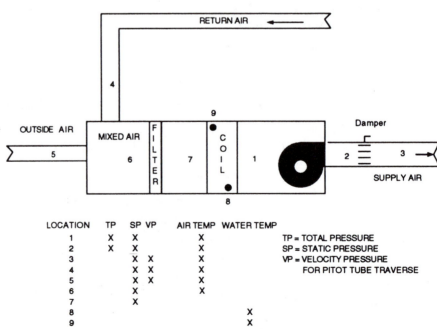

LOCATION	TP	SP	VP	AIR TEMP	WATER TEMP
1	X	X		X	
2	X	X		X	
3		X	X	X	
4		X	X	X	
5		X	X	X	
6		X		X	
7		X			
8					X
9					X

TP = TOTAL PRESSURE
SP = STATIC PRESSURE
VP = VELOCITY PRESSURE
     FOR PITOT TUBE TRAVERSE

is measured at 5,980 cfm. The entering air wet bulb is 66°F and the leaving air wet bulb is 54°F.

- Calculate the heat transfer from the water temperatures:

$$\text{Btuh} = \text{gpm} \times 500 \times \Delta T_w$$
$$\text{Btuh} = 34 \times 500 \times 13$$
$$\text{Btuh} = 221,000$$

- Calculate the heat transfer from the air temperatures:

$$\text{Btuh}_T = \text{cfm} \times 4.5 \times \Delta h$$
$$\text{Btuh}_T = 5,980 \times 4.5 \times 8.21$$
$$\text{Btuh}_T = 220,930$$

$$66°F \ WB = 30.83 \ \text{Btu/lb}$$
$$54°F \ WB = \underline{22.62 \ \text{Btu/lb}}$$
$$8.21 \ \text{Btu/lb}$$

- The readings verify the performance of the coil (221 MBH vs. 220.9 MBH).

Example 1.4: Determine the performance of a heating coil from air temperatures. The entering air temperature of a heating coil is 68°F. The leaving air temperature is 100°F. The air flow across the coil is 2,500 cfm. The calculated heat transfer is 86.4 MBH.

$$\text{Btuh}_S = \text{cfm} \times 1.08 \times \Delta T$$
$$\text{Btuh}_S = 2,500 \times 1.08 \times 32$$
$$\text{Btuh}_S = 86,400$$

CHAPTER 2 Central System Components—Fans

The first section of this chapter is an introduction to fans, fan categories and fan characteristics including multiple fan arrangements. You will learn how to calculate fan power, fan efficiency and fan tip speed. You will learn how to use the fan laws to determine system performance and how to use fan performance curves and fan multirating tables to predict fan performance. At the end of this chapter you will learn how to plot a system curve and to determine the operating point of fans in series and parallel. By using the fan laws, the fan performance curve and the fan system curve, you can calculate and graphically depict any change to the fan speed or the duct system.

HOW FANS OPERATE

As the fan rotates, centrifugal force throws the air outward reducing the pressure at the inlet of the fan wheel. Air is forced through the fan by the greater atmospheric pressure at the inlet. The air leaves the fan wheel at a relatively high velocity. Then, in the fan housing, the velocity is reduced and converted into pressure. The size of the fan wheel and its rotational speed determine the pressure developed by the fan. Most fans in HVAC work are constant volume machines. The exceptions are when the fan is coupled with a variable frequency motor to change the speed of the fan, or automatic dampers are used to restrict the amount of air into or out of the fan. These are variable volume fans. Fans for HVAC applications are, relatively speaking, low pressure machines generally at 10 in. wg pressure or less.

CATEGORIES OF FANS

Types

Heating, ventilating and air-conditioning fans are divided into three general types: axial, centrifugal, and special design fans. Each type is described in detail later in this chapter, but the following paragraphs introduce each type.

Axial fans are fans in which the airflow within the fan wheel is parallel to the fan shaft. Axial fans are further categorized as propeller (*P*), tubeaxial (*TA*), and vaneaxial (*VA*).

Centrifugal fans are fans in which the airflow within the fan wheel is radial or circular to the fan shaft. Five classifications of centrifugal fans are forward curved (*FC*), backward curved (*BC*), backward inclined (*BI*), airfoil (*AF*) and radial (*R*).

Special design fans are fans such as centrifugal power roof ventilators, axial power roof ventilators and tubular centrifugal fans. Tubular centrifugal fans are also known as inline centrifugal fans.

Fans are further categorized by the class of construction (called pressure class), the direction of fan rotation, the arrangement of the drive components, the motor location, the direction of air discharge and for centrifugal fans—the width of the fan wheel.

Pressure Class

Pressure classes (I, II, III and IV) are based on pressure developed by the fan. Each higher class represents heavier fan construction, and higher pressure, speed and performance capabilities of the fan.

Direction of Fan Rotation

Wheel rotation for centrifugal fans is described as either clockwise (*CW*) or counterclockwise (*CCW*). To determine fan rotation, view a centrifugal fan from the drive side, not the inlet side. The drive side is intended to mean the side that is driven by the motor, but it really varies with the fan configuration. On a single inlet, single wide (*SISW*) fan, the drive side is the side opposite the inlet. On a double inlet, double wide (*DIDW*) fan, the drive side is the side that has the drive. On fans with dual drives the side with the higher horsepower rating is considered the drive side. Axial fans will normally have an arrow on the housing which indicates direction of rotation. Rotation should be observed and corrected, if necessary, when the fan motor is wired. Correct fan rotation is important. On a centrifugal fan that's rotating backwards, the airflow can be reduced by 50% or more. An axial fan rotating backwards will not produce airflow in the proper direction.

Width of Fan Wheel

Centrifugal fans are designated either single inlet, single wide or double inlet, double wide. Axial fans do not use width designations. Single inlet, single wide (*SISW*) fans have one fan wheel and a single entry. Since in SISW fans the motor is opposite the inlet, the fan bearings are not in the air stream. This makes SISW fans more common in smaller sizes where inlet duct needs to be attached, and in applications where the air temperature is relatively high or the air is dirty.

Double inlet, double wide (*DIDW*) fans have two single wide fan wheels mounted back to back on a common shaft in a single housing. Air enters both sides of the fan. DIDW fans are generally used for moving large volumes of air in open inlet systems.

Arrangement of the Drive Components

The arrangement describes the drive placement in the fan. Arrangements are numbered from 1 to 10. Fan arrangements are found in the manufacturer's catalog.

Air Discharge Direction

Centrifugal fan discharge (Fig. 2.1), as with fan rotation, is viewed from the drive side (as if the fan was sitting on the ground). Angular discharges, either up or down positions without being specified are assumed to be either 45°, 135°, 225° or 315° from the vertical and in the direction of rotation. An up blast discharge is 360° and is designated simply "UB," while a top angle up blast 45° from vertical with a clockwise rotation is designated TAU CW 45. As you can see, when describing fans it is best to make a complete call out of the fan, such as clockwise top angle upblast 30° from vertical (TAU CW 30). Axial fans do not have discharge direction designations. The centrifugal fan designations are:

- Clockwise (CW)
- Counterclockwise (CCW)
- Top Horizontal Discharge (TH or THD)
- Top Angular Down Discharge (TAD or TADD)
- Down Blast Discharge (DB or DBD)
- Bottom Angular Down Discharge (BAD or BADD)
- Top Angular Up Discharge (TAU or TAUD)
- Up Blast Discharge (UB or UBD)
- Bottom Angular Up Discharge (BAU or BAUD)
- Bottom Horizontal Discharge (BH or BHD)

FIGURE 2.1.   TYPICAL FAN ROTATION AND DISCHARGE DESIGNATION CLOCKWISE ROTATION

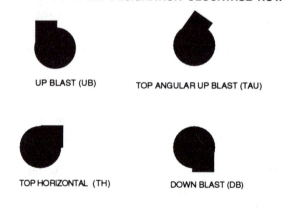

UP BLAST (UB)   TOP ANGULAR UP BLAST (TAU)

TOP HORIZONTAL (TH)   DOWN BLAST (DB)

CHARACTERISTICS OF FANS

Axial Fans

The general classifications of axial fans are: propeller (*P*), tubeaxial (*TA*) and vaneaxial (*VA*). Table 2.1 provides an overview of the differences between these three types of axial fans.

TABLE 2.1.   GENERAL CHARACTERISTICS OF THREE TYPES OF AXIAL FANS

	P	TA	VA
Fan and Wheel Construction			
Simple ring enclosure	x		
Cylindrical tube		x	
Tube with straightening vanes			x
Single thickness blades	x	x	
Airfoil blades		x	x
Long blades and small hub		x	
Short blades and large hub			x
2 or more blades lightweight construction	x		
4 to 8 heavy constructed blades		x	x
Adjustable pitch blades			x
Static Pressure Range			
Low pressure. Generally, three-quarters in. wg or less	x		
Medium pressure. Typically to 3 in. wg		x	
Medium to high pressure			x
Airflow and Discharge			
Airflow is parallel to the shaft	x	x	x
Circular or spiral pattern	x	x	
Vanes straighten out spiral pattern			x
Efficiency			
Low	x		
Medium		x	
High			x
Maximum efficiency near full airflow	x	x	x
Horsepower Characteristics			
Lowest at maximum airflow	x	x	x
Highest at minimum airflow	x	x	x
Horsepower increases as static pressure increases	x	x	x

Typical Applications

Used in low pressure applications. (P)
Used for delivering large volumes of air at low pressures. (P)
Used for general air circulation or exhaust without any attached ductwork. (P)
Used in medium pressure, medium to high air volume applications. (TA)
Used with ducted systems: fume exhausts, paint spray booths and drying ovens. (TA)
Used in applications where good downstream air distribution is needed. (VA)
Used in medium to high pressure, medium to high air volume applications. (VA)

Performance Curve Characteristics

The performance curve has a dip to the left of peak pressure caused by "aerodynamic" stall. Pressures in this area should be avoided. (TA, VA)

> **Key:**
> P = Propeller fan
> TA = Tubeaxial fan
> VA = Vaneaxial fan

Propeller Fans

These produce large volumes of airflow at low pressures. A typical commercial application of propeller fans would be general room air circulation or exhaust ventilation. Very large propeller fans are sometimes used in cooling towers. The housing for a typical propeller fan is normally a simple ring enclosure and the fan will usually have two or more single thickness blades. Propeller fans are generally not very efficient. A characteristic of propeller fans is that the operating horsepower, which is called brake horsepower, is lowest at maximum airflow and highest at minimum airflow. An example of this characteristic is the typical box-type home fan. If you look at this type of fan you'll notice that the first position on the air volume switch is "off." The next position is "high," then "medium," and then "low." This means that when the fan is turned on the electrical current draw and the horsepower will be at its lowest. In the last position, or the "low" position, the current and horsepower are the highest.

Tubeaxial Fans

These are heavy duty propeller fans used in such HVAC applications as fume hood exhaust systems, paint spray booths and drying ovens. The wheel of the tubeaxial fan is enclosed in a cylindrical tube and is similar to the propeller type wheel. The main exception is that the wheel has more blades, usually 4 to 8, and the blades are of much heavier construction. The tubeaxial fan is more efficient than the propeller fan and is most efficient when it's operating at its highest air volume. Like the propeller fan, the tubeaxial fan's operating horsepower is lowest at maximum airflow and highest at minimum airflow.

Vaneaxial Fans

These are basically tubeaxial fans with straightening vanes. They're used in HVAC ducted systems in office buildings or other commercial applications to provide airflow to the conditioned space. The housing is a cylindrical tube similar to the tubeaxial fan with the addition of air straightening vanes. The straightening vanes straighten out the spiral motion of the air and improve the efficiency of the fan. The vaneaxial fan has the highest efficiency of all the axial type fans. The wheel of the vaneaxial fans has shorter blades and a larger hub than the tubeaxial and, like the propeller fan and the tubeaxial fan, the operating horsepower is lowest at maximum airflow and highest at minimum airflow.

Centrifugal Fans

Centrifugal fans are divided into four general categories. The categories are: forward curved, backward curved or backward inclined, airfoil and radial. Table 2.2 provides an overview of the differences among these 4 types of centrifugal fans.

The Forward Curved Fan

Sometimes called a squirrel cage fan, it is generally used in residences and in small commercial and industrial applications. The fan housing is of light weight construction. The fan wheel has 24 to 64 shallow blades that curve toward the direction of rotation. The wheel is usually 24 inches in diameter or smaller. There may also be multiple wheels on a common

TABLE 2.2.   GENERAL CHARACTERISTICS OF FOUR TYPES OF CENTRIFUGAL FANS

	FC	BC	BI	AF	R
Fan and Wheel Construction					
24 to 64 shallow blades	x				
Blades curve in the direction of rotation	x				
Wheel is usually 24 inches in diameter or smaller	x				
Multiple wheels on a common shaft	x				
Lightweight construction	x				
10 to 16 blades (curved blades)		x			
Blades curve away from rotation		x		x	
Medium to heavyweight construction		x	x	x	
10 to 16 flat blades			x		
Blades incline away from rotation			x		
10 to 16 aerodynamically shaped blades				x	
Blades curve away from rotation				x	
6 to 10 blades (flat blades)					x
Blade tips are radial to the center of the wheel					x
Heavyweight construction					x
Static Pressure Range					
Low to medium	x				
Medium to high		x	x	x	x
High					x
Airflow and Discharge					
Airflow is radial to the shaft	x	x	x	x	x
Top or bottom horizontal	x	x	x	x	x
Up or down blast	x	x	x	x	x
Top or bottom angular down	x	x	x	x	x
Top or bottom angular up	x	x	x	x	x
Efficiency					
Low to medium efficiency	x				
Medium efficiency		x	x		
Most efficient of the centrifugal fans				x	
Least efficient of all centrifugal fans					x

Horsepower Characteristics

Horsepower increases continuously as air quantity increases and static pressure decreases. "Overloading." (FC, R)

Horsepower increases with an increase in air quantity but only to a point to the right of maximum efficiency and then gradually decreases. "Nonoverloading." (BC, BI, AF)

Typical Applications

Used in low to medium pressure and volume applications such as residences and package units. (FC)

Used in industrial medium to high pressure and volume applications where dust might cause erosion to airfoil blades. (BC, BI)

Used in commercial and industrial medium to high pressure and volume HVAC applications. (BC, BI, AF)

Used in industrial high pressure and velocity waste collection and material handling applications. (R)

TABLE 2.2. *(Continued)*

Performance Curve Characteristics

Highest efficiency occurs to the right of peak pressure when the fan is delivering 40 to 50% of full volume. (FC)

A dip in the pressure curve left of the peak pressure point which under certain conditions the fan may operate at one point and then another resulting in fan pulsations. (FC)

The pressure curve is less steep and the efficiency is less than backward curved, backward inclined or airfoil fans. (FC)

No dip in the pressure curve left of the peak pressure point and therefore, more stable and predictable operation. (BC, BI, AF)

Highest efficiencies occur 50 to 65% of full volume. (BC, BI, AF)

May have a dip to the left of peak pressure but not as great as with the forward curved fan. Usually not enough of a dip to cause difficulty. (R)

> **Key:**
> FC = Forward Curved fan
> BC = Backward Curved fan
> BI = Backward Inclined fan
> AF = Airfoil fan
> R = Radial fan

shaft. A characteristic of this type of fan is that the operating horsepower is low when the fan's airflow is also low but continues to increase as the airflow increases.

The Backward Curved and Backward Inclined Fans

The wheel on the backward curved or backward inclined fan has 10 to 16 blades that curve or lean away from the direction of rotation. The operating horsepower of this type of fan increases with an increase in airflow (but only to a point), and then gradually decreases. Because of this characteristic, the backward curved and backward inclined fans are called "nonoverloading" fans. In other words, the fan motor, if selected properly, will not draw more electrical current than its nameplate rated current regardless of the airflow. Backward fans are more efficient than forward curved fans but less efficient than airfoil fans.

The Airfoil Fan

It has the best efficiency of all the centrifugal fans. The fan wheel has 10 to 16 aerodynamically shaped blades similar to an airplane wing which curve away from the direction of rotation. As with the backward curved fans, the operating horsepower increases with an increase in airflow (but only to a point), and then gradually decreases. Therefore, airfoil fans are also "nonoverloading" fans.

Radial Fan

It is used in heavy-duty industrial applications such as waste collection and other types of material handling such as sawdust collection. The fan housing is made of heavy construction. The fan wheel is also generally heavily built with the blades being narrow but large in diameter. The fan wheel has 6 to 10 "paddle wheel" blades and is sometimes coated with special materials for protection. The radial fan is the least efficient of all the centrifugal fans. The operating horsepower increases continuously as the airflow increases. Therefore, the radial fan, like the forward curved fan, is an "overloading" type of fan.

Special Design Fans

The classifications of special design fans are tubular centrifugal or inline centrifugal fans, centrifugal power roof ventilators and axial power roof ventilators. The latter two are very small fans used to exhaust air from areas such as restrooms and attic spaces. They operate at low static pressures, horsepowers, and efficiencies.

Tubular or Inline Centrifugal Fans

- **Wheel Construction**—The wheel is housed in a cylindrical tube with backward inclined or airfoil blades.

- **Static Pressure Range**—Low.

- **Air Flow and Discharge**—The air is discharged radially from the wheel, changes directions by 90° to flow through a guide vane section and then flows parallel to the fan shaft.

- **Efficiency**—Lower than backward curved or backward inclined fans.

- **Horsepower Characteristics**—The horsepower curve increases with an increase in air quantity (but only to a point) to the right of maximum efficiency and then gradually decreases. "Nonoverloading".

- **Typical Applications**—Used in return systems where saving space is a consideration.

- **Performance Curve Characteristics**—Generally, no dip in the pressure curve left of the peak pressure point, although some may have a dip similar to axial fan. Generally, stable and predictable operation. Similar to the backward bladed fans except for lower capacities, pressures and efficiencies.

HOW TO CALCULATE FAN AIR HORSEPOWER AND BRAKE HORSEPOWER

The theoretical horsepower required to drive a fan, if the fan were 100% efficient, is called air horsepower (*ahp*). Equation 2.1 gives air horsepower. The actual power required to drive a fan is its brake horsepower (*bhp*). Equations 2.2 and 2.3 are used to calculate brake horsepower. The following terms will be used in the equations:

$$
\begin{aligned}
ahp &= \text{air horsepower} \\
cfm &= \text{airflow volume in cubic feet per minute} \\
P &= \text{fan pressure, in. wg} \\
6{,}356 &= \text{constant for air} \\
bhp &= \text{brake horsepower} \\
SP_F &= \text{fan static pressure} \\
SE_F &= \text{fan static efficiency} \\
TP_F &= \text{fan total pressure} \\
TE_F &= \text{fan total efficiency}
\end{aligned}
$$

Equation 2.1: Air horsepower

$$ahp = \frac{cfm \times TP_F}{6,356}$$

Equation 2.2: Brake horsepower

$$bhp = \frac{cfm \times SP_F}{6,356 \times SE_F}$$

Equation 2.3: Brake horsepower

$$bhp = \frac{cfm \times TP_F}{6,356 \times TE_F}$$

Example 2.1: A fan has a measured airflow of 10,400 cfm. The fan static pressure is 4 inches and the fan static efficiency is 60%. The calculated brake horsepower is 10.9.

$$bhp = \frac{cfm \times SP_F}{6,356 \times SE_F}$$

$$bhp = \frac{10,400 \times 4}{6,356 \times 0.60}$$

$$bhp = 10.9$$

HOW TO CALCULATE FAN EFFICIENCY

Fan efficiency is defined as the useful energy output divided by the power input. If brake horsepower is known, fan efficiencies can be determined from Eq. 2.4 and 2.5. To determine brake horsepower when fan efficiency is unknown, use 0.70 for fan static efficiency.

Equation 2.4: Fan static efficiency

$$SE_F = \frac{cfm \times SP_F}{6,356 \times bhp}$$

Equation 2.5: Fan total efficiency

$$TE_F = \frac{cfm \times TP_F}{6,356 \times bhp}$$

Example 2.2: A fan has a measured airflow of 100,400 cfm. The fan static pressure is 5.5 inches and the brake horsepower is 140.9. The calculated fan static efficiency is about 62%.

$$SE_F = \frac{cfm \times SP_F}{6,356 \times bhp}$$

$$SE_F = \frac{100,400 \times 5.5}{6,356 \times 140.9}$$

$$SE_F = 61.7$$

HOW TO CALCULATE FAN TIP SPEED

Fans that operate at high speeds and pressures are built to withstand the stresses placed on the fan wheel. However, if the fan is rotating too fast, the fan wheel bearings and fan shaft could be damaged. Fan manufacturers list the maximum allowable fan tip speed for each fan class. To calculate the tip speed of any fan at a given speed use Eq. 2.6.

Equation 2.6: Tip speed

$$\text{Tip speed} = \frac{3.14 \times D \times \text{rpm}}{12}$$

Tip speed = speed at the tip of the fan blade in feet per minute
D = fan wheel diameter in inches
rpm = fan speed
3.14 = constant
12 = constant, inches per foot

Example 2.3: A fan wheel is 66 inches in diameter and fan speed is 543 fpm. The fan tip speed is 9,978 fpm.

$$\text{Tip speed} = \frac{\pi(d)(\text{rpm})}{12}$$

$$\text{Tip speed} = \frac{3.14(66)(543)}{12}$$

$$\text{Tip speed} = 9{,}978 \text{ fpm}$$

HOW TO USE THE FAN LAWS TO DETERMINE SYSTEM PERFORMANCE

The performance of a fan can be predicted by the basic fan laws from available fan performance test data. The fan test data is obtained at a constant fan speed and standard air density. The fan laws provide air volume in cubic feet per minute (cfm), total pressure (*TP*), velocity pressure (*VP*), static pressure (*SP*), and brake horsepower (bhp) at varying fan speeds (rpm), motor sheave pitch diameters (*pd*), and air densities (*d*). The following terms will be used in the fan law equations:

cfm_1 = initial air volume
cfm_2 = final air volume
rpm_1 = initial fan speed
rpm_2 = final fan speed
pd_1 = initial pitch diameter
pd_2 = final pitch diameter
SP_1 = initial static pressure
SP_2 = final static pressure
bhp_1 = initial brake horsepower
bhp_2 = final brake horsepower

Fan Laws At Standard Air Density and Constant Fan Speed

1. Air volume (cfm) varies in direct proportion to fan speed (rpm).

   a. Air volume (cfm) varies in direct proportion to pitch diameter (pd) of the motor sheave.

   b. Fan speed (rpm) varies in direct proportion to pitch diameter (pd) of the motor sheave.

$$\frac{cfm_2}{cfm_1} = \frac{rpm_2}{rpm_1}$$

$$\frac{cfm_2}{cfm_1} = \frac{pd_2}{pd_1}$$

$$\frac{rpm_2}{rpm_1} = \frac{pd_2}{pd_1}$$

2. Pressure [for most calculations static pressure (SP) is used] varies as the square of the fan speed (rpm).

   a. Static pressure (SP) varies as the square of the pitch diameter (pd).

   b. Static pressure (SP) varies as the square of the air volume (cfm).

$$\frac{SP_2}{SP_1} = \left[\frac{rpm_2}{rpm_1}\right]^2$$

$$\frac{SP_2}{SP_1} = \left[\frac{pd_2}{pd_1}\right]^2$$

$$\frac{SP_2}{SP_1} = \left[\frac{cfm_2}{cfm_1}\right]^2$$

3. Brake horsepower (bhp) varies as the cube of the fan speed (rpm).

   a. Brake horsepower (bhp) varies as the cube of the pitch diameter (pd) of the motor sheave.

   b. Brake horsepower (bhp) varies as the cube of the air volume (cfm).

   c. Brake horsepower (bhp) varies as the square root of the static pressures (SP) cubed.

$$\frac{bhp_2}{bhp_1} = \left[\frac{rpm_2}{rpm_1}\right]^3$$

$$\frac{bhp_2}{bhp_1} = \left[\frac{pd_2}{pd_1}\right]^3$$

$$\frac{bhp_2}{bhp_1} = \left[\frac{cfm_2}{cfm_1}\right]^3$$

$$\frac{bhp_2}{bhp_1} = \sqrt{\left[\frac{SP_2}{SP_1}\right]^3}$$

Example 2.4: A fan has a measured airflow of 100,400 cfm. The fan static pressure is 5.5 inches and the brake horsepower is 140.9. The fan static efficiency is 62%. The pitch diameter of the motor sheave is 7.25 inches. The fan speed is 645 rpm. The system is air balanced and now has excess airflow. The air can be reduced from 100,400 cfm to 92,350 cfm. The new bhp is 109.7. The new static pressure is 4.65 inches. The new fan speed is 593 rpm.

$$\text{bhp}_2 = \text{bhp}_1 \times \left[\frac{\text{cfm}_2}{\text{cfm}_1}\right]^3$$

$$\text{bhp}_2 = 140.9 \times \left[\frac{92,350}{100,400}\right]^3$$

$$\text{bhp}_2 = 109.7$$

$$SP_2 = SP_1 \times \left[\frac{\text{cfm}_2}{\text{cfm}_1}\right]^2$$

$$SP_2 = 5.5 \times \left[\frac{92,350}{100,400}\right]^2$$

$$SP_2 = 4.65$$

$$\text{rpm}_2 = \text{rpm}_1 \times \frac{\text{cfm}_2}{\text{cfm}_1}$$

$$\text{rpm}_2 = 645 \times \frac{92,350}{100,400}$$

$$\text{rpm}_2 = 593$$

Fan Laws with Changes in Air Density

To compare one fan with another, fans are rated by their manufacturers based on standard air density (0.075 pounds per cubic foot). During the course of a typical fan performance test it is highly unlikely that the ambient air will be at standard air conditions. Therefore, the test results must be corrected for standard air. Essentially, a fan is a constant volume machine and will handle the same airflow regardless of air density for any given speed. However, with changes in density, the fan static pressure (against which the fan can deliver the air quantity) and the power required to drive the fan will change. The reason for the change in fan static pressure and brake horsepower is that as the density of the air changes, the weight of the air delivered by the fan also changes. For example, a motor driving a fan delivering 10,000 cubic feet per minute of air at a density of 0.15 pounds per cubic foot (twice the density of standard air) will have to work twice as hard as the same motor driving the same fan at the same cubic feet per minute but at standard density. The motor will actually use twice as much horsepower to deliver the greater total weight of air (10,000 cubic feet per minute × 0.15 pounds per cubic foot = 1,500 pounds per minute vs. 10,000 cubic feet per minute × 0.075 pounds per cubic foot = 750 pounds per minute).

Fan Laws at Constant Fan Speed and Constant Volume

1. Air volume (cfm) remains constant with changes in air density (d).
2. Pressure [for most calculations static pressure (SP) is used] varies in direct proportion to density (d).

$$\frac{SP_2}{SP_1} = \frac{d_2}{d_1}$$

3. Brake horsepower (bhp) varies in direct proportion to density (d).

$$\frac{\text{bhp}_2}{\text{bhp}_1} = \frac{d_2}{d_1}$$

$$d_1 = \text{initial density in pounds per cubic foot}$$
$$d_2 = \text{final density in pounds per cubic foot}$$

Example 2.5: A fan is selected to deliver 30,000 cfm at 2.0 in. wg SP at standard density. The power required is 12.6 bhp. If the fan is operating at an actual density of 0.060 pounds per cubic foot, the static pressure is 1.6 inches and the brake horsepower is 10.08. The fan speed is unchanged.

$$\frac{SP_2}{SP_1} = \frac{d_2}{d_1}$$

$$SP_2 = SP_1 \times \frac{d_2}{d_1}$$

$$SP_2 = 2.0 \times \frac{0.060}{0.075}$$

$$SP_2 = 1.6$$

Example 2.6: A fan is tested in a density of 0.057 pounds per cubic foot. The tested brake horsepower is 14.8. The tested fan static pressure is 2.5 in. wg. When corrected for standard air the rated static pressure is 3.29 inches and the rated brake horsepower is 19.47.

$$\frac{SP_2}{SP_1} = \frac{d_2}{d_1}$$

$$SP_2 = SP_1 \times \frac{d_2}{d_1}$$

$$SP_2 = 2.5 \times \frac{0.075}{0.057}$$

$$SP_2 = 3.29$$

$$\frac{\text{bhp}_2}{\text{bhp}_1} = \frac{d_2}{d_1}$$

$$bhp_2 = bhp_1 \times \frac{d_2}{d_1}$$

$$bhp_2 = 14.8 \times \frac{0.075}{0.057}$$

$$bhp_2 = 19.47$$

HOW TO USE THE FAN PERFORMANCE CURVE AND FAN MULTIRATING TABLE TO PREDICT SYSTEM PERFORMANCE

Fan performance curves (Fig. 2.2) are developed from actual tests and represent the performance of the fan from full airflow to zero airflow. The tests are performed with a constant fan speed. Measurements and calculations are made for torque, horsepower, dry bulb and wet bulb temperatures, barometric pressures, inlet and outlet fan pressures and airflow. These values are then plotted to develop the fan performance curve. The curve may show static pressure (SP), static efficiency (SE), total pressure (TP), total efficiency (TE) or mechanical efficiency (ME), brake horsepower (bhp), and air volume (cfm).

If the system has good fan inlet and outlet conditions, you can plot the fan air volume by measuring fan static pressure:

- Enter the curve (Fig. 2.3) on the static pressure side.

- Draw a line from the static pressure to the static pressure-rpm fan curve.

- At this point, draw a line down to intersect the cfm. This is the air volume produced by the fan.

Air volume as determined from field static pressure tests (Chap. 1) and the performance curve may be less than anticipated because of poor inlet and outlet measuring conditions.

If measuring fan static pressure directly is not practical because of the inlet and outlet conditions, you can determine the actual fan pressure (which includes any system effect) by measuring fan speed and total air volume, and plotting the results on the performance curve:

FIGURE 2.2.   FAN PERFORMANCE CURVE

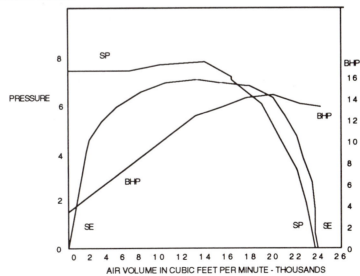

FIGURE 2.3.   DETERMINING CFM FROM STATIC PRESSURE

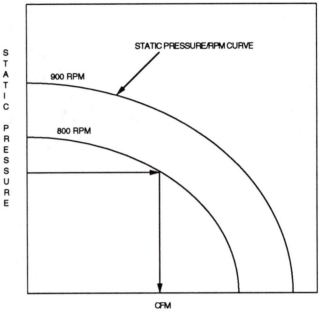

- Take fan speed measurements and total airflow from Pitot tube traverses or totaling the outlets.
- Obtain a fan performance curve for the fan speed measured (Fig. 2.4).
- Enter this curve at the measured cfm and draw a line up to the actual fan speed.
- At this point, draw a line to the left to intersect the fan static pressure.

This is the actual fan static pressure produced by the fan at the measured fan speed and airflow.

FIGURE 2.4.   DETERMINING STATIC PRESSURE FROM CFM AND RPM

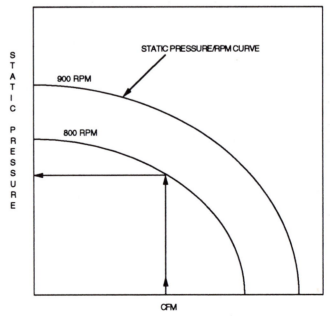

How to Plot the Operating Point for the System and the Fan

The operating point for both the air distribution system and the fan is the intersection of the air distribution "system" curve and the fan performance (static pressure-rpm curve). In Fig. 2.5, a fan is operating on fan curve X. The operating point is found from actual field measurements of air volume and static pressure.

FIGURE 2.5.   THE OPERATING POINT IS THE INTERSECTION OF THE SYSTEM CURVE AND THE FAN CURVE

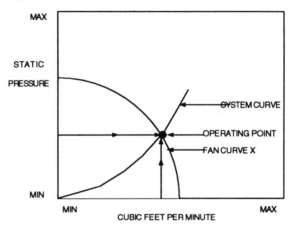

A system curve shows the static pressure required to overcome the pressure losses in the air distribution system in order to deliver various air quantities. An air distribution system operates only along its system curve. After you find the initial operating point from field measurements (airflow and static pressure), you can find any other point on the curve by using the fan law that states that static pressure varies as the square of the cfm. To plot the system curve, you arbitrarily select a number of other points.

Example 2.7: The fan curve is for a backward inclined fan. The solid curved lines indicate rpm. The diagonal broken lines indicate bhp.

- **Fan Curve 1:** The fan is selected for 936 rpm, 22,000 cfm, 4.9 inches SP, 0.75 eff.
- **Fan Curve 1:** From the fan curve or by calculation the fan is designed to operate at 22.6 bhp.
- **Fan Curve 2:** Plot the system curve. The arbitrarily selected points are:

	cfm	
24,000	14,000	6,000
20,000	12,000	4,000
18,000	10,000	0
16,000	8,000	

FAN CURVE 1.
Backward Inclined Fan

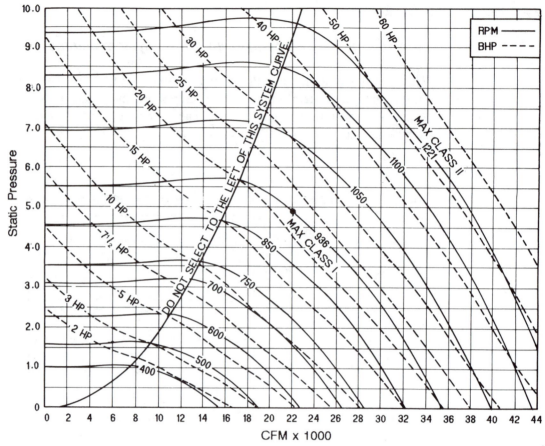

FAN CURVE 2.
Backward Inclined Fan

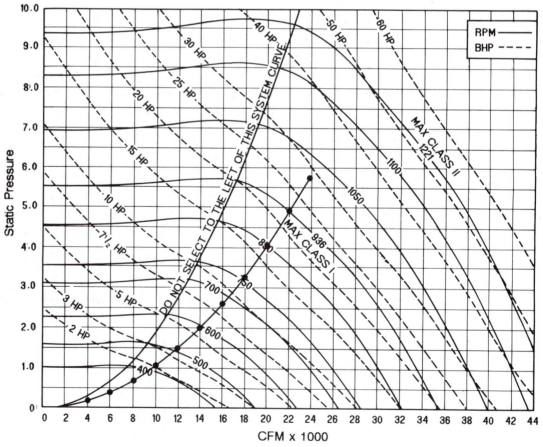

FAN CURVE 3.
Backward Inclined Fan

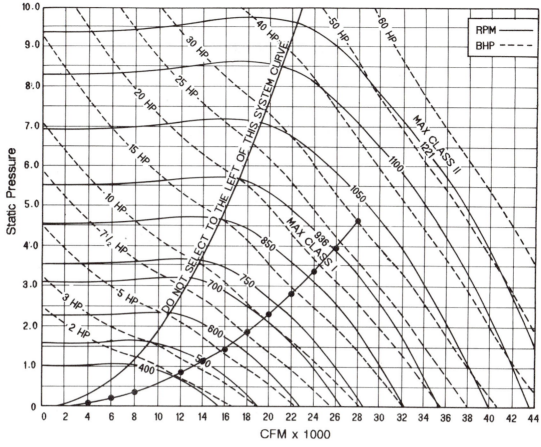

CFM x 1000

FAN CURVE 4.
Backward Inclined Fan

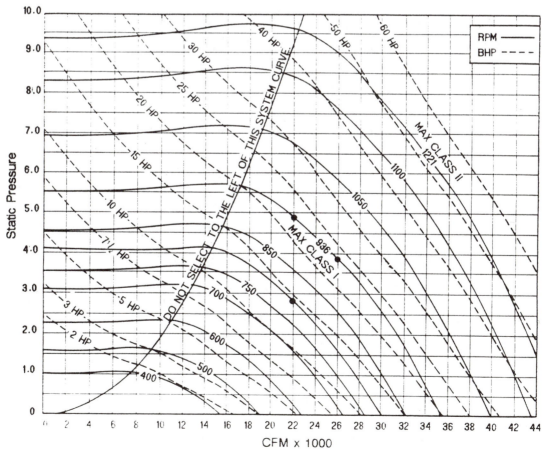

CFM x 1000

- The static pressures for the various cfm are calculated as follows:

$$SP_2 = SP_1 \left[\frac{\text{cfm}_2}{\text{cfm}_1} \right]^2$$

$$SP_2 = 4.9 \left[\frac{24{,}000}{22{,}000} \right]^2$$

$$SP_2 = 5.83$$

	cfm	SP
	24,000	5.83
Design point	22,000	4.90
	20,000	4.05
	18,000	3.28
	16,000	2.59
	14,000	1.98
	12,000	1.46
	10,000	1.01
	8,000	0.65
	6,000	0.36
	4,000	0.16
	0	0.00

- Plot the static pressures and draw the system curve.
- **Fan Curve 3:** The fan is tested and is operating at 936 rpm, 26,000 cfm, 3.9 inches SP, 0.68 eff.
- **Fan Curve 3:** From the fan curve or by calculation the fan is operating at approximately 23.5 bhp.
- **Fan Curve 3:** Plot the system curve. The arbitrarily selected points are

cfm
28,000
24,000
22,000
20,000
18,000
16,000
14,000
12,000
10,000
8,000
6,000
4,000
0

- The static pressures for the various cfm are calculated as follows:

$$SP_2 = SP_1 \left[\frac{cfm_2}{cfm_1} \right]^2$$

$$SP_2 = 3.9 \left[\frac{28,000}{26,000} \right]^2$$

$$SP_2 = 4.52$$

	cfm	SP
	28,000	4.52
Operating point	26,000	3.90
	24,000	3.32
	22,000	2.79
	20,000	2.31
	18,000	1.87
	16,000	1.48
	14,000	1.13
	12,000	0.83
	10,000	0.57
	8,000	0.34
	6,000	0.21
	4,000	0.09
	0	0.00

- Plot the static pressures and draw the system curve.
- Reduce the cfm from 26,000 to design (22,000 cfm).
- *Option 1:* Close the main volume damper and impose 1 inch of additional resistance in the system to drop the cfm to 22,000 at 4.9 inches SP. By imposing this additional resistance the fan is "backed up on its curve" (Fan Curve 3). Closing the main damper will save approximately 0.9 bhp (23.5 bhp − 22.6 bhp).
- *Option 2:* Reduce the fan speed to 792 rpm

$$rpm_2 = rpm_1 \left[\frac{cfm_2}{cfm_1} \right]$$

$$rpm_2 = 936 \left[\frac{22,000}{26,000} \right]$$

$$rpm_2 = 792$$

- **Fan Curve 4:** From the fan curve or by calculation the fan is operating at approximately 2.79 inches SP, 0.68 eff. and 14.25 bhp.

$$SP_2 = SP_1 \left[\frac{cfm_2}{cfm_1} \right]^2$$

$$SP_2 = 3.0 \left[\frac{22{,}000}{26{,}000}\right]^2$$

$$SP_2 = 2.79$$

$$bhp_2 = bhp_1 \left[\frac{cfm_2}{cfm_1}\right]^3$$

$$bhp_2 = 23.5 \left[\frac{22{,}000}{26{,}000}\right]^3$$

$$bhp_2 = 14.25$$

- **Fan Curve 5:** Plot the system curve. Since only the fan speed has changed the system curve remains the same as plotted for the operating condition on Fan Curve 3.

	cfm	*SP*
	28,000	4.52
Operating point	26,000	3.90
	24,000	3.32
	22,000	2.79
	20,000	2.31
	18,000	1.87
	16,000	1.48
	14,000	1.13
	12,000	0.83
	10,000	0.57
	8,000	0.34
	6,000	0.21
	4,000	0.09
	0	0.00

- The savings in energy costs can be calculated.

$$\text{Savings per year} = (bhp_1 - bhp_2)\left[\frac{0.746\ \text{Kw/bhp}}{\text{Motor Eff}}\right] \times \text{hours per year} \times \text{cost per Kwh}$$

This system operates 3,000 hours per year and the cost of electricity is $0.09 per kilowatt hour. The motor efficiency is 0.85.

Closing the main damper:

$$\text{Savings per year} = (23.5 - 22.6)\left[\frac{0.746}{0.85}\right] \times 3{,}000 \times 0.09$$

$$\text{Savings per year} = \$213.00$$

Fan speed change:

$$\text{Savings per year} = (23.5 - 14.25)\left[\frac{(0.746)}{0.85}\right] \times 3{,}000 \times 0.09$$

$$\text{Savings per year} = \$2{,}192.00$$

FAN CURVE 5. **Backward Inclined Fan**

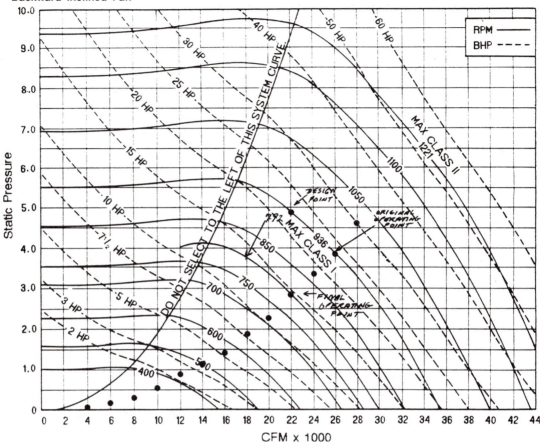

POINTS TO REMEMBER

- To increase or decrease the airflow in a system a physical change must be made in either the duct system or in the fan speed. If the change is to the fan speed, the fan will operate on a new fan performance-rpm curve that runs parallel to the original curve. Since the duct system is unchanged, the system curve remains the same. A decrease in fan speed results in a new fan curve (Fig. 2.6). The fan is operating at a lower cfm and static pressure.

- If the increase or decrease in airflow is the result of reducing or adding resistance to the duct system—for example, opening or closing a main damper—a new system curve (Fig. 2.7) is established, while the fan performance curve remains unchanged. If there are no changes in the system resistance such as closing or opening of dampers, etc., an increase or decrease in system resistance results only from an increase or decrease in cfm. The change in resistance will fall along the system curve. If the system's operating resistance (static pressure) exceeds the design system resistance, the measured cfm and horsepower will be less than design. If, on the other hand, the system's operating resistance is lower than design, the measured cfm and horsepower will be greater than design.

FIGURE 2.6.   A DECREASE IN FAN SPEED RESULTS IN A NEW FAN CURVE

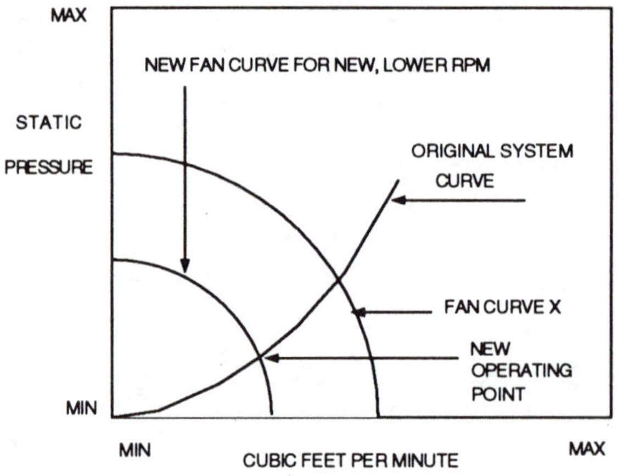

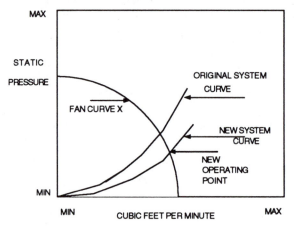

Multiple Fan Characteristics—Fans Operating in Series and Parallel

When fans operate in series, the pressures and horsepowers are additive at equivalent capacities. When they operate in parallel, the fan capacities and horsepowers are additive at equivalent pressures.

Example 2.8: Two fans, each separately handling 20,000 cfm at 3 in. wg at 13.5 bhp, would operate at 20,000 cfm at 6 in. wg and 27 bhp if placed in series and 40,000 cfm at 3 in. wg static pressure and 27 bhp if placed in parallel.

How to Use the Fan and System Curves to Predict Operation of Fans in Multiple Arrangements

To determine the operation of parallel fans, use the performance curve and the system curve (Fig. 2.8). The operating point of each fan in parallel (with all fans on) is on the single fan performance curve (Fig. 2.8, Point 1). The design operating condition for all fans in parallel is the intersection of the system curve and the parallel fan performance curve (Fig.2.8, Point 2). When only one fan is operating, the point of operation shifts to the intersection of the single fan curve with the established system curve (Fig. 2.8, Point 3). The shift occurs be-

FIGURE 2.7.   A DECREASE IN SYSTEM RESISTANCE RESULTS IN A NEW SYSTEM CURVE

FIGURE 2.8.   FANS OPERATING IN PARALLEL

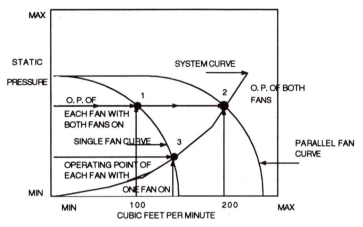

cause one fan is moving air through a duct system capable of handling twice the capacity. If the system goes to single fan operation, the static pressure required decreases (because there is less system resistance) and therefore the airflow and horsepower increase. Parallel fan operation requires that each fan motor be sized for its maximum horsepower, which occurs when only one fan is operating.

Fans in parallel operation should be tested both in normal operation (all fans operating), and individually. Test in each condition for airflow and power requirements. When testing fans individually measure the airflow through the operating fan and then read the pressure differential across the nonoperating fan to ensure that all dampers are working properly and there's no airflow through the nonoperating fans.

To determine the operation of series fans, use the performance curve and the system curve (Fig. 2.9). The design operating condition for all fans in series is the intersection of the system curve and the series fan performance curve (Fig. 2.9, Point 1). The operating point of each fan in series (with all fans on) is on the series fan performance curve (Fig. 2.9,

FIGURE 2.9.   FANS OPERATING IN SERIES

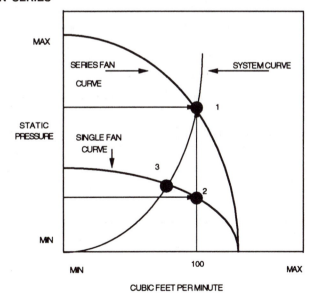

Point 2). When only one fan is operating, the point of operation shifts to the intersection of the single fan curve with the established system curve (Fig 2.9, Point 3). The shift occurs because only one fan is trying to moving air through a duct system designed for twice the static pressure. If the system goes to single fan operation, the static pressure required to overcome the system resistance increases and therefore the airflow and horsepower decrease. Series fan operation requires that each fan motor be sized for its maximum horsepower, which occurs when both fans are operating.

Fans in series operation should be tested both in normal operation (all fans operating), and individually. Test in each condition for airflow and power requirements.

Using Fan Multirating Tables to Select the Appropriate Fan

To make comparisons and selection as easy as possible, fan manufacturers provide multirating performance tables for each wheel diameter manufactured (see Table 2.3). The tables generally provide the following information:

- the air volume (cfm)
- static pressure (SP)
- fan speed (rpm)
- outlet velocity (OV)
- brake horsepower (bhp)
- blade configuration (backward inclined, forward curved, etc.)
- wheel configuration (single or double wide, etc.)
- fan wheel diameter (inches)
- outlet area (square feet)
- tip speed equation
- maximum brake horsepower equation
- pressure class limit

To predict fan air volume, brake horsepower, and outlet velocity from a fan multirating table:

- Enter the table at the measured fan static pressure.
- Go down the column until you find the measured fan speed (rpm).

TABLE 2.3.   FAN MULTIRATING TABLE

Multirating Table		*Fan Wheel Diameter 27 Inches*						*Fan Class I*	
Static Pressure		*1.00 Inch*		*1.25 Inches*		*1.50 Inches*		*1.75 Inches*	
cfm	*OV*	*rpm*	*bhp*	*rpm*	*bhp*	*rpm*	*bhp*	*rpm*	*bhp*
3300	788	600	0.73						
3750	895	616	0.82	673	1.03				
4200	1002	632	0.92	689	1.15	741	1.39		
4650	1110	651	1.03	705	1.28	756	1.53	804	1.79

- At fan rpm go left to find fan cfm and OV, and go right to find fan motor brake horse-power (bhp). In some cases the measured rpm may not be listed and it will be necessary to interpolate to find cfm and bhp.

Example 2.9: A fan is operating at 1.0 inch fan static pressure. The measured fan speed is 632 rpm. Enter the chart at 1.0 inch. Find 632 rpm. The brake horsepower is 0.92. The predicted fan volume is 4,200 cfm and the outlet velocity is 1,002 fpm.

Example 2.10: A fan is operating at 1.0 inch fan static pressure. The measured fan speed is 624 rpm. Enter the chart at 1.0 inch. Find that 624 rpm falls between the listed fan speeds of 616 rpm and 632 rpm. The air volume at 616 rpm is 3,750 cfm, and at 632 rpm the air volume is 4,200 cfm. The respective brake horsepowers are 0.82 and 0.92. Through interpolation the cfm at 624 rpm is 3,975 and the bhp is 0.87.

$$632 - 616 = 16$$
$$624 - 616 = 8$$
$$\frac{8}{16} = 0.5$$
$$4200 - 3750 = 450$$
$$0.5 \times 450 = 225$$
$$3750 \times 225 = 3,975 \text{ cfm}$$
$$0.92 - 0.82 = 0.10$$
$$0.5 \times 0.10 = 0.05$$
$$0.82 + 0.05 = 0.87 \text{ bhp}$$

CHAPTER 3 Verification of System Performance Airflow

In this chapter you will learn how to take velocity pressure readings for a Pitot tube traverse in the duct and how to measure airflow at the outlets. Also included in this chapter are the various reporting forms needed to quantify the airflow. Correction of instrument readings for changes in density is also included.

HOW TO TAKE VELOCITY PRESSURE READINGS

Velocity pressure (VP) is the pressure of the air in motion and is calculated by subtracting static pressure from total pressure. Velocity pressure is always a differential value ($VP = TP - SP$). Use a Pitot tube connected to a liquid filled, or electronic manometer to take velocity pressure readings at designated traverse points in the duct system.

Take velocity pressure in the ductwork on either the inlet of the fan (negative side), or the outlet of the fan (positive side). Use flexible tubing to (a) connect the Pitot tube static pressure port to the low or negative (−) side of the instrument, and (b) to connect the Pitot tube total pressure port to the high or positive (+) side of the instrument (Fig. 3.1).

Determining Traverse Location

To accurately measure velocity pressure readings, locate the test holes far enough down-stream and upstream of any change in the duct system so that the airflow will be as non-

FIGURE 3.1. TAKING VELOCITY PRESSURE READINGS ON THE INLET OR OUTLET SIDE

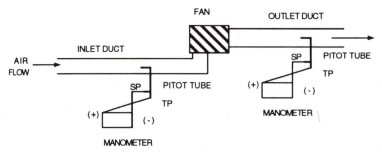

turbulent as possible. The suggested minimums are 8 duct diameters downstream and 2 duct diameters upstream of any changes in the duct such as elbows, transitions, dampers, take-offs, etc. Generally, because of building conditions, this is not possible in many duct systems. The next alternative is to try 4 duct diameters downstream and 2 duct diameters upstream. If 4 duct diameters is not possible and the readings are needed, move the traverse location as far downstream as possible. For any traverse location take a set of velocity pressure readings across the duct. If the readings are uniform, the traverse location is good. Large variations in the readings indicate that there's considerable turbulence in the duct and this may not be a proper location for the traverse. Do not take readings in transitions, elbows, or other fittings. If the duct is rectangular, use Eq. 3.1 to find approximate equivalent round duct size to determine traverse location.

Equation 3.1: Determining equivalent round duct diameter for rectangular duct:

$$D = \sqrt{\frac{4AB}{3.14}}$$

D = equivalent duct diameter in inches
A = length of one side of duct in inches
B = length of adjacent side of duct in inches
3.14 = constant, π

Example 3.1: Find the equivalent duct diameter of a duct that's 18×16 inches.

$$dD = \sqrt{\frac{4AB}{3.14}}$$

$$D = \sqrt{\frac{4 \times 18 \times 16}{3.14}}$$

$$D = 19 \text{ inches.}$$

The location of the holes for the traverse will be approximately 152 inches (8×19 inches) downstream and 38 inches (2×19 inches) upstream from any changes in the duct (Fig. 3.2).

Spacing Traverse Points

The spacing of the traverse holes in the duct are as important as the downstream and upstream location. In order to determine airflow quantity, you need to get the average velocity of the air flowing in the duct. The spacing requirements for square, rectangular, and round duct are listed below.

FIGURE 3.2.   LOCATION FOR PITOT TUBE TRAVERSE

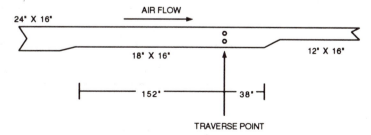

Velocity Pressure Test Points for a Square or Rectangular Duct Traverse.

To make a Pitot tube traverse of a square ore rectangular duct the requirements are:

- A minimum of 16 velocity pressure readings at centers of equal areas.
- The centers of the equal areas are not more than 6 inches apart. (If the number of readings exceeds 64, the distance between the centers may be increased to bring the total number of readings down to 64 or less.
- Areas do not exceed 36 square inches.

Example 3.2: Determine the location of the test holes for a traverse of a rectangular duct. The duct is 30 inches wide by 12 inches deep. The holes will be drilled in the 30 inch side (Fig. 3.3). The size of the hole will normally be ⅜ inches to accommodate the standard Pitot tube.

$$\frac{30 \text{ in.}}{5} = 6 \text{ in.}$$

There will be 5 holes, 6 inches on center. The idea is to drill the least amount of holes but keep the center distance at 6 inches or less.

$$\frac{6 \text{ in.}}{2} = 3 \text{ in.}$$

The first hole always will be one half the center distance. In this example, the first hole will be 3 inches from the side of the duct.

The layout of the holes will be:

Hole No.	Distance from side of duct
1	3 inches
2	9 inches
3	15 inches
4	21 inches
5	27 inches

FIGURE 3.3.   PITOT TUBE TRAVERSE OF A RECTANGULAR DUCT

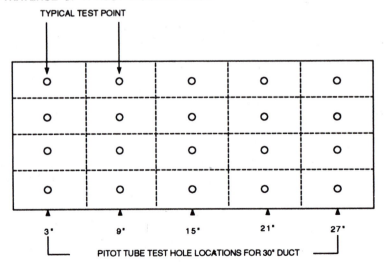

Example 3.3: Determine the markings on the Pitot Tube for the traverse of the 30 inches by 12 inch duct. Since the holes are drilled in the 30 inch side the marks on the Pitot tube will be for the 12 inch side (Fig. 3.4). The tube is commonly marked with a marking pen, electrical tape or duct tape. For convenience in taking a traverse, the standard Pitot tube is divided into inches with a stamped number indicating the even inches and a $\frac{1}{8}$ inch long stamped line indicating odd inches.

$$\frac{12 \text{ in.}}{4} = 3 \text{ in.}$$

There will be 4 marks, 3 inches on center. Mark the Pitot tube for the least amount of test points but enough to have a minimum of 16 readings. This traverse will have 20 test locations (5 holes with 4 readings per test hole).

$$\frac{3 \text{ in.}}{2} = 1.5 \text{ in.}$$

The first mark always will be one half the center distance. In this example, the first mark on the tube will be 1.5 inches.

The layout of the marks will be:

Mark No.	Distance on the tube
1	1.5 inches
2	4.5 inches
3	7.5 inches
4	10.5 inches

FIGURE 3.4.   MARKING THE PITOT TUBE FOR THE 12 INCH SIDE

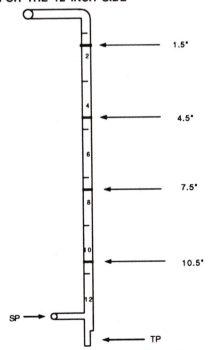

Example 3.4: Determine the location of the holes for a traverse of a duct that is 22 inches wide by 16 inches deep. The holes will be drilled in the 22 inch side.

$$\frac{22 \text{ in.}}{4} = 5.5 \text{ in.}$$

There will be 4 holes 5.5 inches on center.

$$\frac{5.5 \text{ in.}}{2} = 2.75 \text{ in.}$$

The first hole will be 2.75 inches from the side of the duct.

Hole No.	Distance from side of duct
1	2.75 inches
2	8.25 inches
3	13.75 inches
4	19.25 inches

Mark the Pitot tube for the 22 inch by 16 inch duct. The marks on the Pitot tube will be for the 16 inch side.

$$\frac{16 \text{ in.}}{4} = 4 \text{ in.}$$

There will be 4 marks, 4 inches on center. This traverse will have 16 test locations (4 holes with 4 readings per test hole).

$$\frac{4 \text{ in.}}{2} = 2 \text{ in.}$$

The first mark on the tube will be 2 inches.

Mark No.	Distance on the tube
1	2 inches
2	6 inches
3	10 inches
4	14 inches

Example 3.5: Determine the location of the holes for the traverse of a rectangular duct with 1 inch duct liner. The duct is 22 inches wide by 12 inches deep. The holes will be drilled in the 22 inch side. The inside dimensions are 20 inches × 10 inches (Fig. 3.5).

$$\frac{20 \text{ in.}}{4} = 5 \text{ in.}$$

There will be 4 holes, 5 inches on center.

$$\frac{5 \text{ in.}}{2} = 2.5 \text{ in.}$$

The first hole will be 3.5 inches from the side of the duct (2.5 inches + 1 inch).

FIGURE 3.5.   TEST HOLE LOCATIONS AND PITOT TUBE MARKINGS FOR 22 INCH BY 12 INCH LINED DUCT

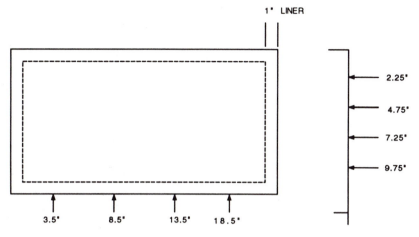

Hole No.	Distance from side of duct including 1 inch liner
1	3.5 inches
2	8.5 inches
3	13.5 inches
4	18.5 inches

Mark the Pitot Tube for the 22 inch by 12 inch lined duct. The marks on the Pitot tube will be for the 12 inch (10 inches inside dimension) side.

$$\frac{10 \text{ in.}}{4} = 2.5 \text{ in.}$$

There will be 4 marks, 2.5 inches on center. This traverse will have 16 test locations (4 holes with 4 readings per test hole).

$$\frac{2.5 \text{ in.}}{2} = 1.25 \text{ in.}$$

The first mark on the tube will be 2.25 inches (1.25 inches + 1 inch).

Mark No.	Distance on the tube plus 1 inch for liner
1	2.25 inches
2	4.75 inches
3	7.25 inches
4	9.75 inches

Velocity Pressure Test Points for a Round Duct Traverse

To make a Pitot tube traverse of a round duct the requirements are:

• Drill two holes in the duct 90° apart.

FIGURE 3.6.   PITOT TUBE TRAVERSE LOCATIONS FOR ROUND DUCT 10 INCH DIAMETER OR SMALLER

1 = 0.088	7 = 0.088
2 = 0.293	8 = 0.293
3 = 0.592	9 = 0.592
4 = 1.408	10 = 1.408
5 = 1.707	11 = 1.707
6 = 1.913	12 = 1.913

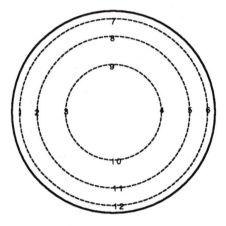

- For duct 10 inches or smaller mark the Pitot tube for the centers of 6 concentric circles of equal areas (Fig. 3.6). This will be a 12 point traverse—6 readings in each of two planes.

- For a duct larger than 10 inches, mark the Pitot tube for the centers of 10 concentric circles of equal areas (Fig. 3.7). This will be a 20 point traverse.

Example 3.6: Determine the location for a traverse of a round duct that is 6 inches in diameter. To determine the markings on the Pitot tube divide the diameter of the duct by 2 to get the radius. Next, multiply the factor for each traverse point times the radius. Convert to the nearest eighth inch.

Test Point No.	Factor	× radius =	Mark (inches)	Nearest eighth inch
1	0.088	3	0.264	¼ inch
2	0.293	3	0.879	⅞ inch
3	0.592	3	1.776	1¾ inches
4	1.408	3	4.224	4¼ inches
5	1.707	3	5.121	5⅛ inches
6	1.913	3	5.739	5¾ inches
7	0.088	3	0.264	¼ inch
8	0.293	3	0.879	⅞ inch
9	0.592	3	1.776	1¾ inches
10	1.408	3	4.224	4¼ inches
11	1.707	3	5.121	5⅛ inches
12	1.913	3	5.739	5¾ inches

FIGURE 3.7. PITOT TUBE TRAVERSE LOCATIONS FOR ROUND DUCT LARGER THAN 10 INCH DIAMETER

1 = 0.052	11 = 0.052
2 = 0.165	12 = 0.165
3 = 0.293	13 = 0.293
4 = 0.454	14 = 0.454
5 = 0.684	15 = 0.684
6 = 1.316	16 = 1.316
7 = 1.547	17 = 1.547
8 = 1.707	18 = 1.707
9 = 1.835	19 = 1.835
10 = 1.948	20 = 1.948

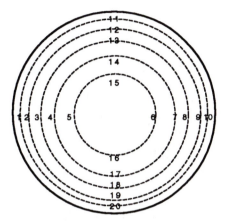

Example 3.7: A duct is 24 inches in diameter. To determine the markings on the Pitot tube divide the diameter of the duct by 2 to get the radius. Next, multiply the factor for each traverse point times the radius. Convert to the nearest eighth inch.

Test Point No.	Factor	× radius =	Mark (inches	Nearest Eighth Inch
1	0.052	12	0.624	⅝ inch
2	0.165	12	1.98	2 inches
3	0.293	12	3.516	3½ inches
4	0.454	12	5.448	5½ inches
5	0.684	12	8.208	8¼ inches
6	1.316	12	15.792	15¾ inches
7	1.547	12	18.564	18½ inches
8	1.707	12	20.484	20½ inches
9	1.835	12	22.02	22 inches
10	1.948	12	23.376	23⅜ inches
11	0.052	12	0.624	⅝ inch
12	0.165	12	1.98	2 inches
13	0.293	12	3.516	3½ inches
14	0.454	12	5.448	5½ inches
15	0.684	12	8.208	8¼ inches
16	1.316	12	15.792	15¾ inches
17	1.547	12	18.564	18½ inches
18	1.707	12	20.484	20½ inches
19	1.835	12	22.02	22 inches
20	1.948	12	23.376	23⅜ inches

Pitot Tube Traverse Procedure to Verify Operating Capacity

After the test holes have been drilled in the duct, set up the instruments and take readings using the following guidelines:

1. Correctly set up the instrument following manufacturer's instructions. For liquid filled manometers, make sure that:

   - Both the left and right tubing connectors are open to atmosphere.
   - The instrument is level.
   - The instrument has been adjusted to ambient temperature.
   - The meniscus is zeroed.
   - Only one person reads through the entire traverse. Different people read the meniscus differently so it is best for one person to complete the traverse. Another person can read the traverse the second time through if necessary. Consistency is the key.
   - Avoid parallax in reading the instrument. Most test quality analog instruments will have a mirrored scale to help with the readings. Read the instrument straight on so that only one image is seen.

2. Check the tubing for cracks or holes, especially near the ends. Repair or replace the tubing as needed.

3. Make sure that the Pitot tube is clean. Dirt and insulation can clog the static and impact holes.

4. Properly connect the Pitot tube to the instrument. Connect the Pitot tube's static pressure port to the low or negative (−) side of the instrument and the total pressure port to the high or positive (+) side of the instrument.

5. Insert the Pitot tube in the test hole and take the required number of velocity pressure readings. Record the velocity pressures on the data sheet.

6. Convert each velocity pressure reading to velocity using Eq. 3.2. Electronic manometers will make this conversion for you and read directly in velocity (feet per minute).

Equation 3.2: Converting velocity pressure (*VP*) to velocity (fpm).

$$\text{where} \quad V = 4{,}005 \sqrt{VP}$$

V = velocity in feet per minute (fpm)

$4{,}005$ = constant for standard air conditions

$\sqrt{VP}$ = square root of the velocity pressure in inches of water column

- Total and average the velocities. Some electronic manometers will do the total and average for you.

- Multiply the average velocity in the duct by the area of the duct in square feet. This is the quantity of airflow at this point in the duct in cubic feet per minute. Use Eq. 3.3.

Equation 3.3: Airflow quantity

$$Q = AV$$
Q = quantity of airflow in cubic feet per minute (cfm)
A = cross sectional area of the duct in square feet (sf)
V = velocity in feet per minute (fpm)

- Plug or otherwise seal the test holes after completing the entire traverse.

Example 3.8: Traverse a rectangular duct and calculate total cfm. The duct is 40 inches wide by 40 inches deep. The total number of readings is 64. The required capacity is 21,000 cfm. The location of the test holes and the marking of the Pitot tube is shown in Fig. 3.8. The traverse sheet is shown in Fig. 3.9.

Example 3.9: Traverse a round duct and calculate total cfm. The duct is 32 inches in diameter. The total number of readings will be 20. The required capacity is 10,000 cfm. The location of the test holes and the marking of the Pitot tube is shown in Fig. 3.10. The traverse sheet is shown in Fig. 3.11. Figure 3.12 shows a traverse sheet for round duct 10 inches or smaller. Figure 3.13 shows a typical traverse summary sheet.

How to Correct Instrument Readings for Air Density Changes

Liquid filled and dry manometers are calibrated to standard air conditions (70°F, 29.92 inches Hg, 0.075 pounds per cubic foot). Taking velocity pressure readings in conditions that vary greatly from standard air conditions require a correction for the instrument. (Note: Some electronic manometers have a density correction component built into the instrument.) To mathematically correct for the instrument reading use Eq. 3.4, 3.5, 3.6 and 3.7.

Equation 3.4: Determining new density.

$$D = 1.325 \frac{P_B}{T_A}$$

Equation 3.5: Determining correction factor for velocity with a change in air density.

$$CF = \sqrt{\frac{0.075}{D}}$$

FIGURE 3.8.   LOCATION OF TEST HOLES AND MARKING THE PITOT TUBE

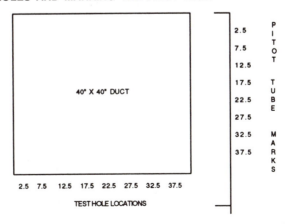

FIGURE 3.9. **RECTANGULAR DUCT TRAVERSE SHEET**

Project: _____

Fan Designation:_____

Traverse Designation:_____

	Specified	Actual
Duct Size (inches)	40″ × 40″	40″ × 40″
Duct Area (Square Feet)	11.1 sq. ft.	11.1 sq. ft.
Volume (Cubic Feet per Minute)	21,000 cfm	20,845 cfm
Average Velocity (Feet Per Minute)	1,892 fpm	1,878 fpm
Center Line Static Pressure	Not Specified	1.76″ SP
Density (pounds per cubic foot)	0.075 lbs/cf	0.075 lbs/cf
Instrument Correction for Density	None	None

No	VP	Vel	No	VP	Vel	No	VP	Vel	No	VP	Vel
1	0.17	1,651	18	0.20	1,791	35	0.19	1,746	52	0.22	1,879
2	0.19	1,746	19	0.21	1,835	36	0.20	1,791	53	0.19	1,746
3	0.20	1,791	20	0.22	1,879	37	0.21	1,835	54	0.17	1,651
4	0.21	1,835	21	0.22	1,879	38	0.21	1,835	55	0.16	1,602
5	0.22	1,879	22	0.23	1,921	39	0.22	1,879	56	0.17	1,651
6	0.19	1,746	23	0.24	1,962	40	0.22	1,879	57	0.18	1,699
7	0.21	1,835	24	0.24	1,962	41	0.20	1,791	58	0.17	1,651
8	0.23	1,921	25	0.26	2,042	42	0.23	1,921	59	0.18	1,699
9	0.24	1,962	26	0.26	2,042	43	0.23	1,921	60	0.19	1,746
10	0.22	1,879	27	0.25	2,003	44	0.24	1,962	61	0.20	1,791
11	0.23	1,921	28	0.24	1,962	45	0.26	2,042	62	0.19	1,746
12	0.24	1,962	29	0.23	1,921	46	0.26	2,042	63	0.19	1,746
13	0.21	1,835	30	0.22	1,879	47	0.27	2,081	64	0.19	1,746
14	0.20	1,791	31	0.21	1,835	48	0.25	2,003			
15	0.19	1,746	32	0.22	1,879	49	0.24	1,962			
16	0.18	1,699	33	0.20	1,791	50	0.23	1,921			
17	0.19	1,746	34	0.21	1,835	51	0.23	1,921			
SUBTOTAL 30,945			SUBTOTAL 34,380			SUBTOTAL 32,532			SUBTOTAL 22,353		

Total 120,210

Total velocity (fpm) divided by no. of readings = average velocity (fpm)

Total Velocity (120,210) fpm divided by (64) readings = 1,878 fpm

Average fpm × area = total cfm (nearest 5 cfm)

1,878 avg. fpm × 11.1 sf area = 20,845 total cfm

Notes:

_____

_____

_____

_____

_____

FIGURE 3.10. PITOT TUBE TRAVERSE FOR 32 INCH ROUND DUCT

1 = 0.052	11 = 0.052
2 = 0.165	12 = 0.165
3 = 0.293	13 = 0.293
4 = 0.454	14 = 0.454
5 = 0.684	15 = 0.684
6 = 1.316	16 = 1.316
7 = 1.547	17 = 1.547
8 = 1.707	18 = 1.707
9 = 1.835	19 = 1.835
10 = 1.948	20 = 1.948

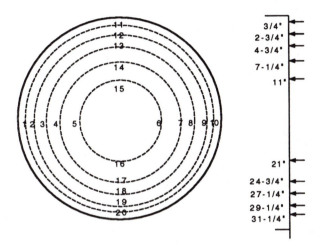

Equation 3.6: Determine average velocity using correction factor.

$$V_C = V_M \times CF$$

Equation 3.7: Determine actual air volume from corrected velocity.

$$Q = A \times V_C$$

D = air density in pounds per cubic foot
1.325 = constant, 0.075 divided by 29.92/530
P_B = barometric pressure in inches of mercury
T_A = absolute temperature in degrees Rankine
(duct air temperature in degrees Fahrenheit plus 460)
CF = correction factor
0.075 = density of standard air in pounds per cubic foot
V_C = corrected velocity
V_M = measured velocity
Q = quantity of airflow in cubic feet per minute
A = area in square feet

Example 3.10: Find new density and determine the actual air volume of the air in the duct in Ex. 3.8 if the air temperature in the duct is 170°F. The velocity pressures were taken using a liquid filled manometer. The duct is exhausting air from a furnace hood in a lab in San Diego, California. The measured average velocity is 1,878 fpm and the measured air volume is 20,845 cfm. The corrected velocity is 2,047 fpm and the corrected air volume is 22,722 cfm.

FIGURE 3.11.   ROUND DUCT TRAVERSE SHEET

Project: _____

System Designation: _____

Traverse Designation: _____

	Specified	Actual
Duct Size (inches)	32″ Diameter	32″ Diameter
Duct Area (Square Feet)	5.58 Sq. Ft.	5.58 Sq. Ft.
Volume (Cubic Feet Per Minute)	10,000 cfm	11,215 cfm
Average Velocity (Feet Per Minute)	1,792 fpm	2,010 fpm
Center Line Static Pressure	Not Specified	1.20″
Density (pounds per cubic foot)	0.075 lbs./cf	0.075 lbs./cf
Instrument Correction for Density	None	None

No	Factor	Inches	VP	Vel	No	Factor	Inches	VP	Vel
1	0.052	3/4	0.22	1,879	11	0.052	3/4	0.23	1,921
2	0.165	2-3/4	0.23	1,921	12	0.165	2-3/4	0.24	1.962
3	0.293	4-3/4	0.24	1,962	13	0.293	4-3/4	0.25	2,003
4	0.454	7-1/4	0.25	2,003	14	0.454	7-1/4	0.26	2,042
5	0.684	11	0.25	2,003	15	0.684	11	0.27	2,081
6	1.316	21	0.26	2,042	16	1.316	21	0.29	2,157
7	1.547	24-3/4	0.27	2,081	17	1.547	24-3/4	0.28	2,119
8	1.707	27-1/4	0.25	2,003	18	1.707	27-1/4	0.25	2,003
9	1.835	29-1/4	0.25	2,003	19	1.835	29-1/4	0.25	2,003
10	1.948	31-1/4	0.25	2,003	20	1.948	31-1/4	0.25	2,003

SUBTOTAL 19,900   SUBTOTAL 20,294

TOTAL 40,194

Total velocity (fpm) divided by 20 readings = average velocity (fpm)

40,194 fpm/20 readings = 2,010 average fpm

Average fpm × area = total cfm

2,010 fpm × 5.58 sf = 11,215 cfm

Notes:

_____

_____

_____

_____

_____

_____

_____

_____

_____

_____

FIGURE 3.12.   ROUND DUCT TRAVERSE SHEET, 10 INCHES OR SMALLER

Project: _____

Fan Designation: _____

Traverse Designation: _____

	Specified	*Actual*

Duct Size (inches)
Duct Area (square feet)
Average Velocity (feet per minute)
Volume (cubic feet per minute)
Center Line Static Pressure
Density (pounds per cubic foot)
Instrument Correction for Density

No.	Factor Inches	VP	Vel
1	0.088		
2	0.293		
3	0.592		
4	1.408		
5	1.707		
6	1.913		
7	0.088		
8	0.293		
9	0.592		
10	1.408		
11	1.707		
12	1.913		
	TOTAL		

Total velocity (fpm) divided by 12 Readings = Average velocity (fpm)

_____ fpm/12 readings = _____ average fpm

Average fpm × Area = Total cfm

_____ avg. fpm × _____ area = _____ total cfm

Notes:

_____
_____
_____
_____
_____
_____
_____
_____
_____
_____
_____

FIGURE 3.13.   TRAVERSE SUMMARY SHEET

Project:_____   **Engineer/Contact:**_____

System:

No.	Designation	Duct Size	Sq. Ft.	SP	Cubic feet/minute			
					Design	Actual	%D	Notes

Notes:

$$D = 1.325 \frac{P_B}{T_A}$$

$$D = 1.325 \frac{29.92}{630}$$

$$D = 0.063 \text{ pounds per cubic foot}$$

$$CF = \sqrt{\frac{0.075}{D}}$$

$$CF = \sqrt{\frac{0.075}{0.063}}$$

$$CF = 1.09$$
$$V_C = V_M \times CF$$
$$V_C = 1{,}878 \times 1.09$$
$$V_C = 2{,}047 \text{ fpm}$$
$$Q = A \times V_C$$
$$Q = 11.1 \times 2{,}047$$
$$Q = 22{,}722 \text{ cfm}$$

HOW TO TAKE AIRFLOW MEASUREMENTS AT THE OUTLET

Manufacturers of supply air outlets conduct various airflow tests on their products. From these tests, they establish an effective area of the outlet. The effective area of an outlet is defined as the sum of the areas of all the *vena contractas* (the smallest area of an air stream leaving an orifice) existing at the outlet. The effective area is based on the number of orifices and the exact location of the vena contractas, along with the size and shape of the grille bars, diffuser rings, etc.

Based on their findings, the manufacturers publish area correction factors or flow factors for their products. These flow factors apply to a specific type and size of grille, register or diffuser, a specific air measuring instrument, and the correct positioning of that particular instrument. The flow factors are called "K factor" or "Ak factor" (Table 3.1).

TABLE 3.1.   FLOW FACTOR TABLE FROM OUTLET MANUFACTURER

	Area Factor—Ak								
Neck Size Model	6 in.	8 in.	10 in.	12 in.	14 in.	16 in.	18 in.	20 in.	24 in.
278H	0.03	0.06	0.12	0.17	0.26	0.32	0.42	0.58	0.87
278V	0.09	0.17	0.24	0.32	0.43	0.54	0.64	0.78	1.01

TEST PROCEDURE:

Position Alnor® tip (2220-A) as shown.

Take a minimum of six equispaced velocity readings.

Average the readings.

Find Ak for neck size (diameter)

cfm = Average velocity (fpm) × Ak Factor

FIGURE 3.14.

FIGURE 3.15.

FIGURE 3.16.   AIR OUTLET TEST SHEET.

AIR DISTRIBUTION TEST SHEET

Project:

Area Served	Terminal No.	Terminal Type	Terminal Size	Ak	CFM	Notes
ROOM 3-41	1	Supply Air CD	24 × 24	1	183	
ROOM 3-41	2	Supply Air CD	24 × 24	1	318	
ROOM 3-41	3	Supply Air CD	24 × 24	1	320	
ROOM 3-41	4	Supply Air CD	24 × 24	1	254	
ROOM 3-41	5	Supply Air CD	24 × 24	1	283	
ROOM 3-41	6	Supply Air CD	24 × 24	1	266	
ROOM 3-41	7	Supply Air CD	24 × 24	1	154	
ROOM 3-41	8	Supply Air CD	24 × 24	1	297	
ROOM 3-41	9	Supply Air CD	24 × 24	1	318	
ROOM 3-41	10	Supply Air CD	24 × 24	1	266	
					2,659	

Area Served	Terminal No.	Terminal Type	Terminal Size	Ak	CFM	
ROOM 3-41	11	Return Air CD	24 × 24	1	361	
ROOM 3-41	12	Return Air CD	24 × 24	1	360	
ROOM 3-41	13	Return Air CD	24 × 24	1	316	
ROOM 3-41	14	Return Air CD	24 × 24	1	359	
ROOM 3-41	15	Return Air CD	24 × 24	1	381	
ROOM 3-41	16	Return Air CD	24 × 24	1	267	
					2,044	

Area Served	Duct No.	Duct Type	Duct Size	Ak	CFM	
ROOM 3-41		Outside Air OP	10″ø	0.545	380	

Notes: CD Ceiling Diffuser

OP Opening

The velocity of the air from the outlet is read and averaged. In Eq. 3.8, the average velocity is multiplied by the Ak factor to determine the air volume from the outlet.

Equation 3.8: Air volume:

$$cfm = V \times A_k$$

cfm = air volume in cubic feet per minute
V = average velocity in feet per minute
A_k = manufacturer's area correction or flow factor

Example 3.11: A sidewall grille is set for a 0° pattern deflection and read with the anemometer. A correction is applied to the anemometer and an average velocity of 600 fpm is calculated. The grille is 16 inches × 12 inches. The manufacturer's flow factor is 1.13. The total airflow out of the grille is 678 cfm.

$$
\begin{aligned}
\text{cfm} &= V \times A_k \\
\text{cfm} &= 600 \times 1.13 \\
\text{cfm} &= 678
\end{aligned}
$$

Another way to measure air from outlets is to use a capture hood such as the Shortridge FlowHood® (Fig. 3.14) or Alnor Balometer® (Fig. 3.15). Capture hoods collect the air from the outlet and give a direct air volume reading in cubic feet per minute (cfm) so that only one reading is required. Capture hoods eliminate the need for an averaged reading and the Ak factor. Readings with the capture hood are faster and repeatable. Figure 3.16 shows a typical air outlet test sheet.

CHAPTER 4 Air Distribution Components

In this chapter you will learn about various air distribution components and airflow control devices such as ductwork, dampers, air valves, diverters, terminal boxes and outlets.

DUCTWORK

The ductwork is a passageway used for conveying air and is typically made of various thicknesses of galvanized sheet metal. The thickness of the metal is called its **gage.** Typical sheet metal duct is 26 to 22 gage. The higher the gage number the thinner the metal. Depending on the application the duct may also be:

- Galvanized sheet metal wrapped with insulation to reduce heat transfer.
- Galvanized sheet metal lined with insulation to reduce heat transfer and noise transfer.
- Fiberglass or fiberboard duct to reduce heat transfer and noise transfer.
- Stainless steel
- Black iron
- Aluminum
- Aluminum flex
- Plastic-wrapped insulated wire flex
- Polyvinyl chloride (PVC)
- Plastic
- Resin

Table 4.1 lists a "rule of thumb" correction factor for pressure losses through various materials used in ductwork.

Example 4.1: At a certain velocity the friction loss through 100 feet of a certain sized galvanized duct is 0.1 inches. The friction loss through the same size and length of plastic

TABLE 4.1.

Material	Correction Factor
Galvanized duct	1.00
Fiberglass duct	1.35
Lined duct	1.08–1.42
Plastic flex duct, fully extended	1.85
Plastic flex duct, compressed 10%	3.65

flex duct with the same velocity would be approximately 0.185 inches. If the 100 feet of flex duct is compressed to 90 feet, the loss will be approximately 0.365 inches.

Shapes and Sizes of Ducts

Duct is formed to fit the architecture and construction of the building. Duct may be round (the most efficient), rectangular, square, or flat-oval (a round duct that is spread to form essentially a rectangular duct with a semicircle at either side Fig. 4.1).

Duct sizes are normally given in inches. Generally, HVAC round duct is manufactured in one inch diameter increments from 3 inches to 10 inches. Above ten inches, standard round duct is made in two inch increments (10, 12, 14, etc.). Rectangular ducts are generally made in even sizes such as 14 inches × 12 inches, 24 inches × 20 inches, 46 inches × 30 inches, etc. For rectangular ducts, the first number given is the side of the duct that is being viewed. For example, a rectangular duct being viewed from the top (top view is assumed unless otherwise stated) has a callout of 24 inches × 18 inches. A side or elevation view of the same duct would be designated as 18 inches × 24 inches. Duct lengths may be given in inches or feet. For example, a call may be made for either a 48 inch length (or joint) of duct or a 4 foot joint of duct.

Duct Friction Loss and Aspect Ratio

Round duct has less friction loss than other duct configurations because it has less material in contact with the air.

Example 4.2

	Round	Square	Rectangular
Area, sf.	1.0	1.0	1.0
Duct size	13.5 in.	12 in. × 12 in.	24 in. × 6 in.
Perimeter, in.	42.4 in.	48 in.	60 in.
Aspect ratio*	NA	1:1	4:1
Friction Loss/100 feet	0.12	0.13	0.16

*The aspect ratio in rectangular ducts is the ratio of the adjacent sides. A duct that is 24 inches × 18 inches has an aspect ratio of 1.33:1 (24/18 = 1.33). For energy conservation the aspect ratio should not exceed 3:1.

FIGURE 4.1.   FLAT OVAL DUCT

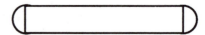

TABLE 4.2.

Pressure Class	Static Pressure (inches of water)	Velocity
Low	To 2 inches	To 2,500 fpm
Medium	Between 2 and 6 inches	Between 2,000 and 4,000 fpm
High	Above 6 inches	Above 2,000 fpm

Duct Pressures

For rating purposes, ductwork is designed, fabricated and installed as either low, medium, or high pressure. Medium and high pressure ductwork should be pressure leak tested. All ductwork should be sealed. Table 4.2 shows static pressure (inches of water) and velocity for each pressure class.

TYPES OF DUCT SYSTEMS

Duct systems are either single path or dual path.

A **single path system** (Fig. 4.2) is one in which the airflows through coils which are located essentially in series to each other. Single zone heating and cooling units and multizone terminal reheat units are single path duct systems.

A **dual path system** is one in which the airflows through heating and cooling coils which are located essentially parallel to each other. The coils may be side-by-side or stacked. The heating coil is located in the hot deck (*HD*) and the cooling coil is located in the cold deck (*CD*). Some systems may not have a heating coil, but instead bypass return air or mixed air into the hot deck. Multizone and dual duct systems (Figs. 4.3 and 4.4) are dual path duct systems.

FIGURE 4.2. SINGLE PATH SYSTEM

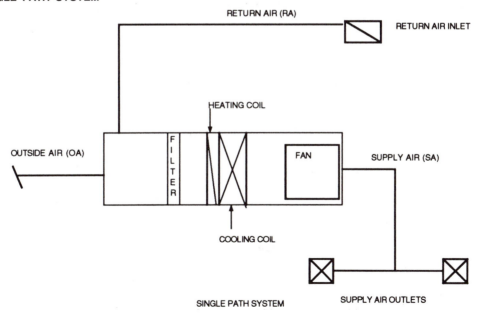

FIGURE 4.3.   DUAL PATH MULTIZONE MIXING DAMPER SYSTEM

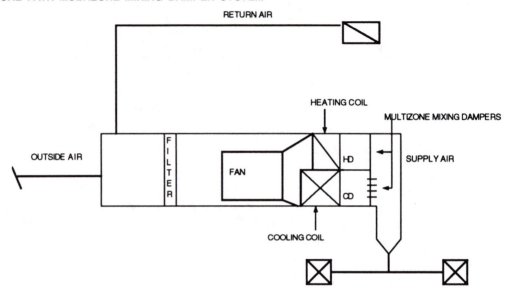

FIGURE 4.4.   DUAL PATH DUAL DUCT MIXING BOX SYSTEM

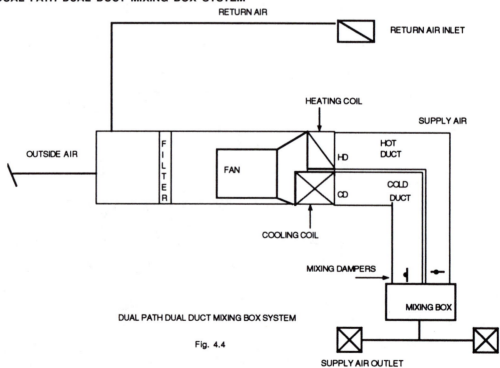

DUAL PATH DUAL DUCT MIXING BOX SYSTEM

Fig. 4.4

AIRFLOW CONTROL

Airflow is controlled by using devices such as dampers, air valves, diverters, terminal boxes and outlets.

Dampers

A damper is a device used to control the volume of airflow. Flow characteristics of dampers aren't consistent and may vary from manufacturer to manufacturer, and from one system to another. The actual effect of closing a particular damper in a particular system can only be determined in the field by measurement. There are three types of dampers:

- Automatic—automatic temperature control dampers (single or multibladed)
- Manual—volume or balancing dampers (single or multibladed)
- Gravity controlled—backdraft dampers

Single Blade Dampers

The two common types of single blade damper are butterfly and blast gate. The butterfly damper has a flat blade attached to a rod. The damper is placed in the center of the duct with the rod protruding through the duct. A locking handle on one end of the rod controls the position of the damper. The blast gate damper protrudes through the duct and is moved into the duct to control air volume.

Multibladed Dampers

Parallel Blade Damper (PBD)

The blades of parallel bladed dampers (Figs. 4.5A and 4.5B) rotate parallel to each other, producing a "diverting" air pattern when partially closed. This diverting pattern throws the air to the side of the duct, or to the top or bottom of the duct. If the damper is placed too close upstream to a coil, the diverting flow pattern may adversely affect the heat transfer of the coil. If the damper is placed too close upstream to a branch duct takeoff, the diverting flow pattern may reduce airflow into the branch duct. Parallel blade dampers are best used in mixing applications but may be used for volume control.

Opposed Blade Damper (OBD)

Opposed blade dampers (Fig. 4.6A and 4.6B) operate so that the adjacent blades rotate in opposite directions to each other. This means that the blade openings become increasingly narrow as the damper closes, resulting in a uniform, "nondiverting" airflow pattern. Opposed blade dampers have a better flow characteristic than parallel blade dampers. They are generally recommended for large duct systems for volume control, but may also be used in mixing applications.

Automatic Dampers

Automatic Temperature Control Dampers (ATCD)

Dampers controlled by the temperature requirements of the system are called automatic temperature control dampers. Automatic temperature control dampers are usually multibladed parallel blade or multibladed opposed blade dampers. They are controlled for

FIGURE 4.5A.   PARALLEL BLADE DAMPER

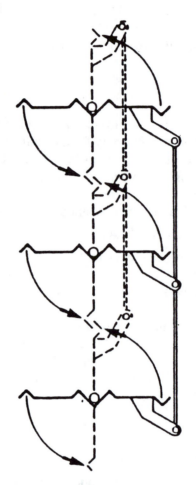

FIGURE 4.5B.   PARALLEL BLADE DAMPER WITH DIVERTING AIRFLOW PATTERN

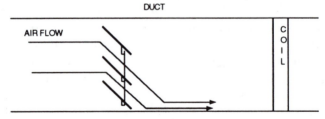

either two-position or modulating control action. The control source may be electricity (electrical or electronic) or compressed air (pneumatic).

Manual Dampers

Volume Dampers

Manual volume dampers are used to control the quantity of airflow in the air distribution duct by adding resistance to the flow (Fig. 4.7). The single blade butterfly damper is the most common type of manual damper used for volume control. The proper selection and placement of volume dampers equalizes the pressure drops in the different air paths allowing the system to be balanced in the least amount of time with the least amount of resistance and air noise. If volume dampers are not properly selected, placed, installed and adjusted, they may not control the air as intended. They may also add unnecessary resistance to the

FIGURE 4.6A.   OPPOSED BLADE DAMPER

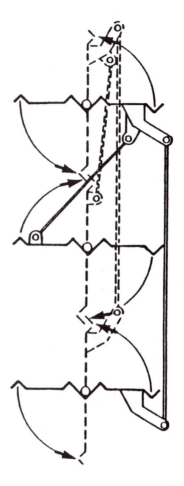

FIGURE 4.6B.   OPPOSED BLADE DAMPER WITH NONDIVERTING AIRFLOW PATTERN

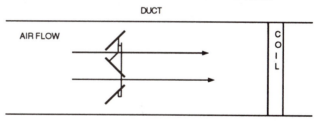

system and create noise problems. Manual volume dampers should be provided in the following locations:

- Mains
- Submains
- Branches
- Subbranches
- Takeoffs
- Zones of a multizone unit, at the unit

Volume dampers should not be installed immediately behind diffusers and grilles. When throttled, the dampers create noise at the outlet and change the effective area of the outlet so the manufacturer's flow factor (Chap. 3) is no longer valid. Proper selection and installation of manual volume dampers in the takeoffs eliminates the need for volume controls at grilles and diffusers.

FIGURE 4.7.   MANUAL VOLUME DAMPERS

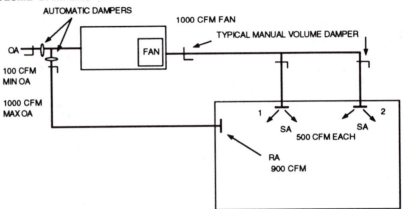

THE PRESSURE LOSS FROM THE FAN THROUGH OUTLET 1 IS 1.25"

THE PRESSURE LOSS FROM THE FAN THROUGH OUTLET 2 IS 1.50"

THE VOLUME DAMPER TO OUTLET 1 MUST BE CLOSED TO IMPOSE 0.25"
OF ADDITIONAL PRESSURE SO THAT THE AIR IS EQUAL TO EACH OUTLET.

MANUAL VOLUME DAMPERS ARE INSTALLED IN THE RETURN AND
OUTSIDE AIR DUCTS SO THAT THEY CAN BE ADJUSTED SO THAT
THE FAN "SEES" THE SAME PRESSURE LOSS THROUGH EACH RUN
WHEN IT IS FULL OPEN - EITHER FULL RA OR FULL OA ECONOMIZER.

Manual volume dampers may need to be installed in the outside, relief and return air connections to the mixed air plenum in addition to any automatic dampers (Fig. 4.7). These volume control dampers balance the pressure drops in the various flow paths so the pressure drop in the entire system stays constant as the proportions of return air and outside air vary to satisfy the temperature requirements.

Another type of manual volume damper is a **blast gate damper** (Fig. 4.8). This is a single blade damper normally used in exhaust applications such as in the exhaust duct from a laboratory fume hood.

FIGURE 4.8.   BLAST GATE DAMPER

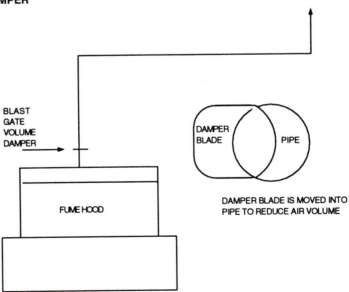

Gravity Controlled Dampers Backdraft Dampers (BDD)

Gravity backdraft dampers open when the air pressure on the upstream (entering air) side of the damper is greater than the downstream (leaving air) pressure. They close from gravitational force when there's no airflow.

AIR VALVES

Air valves are automatically controlled air volume devices. They are used in higher pressure systems and terminal boxes. Air valves have better flow characteristics than dampers and are used to give a more positive control in the system or terminal. The control source for the air valve may be electricity (electrical or electronic) or compressed air (pneumatic).

DIVERTERS

Low pressure systems may also have movable diverting devices such as extractors (Fig. 4.9) and splitters (Fig. 4.10) installed in the duct. These devices are moved to divert air into branch ducts or takeoffs. An extractor is sometimes called a "pickup." A splitter is often

FIGURE 4.9. EXTRACTOR

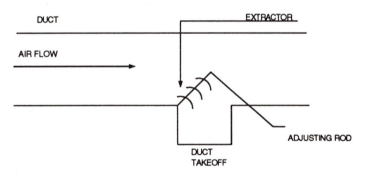

FIGURE 4.10. SPLITTER

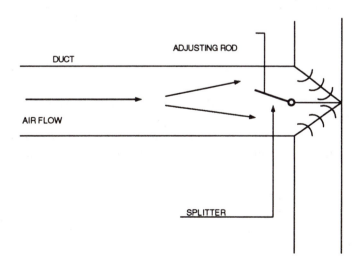

incorrectly called a "splitter damper". A splitter is a diverting device, not a damper. A damper regulates volume of air while a diverter directs the air.

TERMINAL BOXES

A terminal box is a unit which controls supply airflow, temperature and humidity to the conditioned space. The box may also have dampers, sound attenuation, and heating or cooling coils. The air volume through the box may be factory set but should also be adjustable in the field. Terminal boxes reduce the inlet pressures to a level consistent with the low pressure, low velocity duct connected to the discharge of the box. Any noise that's generated within the box in the reduction of the pressure is attenuated. Baffles or other devices are installed which reflect the sound back into the box where it can be absorbed by the box lining. Commonly, the boxes are lined with fiberglass which also provides thermal insulation so the conditioned air within the box won't be heated or cooled by the ambient air in the spaces surrounding the box. Terminal boxes work off static pressure in the duct system. Each box has a minimum inlet static pressure requirement to overcome the pressure losses through the box plus any losses through the discharge duct, volume dampers and outlets. The general classifications for terminal boxes are:

- Constant air volume
- Variable air volume
- Single inlet
- Dual inlet
- Medium pressure
- High pressure
- Pressure dependent
- Pressure independent
- System powered
- Fan powered
- Induction
- Reheat
- Recool
- Bypass

Constant Air Volume (CAV) Terminal Box

A terminal box which delivers a constant quantity of air is called a constant volume terminal box. Constant volume boxes may be single duct, dual duct or induction type.

Single Duct

A single duct terminal box (Fig. 4.11) is usually supplied with cool, conditioned air through a single inlet duct. Air flowing through the box is controlled to maintain a constant

FIGURE 4.11.   SINGLE DUCT CONSTANT VOLUME TERMINAL BOX

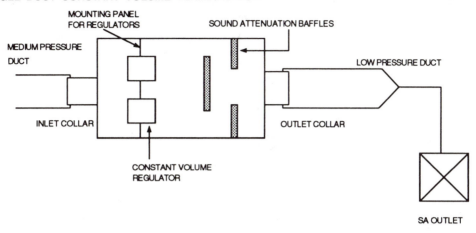

volume to the conditioned space. A reheat coil (water, steam or electric) or recool coil (refrigerant or water) may be installed in the box or immediately downstream from it. A room thermostat controls the coil.

Dual Duct

Conditioned air is supplied to the box through separate hot and cold inlet ducts. The hot duct supplies warm air which may be either heated air or return air from the conditioned space. The cold duct supplies cool air which may be either cooled and dehumidified when the refrigeration unit is operating, or simply cool outside air brought in by the economizer cycle. Dampers in the inlet ducts mix warm and cool air as needed to properly condition the space and maintain a constant volume of discharge air. The mixing dampers are controlled by the room thermostat.

Induction

Constant air volume induction boxes (Fig. 4.12) are supplied with a constant temperature supply air. The supply air is forced through a discharge nozzle at relatively high velocities. The high velocity of the supply air creates a low pressure region in the box which induces the higher pressure room or return air. The induced air mixes with the supply air. This mixed air is supplied to the conditioned space. Some induction boxes have heating or cooling coils. The room or return air is induced through the coil.

Variable Air Volume (VAV) Terminal Box

Variable air volume boxes vary the amount of air delivered to the conditioned space as the heat load varies. The operating costs of a VAV system are reduced when the total volume of air is reduced throughout the system. When the supply fan reduces its volume output, the brake horsepower and electrical cost to operate the fan go down by as much as the cube of the reduction in air volume. The exception to this is the VAV bypass box system which is a constant volume system at the central fan.

Variable air volume systems generally will have a diversity. A VAV diversity is when the total volume of all the VAV boxes is greater than the maximum output of the fan. For

FIGURE 4.12.   CONSTANT AIR VOLUME INDUCTION BOX

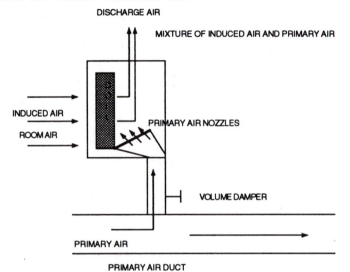

example, if the boxes total 10,000 cfm and the fan output is 7,000 cfm, the diversity is 0.70. VAV boxes are available in many combinations that include:

- Single duct
- Dual duct
- Pressure dependent
- Pressure independent
- Cooling only
- Cooling with reheat
- Induction
- Bypass
- Fan powered
- Throttling
- Pneumatic
- Electric
- Electronic
- System powered
- Direct acting
- Reverse acting
- Normally open
- Normally closed

Single Duct

Conditioned air is supplied to the VAV box through a single inlet duct (Fig. 4.13). The volume of air through the box is varied by the throttling action of an internal damper or air valve.

FIGURE 4.13.   SINGLE DUCT VARIABLE AIR VOLUME PRESSURE INDEPENDENT TERMINAL BOX

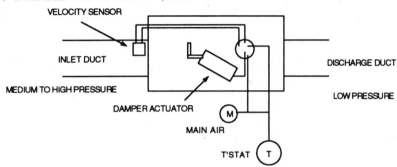

Dual Duct

Conditioned air is supplied to the VAV box through separate hot and cold inlet ducts. There are a variety of control schemes to vary the air volume and discharge air temperature out of dual duct boxes. It is always best to consult the manufacturer's specifications for setting and operating instructions.

Pressure Dependent

Pressure dependent means that the quantity of air passing through the box is dependent on the inlet static pressure. When a variable air volume box is classified as pressure dependent it means that the controls only position the motorized volume damper or air valve in response to a signal from the conditioned space thermostat. The box is essentially only a pressure reducing and sound attenuation device. Pressure dependent boxes do not control airflow as well as pressure independent boxes. In Fig. 4.14, the pressure dependent box varies more than 100 cfm, from 0.03 inches of inlet pressure to only 0.3 inches of pressure. However, the pressure independent box in Fig. 4.15 only varies about 50 cfm, between 0.3 and 4.0 inches of pressure.

FIGURE 4.14.   PRESSURE DEPENDENT BOX

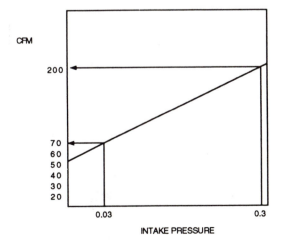

FIGURE 4.15.   PRESSURE INDEPENDENT BOX

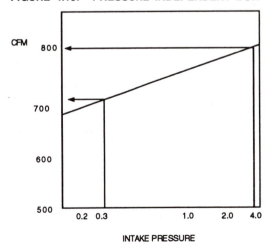

Pressure Independent

Pressure independent means that the quantity of air passing through the box is independent (within design limits) of the inlet static pressure. Pressure independent variable air volume boxes, in addition to the motorized volume damper or air valve, have flow sensing devices inside the boxes. The sensing device and its controller regulate the air volume through the boxes in response to the room thermostat. The controller is mounted on the outside of the box with connections to the sensing device, volume damper or air valve and the room thermostat. The controller may be electric, electronic or pneumatic. The sensing devices in these boxes maintain air volume at any point between maximum and minimum settings, regardless of the box's inlet static pressure, as long as the pressure is within the design operating range. The air volume will vary from a design maximum airflow down to a minimum airflow. Minimum airflows are generally around 50% to 25% of maximum, but may be as low as zero (shutoff) airflow.

Example 4.3: The temperature rises in the space and the room thermostat (responding to the load conditions in the space) sends a signal to the controller. The controller responds by actuating the volume damper or air valve (volume control device) to open for more cooling. The sensing device determines the differential pressure in the box and transmits a signal to the box controller which regulates the airflow within the preset maximum and minimum range. The controller opens the volume control device more. As the temperature in the space drops, the volume control device closes. If the box also has a reheat coil, the volume control device, on a call for heating, would close to its minimum position (usually not less than 50% of maximum) and the reheat coil would be activated. Because of its pressure independence, assuming the box is working properly, the airflow through the box is not affected by pressure changes in the system as other VAV boxes modulate open or closed.

Induction

A variable air volume ceiling induction box (Fig. 4.16) has a volume damper or air valve. It is located in the inlet duct to control the flow of primary supply air into the box. An induction damper allows return air from the ceiling plenum into the box to mix with the primary supply air. When the thermostat in the conditioned space calls for cooling, the volume damper opens and the induction damper closes. As the space cools down, the volume damper throttles back and the induction damper opens to maintain a relatively constant mixed airflow into the conditioned space. When the induction damper is wide open, the volume damper is throttled to a maximum of about 75% to allow for the proper induction/primary air ratio.

FIGURE 4.16.   VARIABLE AIR VOLUME INDUCTION BOX

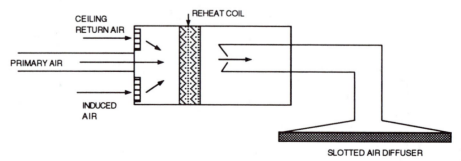

Bypass

A VAV bypass box (Fig. 4.17) is supplied with a constant air volume but supplies a variable air volume to the conditioned space. The supply air enters the box and can exit either into the conditioned space through the discharge ductwork or back to the return system through a bypass damper. A room thermostat regulates the bypass damper so that the conditioned space receives either all the cool supply air entering the box or only a part of it. There is no reduction in the main supply air volume feeding the box. This type of system has no central fan energy savings.

FIGURE 4.17.   VARIABLE AIR VOLUME BYPASS BOX

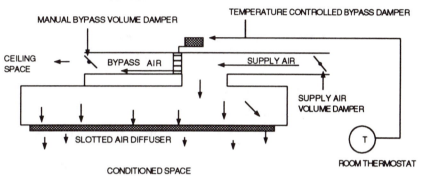

TABLE 4.3.   COMPARISON OF SERIES AND PARALLEL FAN POWERED BOX

Series	*Low Cooling Loads*	*Mid- to High Cooling Loads*	*Heating Loads*
Secondary fan:	on	on	on
Airflow:	constant	constant	constant
Secondary fan sound:	constant	constant	constant
Discharge air sound:	constant	constant	constant
Discharge air temperature:	variable	variable	variable
Secondary fan selection:	Selected for cooling load (typically 100% of cooling).		
Secondary fan control:	Interlocked with central fan.		
Secondary fan energy usage:	Operates continuously under greater loads.		
Central fan static pressure:	Must provide static pressure to overcome losses in air valve.		
Box inlet static pressure:	Adequate to overcome air valve only.		
Air valve selection:	Selected for cooling load.		
Parallel	*Low Cooling Loads*	*Mid- to High Cooling Loads*	*Heating Loads*
Secondary fan:	on	off	on
Air flow:	constant	variable	constant
Secondary fan sound:	audible	none	audible
Discharge air sound:	audible	none	audible
Discharge air temperature:	variable	constant	variable
Secondary fan selection:	Selected for heating load (typically 60% of cooling).		
Secondary fan control:	From T-stat signal. Not interlocked with central fan.		
Secondary fan energy usage:	Operates fewer hours under lighter loads.		
Central fan static pressure:	Adequate to overcome losses in air valve, low pressure distribution system.		
Box inlet static pressure:	Adequate to overcome air valve plus low pressure distribution system.		
Air vlave selection:	Selected for cooling load.		

FIGURE 4.18.   VARIABLE AIR VOLUME SERIES FAN POWERED BOX

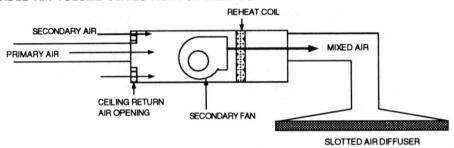

Fan Powered

A variable air volume fan powered box (Table 4.3) combines some of the options already discussed and has several advantages over the standard VAV box.

- Energy savings of a conventional VAV system.
- Heating capabilities.
- A relatively constant airflow to the conditioned space.

The typical fan powered box has a small centrifugal secondary fan and a return air opening from the ceiling plenum. The secondary fan may operate continuously (series system, Fig. 4.18), it may shut off (intermittent, parallel system, Fig. 4.19) or it may operate in a variable volume mode. When the room thermostat signals for cooling, the box operates as would the standard VAV box. However, when the room thermostat signals for heat, the primary air volume is reduced and the secondary fan draws warm air from the ceiling plenum. The cool primary air from the main system mixes with the ceiling return air at either the inlet side (series system) or discharge side (parallel) of the secondary fan. A system of dampers regulates the air volume, direction of airflow, and mixing of the air streams. If the room thermostat continues to signal for heat, the primary air volume damper continues to close and more ceiling air is drawn into the box. This maintains the air volume at a relatively constant rate but varies the temperature of the air into the conditioned space. In areas where there is

FIGURE 4.19.   VARIABLE AIR VOLUME PARALLEL FAN POWERED BOX

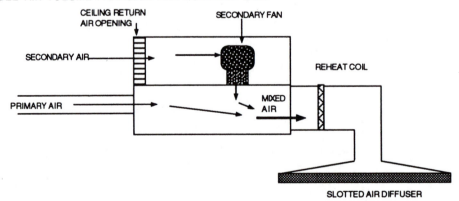

a greater heating load, reheat coils are installed in the boxes. Fan powered VAV boxes are used around the space perimeter or other areas where:

Air stagnation is a problem when the main supply air is reduced.

There are seasonal heating and cooling requirements.

Heat is required during the unoccupied hours when the central fan is off.

The heating load can be satisfied with recirculated ceiling return air.

Fan Powered Bypass

A variable air volume fan powered bypass box acts the same as the conventional by-pass box, with the addition of a secondary fan in the box. The bypass box uses a constant volume supply primary fan but provides variable air volume to the conditioned space. The supply air comes into the box and can exit either into the conditioned space through the secondary fan and the discharge ductwork or back to the return system through a bypass damper. The fan in the box circulates the primary air or return air into the room. The conditioned space receives either all primary air, all return air or a mixture of the two, depending on the signal from the room thermostat. Since there's no reduction in the main supply air volume feeding the box this type of system has no savings of primary fan energy.

System Powered

System powered boxes use, static pressure from the primary supply air duct to power the variable air volume controls. The minimum inlet static pressure required with this type of box is usually higher than other variable air volume systems in order to operate the VAV controls and still provide the proper airflow volume throughout the distribution system.

SUPPLY AIR OUTLETS

The openings in the supply duct that allow the conditioned air into the space are called supply air outlets. Supply air outlets are generally categorized as:

• Ceiling diffusers
• Ceiling grilles or ceiling registers
• Sidewall grilles or sidewall registers
• Floor grilles or floor registers
• Supply openings

Ceiling Diffusers

A ceiling diffuser is a supply air outlet located in the ceiling. Ceiling diffusers will typically have pattern deflectors arranged to promote the mixing of the supply air with the room air to produce a horizontal air pattern. This horizontal flow pattern, called surface effect, is caused by the inducement or entrainment of room air when the outlet discharges air directly parallel with and against a ceiling. The air then tends to flow along the ceiling. A high degree of surface effect is required for cooling applications, especially variable air volume systems, because it helps to reduce the dumping of cold air. Dumping is defined as the rapidly falling action of cold air caused by a variable air volume box or other device reducing its air velocity. Surface effect, however, contributes to "smudging." Smudging is the term used for the black markings on ceilings and supply air outlets made by suspended dirt particles in the room air. The dirt particles are entrained in the mixed air stream and then deposited on the ceilings and outlets. If smudging is a concern, "antismudge" devices are available. These devices physically lower the outlet away from the ceiling and cover the ceiling area a few inches beyond the diffuser.

Ceiling diffusers are generally categorized as:

- Rectangular
- Square
- Round
- Perforated face
- Light troffer
- Linear slot

Rectangular, Square and Round Diffusers

Rectangular or square ceiling diffusers typically supply air in a one-, two-, three- or four-way pattern. Round ceiling diffusers supply air in all directions.

Perforated Face Diffuser

Perforated face diffusers are used with lay-in ceilings and are similar in construction to the standard square ceiling diffuser with an added perforated face plate. They're generally equipped with adjustable pattern vanes to change the flow pattern to a one-, two-, three- or four-way throw.

Light Troffer

A light troffer is a type of ceiling diffuser which fits over a fluorescent lamp fixture and supplies air through a slot along the edge of the fixture. Some troffers supply air to only one side of the fixture, while other types supply air to both sides of the fixture.

Linear Slot

Linear slot diffusers are manufactured in various lengths (1, 2, 4 and 5 foot linears are common) and numbers of slots (1, 2, 3 and 4 slots are common). Linear slots are usually adjustable for different throw patterns (horizontal right or left or both, or vertical).

Grilles and Registers

A grille is a wall, ceiling or floor mounted louvered covering for an air opening. A register is a grille with a built-in or attached damper assembly. The damper is used to control air volume. To control airflow pattern, some grilles have a removable louver. Reversing or rotating the louver changes the air direction. Grilles are also available with adjustable horizontal and/or vertical bars so the direction, throw and spread of the supply air stream can be controlled.

Supply Openings

Supply openings in the duct are used in applications such as pressurized ceiling plenums for a laboratory perforated ceiling air distribution system.

Flow Patterns of Air Leaving Supply Outlets

The flow pattern of air leaving the supply outlet is important for the proper mixing of supply air and room air. As the supply air leaves the outlet it induces room air and creates a mixed airflow into the occupied space. The flow pattern should be such that the supply air mixes with the room air to maintain a uniform temperature and humidity throughout the occupied zone. The occupied zone is defined as the conditioned space from the floor to about six feet above the floor. Studies have shown that the majority of adults feel comfortable when the mixed air is between 68°F and 79°F, and between 20% and 60% relative humidity.

For the proper airflow pattern, ceiling diffusers or high sidewall grilles should be used for cooled air. Heated air should be distributed from low sidewall grilles or floor grilles. If low sidewall grilles or floor grilles are used for cool air distribution, they should be adjusted to direct the air up. The airflow pattern for heated air distributed from ceilings diffusers or high sidewall outlets should be directed down. Generally, since most outlets in air distribution systems are designed for year round operation, the outlets should be adjusted for horizontal patterns. However, for high ceilings or outlets with low velocities, it may be necessary to adjust the outlet for a vertical pattern to force the air down into the occupied zone. Supply air outlets are usually selected to complement the architectural design of the building as well as to control airflow patterns. The proper selection, installation and setting of flow patterns of supply air outlets will help to:

Avoid drafts caused by high air velocity or direction of airflow.

Avoid air stagnation caused by low air velocity or direction of airflow.

RETURN AIR INLETS

The openings in return ducts that allow the air from the conditioned space are called return air inlets. Return air inlets are generally categorized as:

Ceiling returns

Ceiling eggcrate returns

Ceiling grilles or ceiling registers

Sidewall grilles or sidewall registers

Floor grilles or floor registers

Return openings

Flow Patterns of Air Entering Return Air Inlets

The flow pattern of the air entering the return inlet is important for proper room temperature and pressurization. The flow pattern should be such that the supply air does not "short cycle." Short cycling means that the supply air comes from the outlet and goes into the return air inlet without mixing properly with the room air. Short cycling can happen if the return inlet is placed too close to the supply outlet. Return air inlets are usually selected to complement the architectural design of the building as well as to control air flow patterns. The proper selection, installation and placement of return air inlets will help to avoid drafts and air stagnation.

CHAPTER 5 Verification of System Performance Water Side

In this chapter you will learn how to determine the operating capacity and performance of the water side of the heating, ventilating and air conditioning system. You will learn how to set up an agenda to verify system performance, including what documents and forms to use. You will also learn how to measure water pressures at the pump and in the piping to determine pump and system performance.

HOW TO SET UP AN AGENDA

Before beginning the actual field measurements of the hydronic systems, you need documentation.

Get the following from the mechanical contractors:

- Mechanical plans
- Equipment specifications
- Engineering drawings
- Shop drawings

- "As-built" drawings
- Schematics
- Previous water balance reports

Get the following from the equipment manufacturers:

- Equipment catalogs
- Pump description and capacities
- Coil description and capacities

- Recommendations for testing equipment
- Operation and maintenance instructions
- Pump performance curves

Study these documents to become familiar with the hydronic system and its design intent. For clarity, label the pumping equipment and water distribution on the drawings. Make note of any piece of equipment, system component or any other condition that specifically should be investigated during the field verification of performance.

As applicable, prepare the following reporting forms for each water system:

- Pump data and test sheet
- Motor data and test sheet
- Coil data and test sheet
- Schematic of system
- Summary sheet

HOW TO VERIFY THE SYSTEM PERFORMANCE

To determine the operating capacity and performance of the water system, you must do the following:

- Verify the operation of the motor.
- As applicable, take drive information and speed measurements of the pump.
- Take pressure measurements of the pump.
- As applicable, take water temperatures in the system.

Verifying Motor Operation

Take voltage, current and power factor readings. Voltage and current readings can be made using a portable clamp-on volt-ammeter. Generally, all electrical readings are taken at the motor control center or at the disconnect box. The measured voltage should be plus or minus 10% of the motor nameplate voltage. If it's not, note it on the report. The measured current should not be over nameplate amperage. If it is, close the pump discharge valve until the current reading is down to maximum rated amperage.

Record nameplate motor speed. The motor operating speed is generally not measured. Unless the motor has a variable frequency drive, the nameplate rpm is recorded on the report sheet as constant operating speed. If the motor is VFD, note it on the report.

Check the installed motor thermal overload protection devices if the system is new or if changes have been made recently.

Record the following motor and starter nameplate information on the motor data and test sheet (Fig. 5.1).

- Manufacturer
- Frame size
- Horsepower
- Phase
- Hertz
- Motor speed, rpm
- Service factor

- Voltage
- Amperage
- Power factor
- Efficiency
- Starter size
- Overload protection

Taking Pump Drive and Impeller Information

Most HVAC pumps are directly coupled to the motor. If you do encounter a belt driven pump refer to Chap. 1, Fig. 1.2 and record the applicable drive information. Record the nameplate pump impeller size on the pump data sheet (Fig. 5.2).

FIGURE 5.1. MOTOR DATA AND TEST SHEET

Project: _____ **Engineer/Contact:** _____

	Specified	Actual	Specified	Actual
Pump Designation				
Motor Information				
Manufacturer	_____	_____	_____	_____
Frame Size	_____	_____	_____	_____
Horsepower	_____	_____	_____	_____
Phase	_____	_____	_____	_____
Hertz	_____	_____	_____	_____
Speed RPM	_____	_____	_____	_____
Service Factor	_____	_____	_____	_____
Voltage	_____	_____	_____	_____
Amperage	_____	_____	_____	_____
Power Factor	_____	_____	_____	_____
Efficiency	_____	_____	_____	_____
Brake Horsepower	_____	_____	_____	_____
Starter Size	_____	_____	_____	_____
Thermal OLP	_____	_____	_____	_____

Notes: _____

FIGURE 5.2.   PUMP DATA AND TEST SHEET

Project:_____   **Engineer/Contact:** _____

Pump Information	Specified	Actual	Specified	Actual
Designation	_____	_____	_____	_____
Location	_____	_____	_____	_____
Service	_____	_____	_____	_____
Manufacturer	_____	_____	_____	_____
Serial #	_____	_____	_____	_____
Model #	_____	_____	_____	_____
Impeller Diameter	_____	_____	_____	_____
Gallons per Minute	_____	_____	_____	_____
Total Dynamic Head	_____	_____	_____	_____
Pump Speed	_____	_____	_____	_____
Rotation	_____	_____	_____	_____

	Shutoff	Operating	Shutoff	Operating
Gallons per Minute	*******0*******	_____	*******0*******	_____
Discharge Pressure (psig)	_____	_____	_____	_____
Suction Pressure (psig)	_____	_____	_____	_____
PSI Rise	_____	_____	_____	_____
Static Head	_____	_____	_____	_____
Velocity Head	_____	_____	_____	_____
Total Dynamic Head	_____	_____	_____	_____
Impeller Diameter	_____	**********	_____	**********
Brake Horsepower	**********	_____	**********	_____
Efficiency	**********	_____	**********	_____
	_____	_____	_____	_____
	_____	_____	_____	_____

System Condition				
Pump	_____	_____	_____	_____
Pipe	_____	_____	_____	_____
	_____	_____	_____	_____
	_____	_____	_____	_____
	_____	_____	_____	_____
	_____	_____	_____	_____
	_____	_____	_____	_____
	_____	_____	_____	_____
	_____	_____	_____	_____

Notes: _____

_____

_____

_____

_____

_____

_____

_____

Verifying Pump Operation

Test multiple pumps (series, parallel or combinatin series-parallel operation) both in normal operation—all pumps operating except the "standby" pump(s)—and individually. Test in each condition for flow and power requirements. When testing pumps individually, measure the flow through the operating pump and then read the pressure differential across the nonoperating pump to ensure that all valves are working properly and there's no flow through nonoperating pumps. After checking the "normal" pumps, put the standby pump(s) in operation and test. Record the following nameplate and actual information on the pump data and test sheet (Fig. 5.2).

- Manufacturer
- Serial number
- Model number
- Impeller diameter
- Rotation

From the prints or the specifications, total the water quantities for all terminals and compare them with the design capacity for the pump. Reconcile any differences. Determine if rotation is correct by starting and stopping the pump and observing direction of movement. Pumps will normally have an arrow on the housing which indicates direction of rotation. You should be able to see rotation. If the pump is rotating backwards, rewire the motor for proper direction. Verify impeller size. Take a pressure reading at the pump and read the pressure drop across the strainer. Check the pump(s) for proper alignment, water leakage, etc.

HOW TO VERIFY PUMP IMPELLER SIZE

To verify pump performance from the pump curve, or to use the pump as a flow meter to get approximate water flow quantities, you must first verify impeller size. Do the following:

- Read amperage and voltage draw with the pump in normal operation.
- Turn the pump off.
- Attach a test gauge to the pump's suction and discharge pressure taps. Pressure readings at the pump are best taken with a Bourdon tube test gauge and hose manifold (Fig. 5.3).
- Mark the normal position of the discharge valve.
- Close the discharge valve.
- Open the suction valve fully if it is not already open. The suction valve should remain in the full open position.
- Turn the pump on.
- Read discharge static head (psig). Keep the test gauge at the same height for both discharge and suction readings.
- Read suction static head (psig).

FIGURE 5.3.   BOURDON TUBE TEST GAUGE AND MANIFOLD

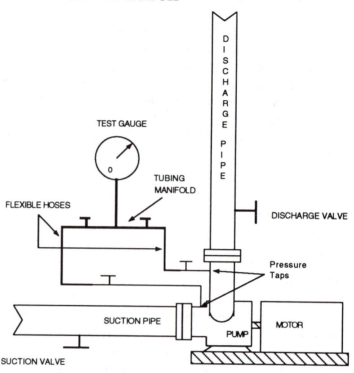

- Read amperage and voltage draw.
- Turn the pump off.
- Open the discharge valve to its normal position.
- Turn the pump on.
- Calculate the psi rise across the pump by subtracting the suction pressure from the discharge pressure. If the suction is in a vacuum, convert the reading to psi and add the result to the discharge pressure.
- Multiply the pump psi rise by 2.31 feet per psi to get rise in feet of water.
- Get the correct pump performance curve.
- Plot the pump rise on the vertical axis of the pump curve at zero flow. This is the shut-off head in feet of water (also called block-tight or no-flow head). The plot will fall on or near one of the impeller curves which will be the installed impeller size.

In the following example problems, the pump discharge valve is closed and readings are taken at the pump to determine impeller size. Use the pump curve in Fig. 5.4.

Example 5.1

Suction pressure = 10 psig
Discharge pressure = 68.5 psig

1. Calculate psi rise. 68.5 − 10 = 58.5
2. Multiply 58.5 psi × 2.31 feet/psi = 135 feet shut-off head.
3. Plot 135 feet on the vertical axis. The impeller size is 12 inches.

Example 5.2

Suction pressure = 10 psig
Discharge pressure = 75 psig

1. Calculate psi rise. $75 - 10 = 65$

2. Multiply 65 psi × 2.31 feet/psi = 150 feet shut-off head.

3. Plot 150 feet on the vertical axis. The impeller size is 12.5 inches.

HOW TO DETERMINE WATER FLOW USING THE PUMP AS A FLOW METER

Follow this procedure for using the pump as a flow meter.

- Verify impeller size.

- Attach a test gauge with manifold to the pump's suction and discharge pressure taps.

- Turn the pump on.

- Read amperage and voltage draw with the pump in normal operation. If the motor is overloaded, close the pump discharge valve until the amperage is down to maximum rated amperage.

- Read discharge static head (psig). Keep the test gauge at the same height for both discharge and suction readings.

- Read suction static head (psig). The suction valve should be fully open.

- Calculate the psi rise across the pump by subtracting the suction pressure from the discharge pressure. If the suction is in a vacuum, convert the reading to psi and add the result to the discharge pressure.

- Multiply the pump psi rise by 2.31 feet per psi to get rise in feet of water.

- Get the correct pump performance curve.

- Plot the pump rise on the vertical axis of the pump curve at zero flow. This is the rise across the pump in total dynamic head (TDH) in feet of water. [This is the total dynamic head when both the inlet and outlet pipe diameters are the same. When the inlet and outlet pipe sizes are different this value is actually static head rise across the pump and a velocity head is added to get actual total dynamic head (TDH = static head + velocity head)].

- Draw a line from the total dynamic head (TDH) horizontally to the right to intersect the pump impeller curve.

- At the intersection of the pump impeller curve and the TDH, draw a line vertically to the bottom of the chart to read the flow in gallons per minute.

- At the intersection of the TDH and the impeller curve, read the brake horsepower and the pump efficiency.

- Pump brake horsepower can also be calculated using Eq. 5.1.

Equation 5.1: Pump brake horsepower

$$\text{bhp} = \left[\frac{\text{gpm} \times TDH}{3960 \times \textit{eff}_p} \right]$$

bhp = brake horsepower
gpm = water flow in gallons per minute
TDH = total dynamic head in feet of water
3,960 = constant for water
eff_p = pump efficiency

In the following example problems the pump impeller size has been verified to be 12 inches (Fig. 5.4).

Example 5.3:

Suction pressure = 11 psig
Discharge pressure = 64.5 psig

- Calculate psi rise. 64.5 − 11 = 53.5

- Multiply 53.5 psi × 2.31 feet/psi = 124 feet total dynamic head.

- Plot 124 feet on the verticle axis.

- Draw a line from 124 feet TDH to intersect the 12 inch impeller curve.

- At the intersection of the pump impeller curve and the TDH draw a line vertically to the bottom of the chart to read 1,500 gallons per minute.

FIGURE 5.4.

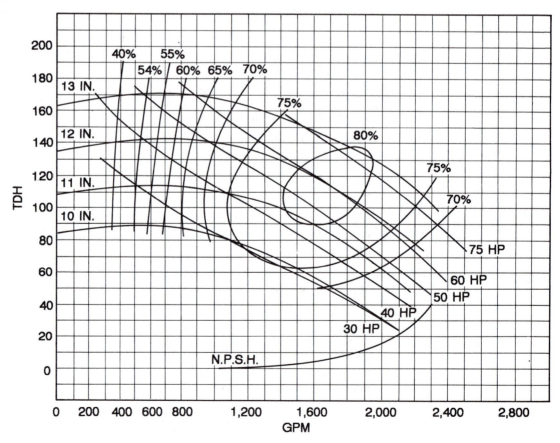

- At the intersection of the TDH and the impeller curve, the load on the motor is approximately 59 brake horsepower.
- At the intersection of the TDH and the impeller curve, the pump efficiency is 80%.
- Calculate brake horsepower.

$$bhp = \left[\frac{gpm \times TDH}{3,960 \times eff_p} \right]$$

$$bhp = \left[\frac{1,500 \times 124 \text{ feet}}{3,960 \times 0.80} \right]$$

$$bhp = 58.7$$

How to Make Velocity Head Corrections

If the inlet and outlet pipes are the same size as they are in the examples above, the velocity head is equal across the pump and there is no correction required. Even with different size piping, if the pump is properly sized in the normal operating range, the velocity head value is usually very small as compared to the static head and it makes little difference in the calculations. Therefore, when using the pump as a flow meter, the suction and discharge static heads are read and used as the only value to calculate total dynamic head. If however, corrections are required, some pump manufacturers have a table (Table 5.1) for velocity corrections. If this table isn't available, use Eq. 5.2 and 5.3:

Equation 5.2: Velocity head

$$H_V = \frac{V_0^2 - V_1^2}{2\,g}$$

Equation 5.3: Velocity

$$V = \frac{0.408\ gpm}{d^2}$$

TABLE 5.1.   VELOCITY CORRECTION TABLE

Disch size	Suct size	Velocity correction in feet of head versus gpm														
		0.1 ft.	0.5 ft.	1 ft.	2 ft.	3 ft.	4 ft.	5 ft.	6 ft.	7 ft.	8 ft.	9 ft.	10 ft.	15 ft.	20 ft.	25 ft.
2.5 in.	2.5 in.						No Correction									
2.5 in.	3 in.	50	110	160	220	270	310	350	390	420	445	475	500	620	720	800
2.5 in.	4 in.	42	90	130	180	220	260	285	315	340	360	385	405	500	580	650
2.5 in.	5 in.	39	85	120	170	210	240	270	295	315	340	360	380	460	540	600

Disch size	Suct size	Velocity Correction in Feet of Head Versus gpm							
		0.1 ft.	0.5 ft.	1 ft.	2 ft.	4 ft.	10 ft.	15 ft.	25 ft.
8 in.	8 in.				No Correction				
8 in.	10 in.	520	1150	1600	2300	3300	5200	6400	8400
8 in.	12 in.	440	1000	1400	2000	2800	4400	5400	7000
8 in.	14 in.	420	950	1320	1900	2700	4200	5200	6800

H_V = velocity head in feet of water
V_o = outlet pipe velocity of water in feet per second
V_i = inlet pipe velocity of water in feet per second
g = acceleration due to gravity, 32.2 feet per second squared
V = velocity in feet per second
0.408 = constant
gpm = water flow in gallons per minute
d = inside diameter of the water pipe

Example 5.4

1,500 gpm
inlet pipe diameter = 10 in.
outlet pipe diameter = 8 in.
Suction pressure = 63 psig
Discharge pressure = 22 psig
Static rise = 41 psi (63 − 22)
Static head = 94.71 ft. (41 × 2.31)
Velocity head = 0.84 ft.
Total dynamic head = 95.55 ft.

$$V = \frac{0.408 \text{ gpm}}{d^2}$$

$$V = \frac{0.408(1,500)}{10^2}$$

$$V = 6.12 \, fps$$

$$V = \frac{0.408 \text{gpm}}{d^2}$$

$$V = \frac{0.408(1,500)}{8^2}$$

$$V = 9.56 \, fps$$

$$H_V = \frac{V_o^2 - V_i^2}{2g}$$

$$H_V = \frac{(9.56)^2 - (6.12)^2}{64.4}$$

$$H_V = \frac{91.39 - 37.45}{64.4}$$

$$H_V = 0.84 \text{ ft.}$$

HOW TO VERIFY WATER FLOW PERFORMANCE

How to Determine Water Flow Using Temperatures

Using temperatures to verify system fluid flow performance is at best only an approximation. Extreme care must be used when taking temperature readings:

- Instruments must be calibrated and used according to the manufacturer's instructions.
- Metal surfaces to be measured must be clean. Water temperatures measured in insertion wells are more accurate than surface temperatures.
- Measure temperatures as close to the heat exchanger as possible.
- Allow enough time for the instruments to register properly. Be patient.
- To determine the water flow, use Eq. 5.4.

Equation 5.4

$$\text{gpm} = \frac{\text{Btuh}}{500 \times \Delta T_w}$$

gpm = water volume in gallons per minute
Btuh = Btu per hour
500 = constant, 60 min/hour × 8.33 lbs/gallon × 1 Btu/lb °F
ΔT_w = temperature difference between the entering and leaving water

Example 5.5: A chilled water coil has a measured 5,500 cfm of air flowing through it. The entering air temperature is 76°F dry bulb, 65°F wet bulb and the leaving air temperature is 60°F dry bulb, 58°F wet bulb. The entering water temperature is measured at 45°F and the leaving water temperature is 57°F. The water flow through the coil is calculated at 20.4 gallons per minute.

$$\text{Btuh} = \text{cfm} \times 4.5 \times \Delta h$$
$$65 \text{ FWB} = 30.06 \text{ Btu per pound}$$
$$58 \text{ FWB} = 25.12 \text{ Btu per pound}$$
$$\text{Btuh} = 5,500 \times 4.5 \times 4.94$$
$$\text{Btuh} = 122,265$$

$$\text{gpm} = \frac{\text{Btuh}}{500 \times \Delta T_w}$$

$$\text{gpm} = \frac{122,265}{500 \times 12}$$

$$\text{gpm} = 20.4$$

Example 5.6: A hot water coil has a measured 5,500 cfm of air flowing through it. The entering air temperature is 76°F dry bulb and the leaving air temperature is 90°F dry bulb. The entering water temperature is measured at 180° F and the leaving water temperature is 160°F. The water flow through the coil is calculated at 8.3 gallons per minute.

$$\text{Btuh} = \text{cfm} \times 1.08 \times \Delta T$$
$$\text{Btuh} = 5,500 \times 1.08 \times 14$$
$$\text{Btuh} = 83,160$$

$$\text{gpm} = \frac{\text{Btuh}}{500 \times \Delta T_w}$$

$$\text{gpm} = \frac{83,160}{500 \times 20}$$

$$\text{gpm} = 8.3$$

How to Determine Water Flow Using Meters

Flow meters such as venturis, orifice plates, annular flow meters and calibrated balancing valves are permanently installed devices used to measure water flow through pumps, heat exchangers and piping. In order for flow meters to provide accurate and reliable readings, they must be installed away from any source of water flow disturbance. The manufacturers of the various flow meters specify the lengths of straight pipe upstream and downstream of their flow meter needed to obtain reliable readings. The pipe lengths will vary with the type and size of flow meter, but typical specifications are between 5 to 25 pipe diameters upstream and 2 to 5 pipe diameters downstream of the flow meter.

A pressure drop across the flow meter is used along a capacity curve to determine water flow. The pressure drop is usually measured with a differential pressure gauge. A flow capacity curve and/or slide rule is supplied with each flow meter. Care must be taken when reading capacity curves as they are logarithmic and can be misleading if close attention is not paid to the numbering on the curves. With most types of flow meters (venturis, orifices plates and annular flow meters) a balancing valve is also required to set the flow. Some other types of water flow sensors and flow meters are: Doppler effect meters, Pitot tubes and manometers, magnetic flow meters, vortex shedding meters and turbine meters.

The Venturi Flow Meter

As water passes through the venturi flow meter (Fig. 5.5) the venturi creates turbulence and a measurable amount of friction loss. A pressure drop is created as a result of a change in the water's velocity. The pressure drop is measured with a differential pressure gauge. A capacity curve (Fig. 5.6) shows flow rate in gallons per minute (gpm) versus measured pressure drop in inches of water column (in. wc).

FIGURE 5.5.   THE VENTURI FLOW METER

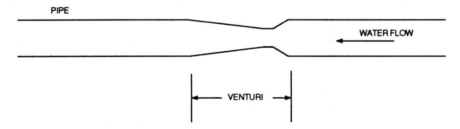

Example 5.6: A 2 inch water branch has a venturi installed. The nameplate on the venturi is stamped 2 inches - 400. The pressure drop across the venturi is measured at 50 inches with a differential gauge.

- Enter the curve (Fig. 5.6) on the left side at 50 inches.
- Draw a line to the right to intersect the correct venturi size (2 inches − 400).
- From this intersection draw a line down to the gallons per minute scale at the bottom of the curve.
- The flow through the venturi is 40 gpm.

FIGURE 5.6.   VENTURI CAPACITY CURVE

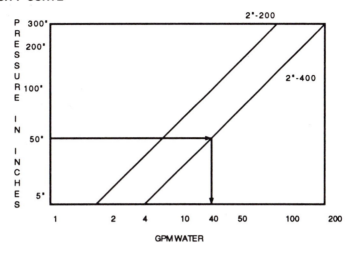

The Orifice Plate

An orifice plate is a fixed circular opening in the pipe. A measurable pressure loss is created as the water passes from the larger diameter pipe through the smaller diameter orifice. A capacity curve which shows flow rate in gallons per minute versus measured pressure drop is furnished with the orifice plate. A water gauge, such as a differential pressure gauge, is connected to the pressure taps and flow is read. The capacity curve is read in the same manner as the venturi capacity curve.

The Annular Flow Meters

The annular flow meter is a multiported flow sensor installed in a housing in the water pipe. The ports in the sensor are spaced to represent equal annular areas of the pipe in the same manner as the Pitot tube traverse is spaced for a round air duct. The upstream ports sense high pressure and the downstream ports sense low pressure. The resulting differential pressure is measured with a differential pressure gauge. A capacity curve (Fig. 5.7) which

FIGURE 5.7.   ANNULAR CAPACITY CURVE

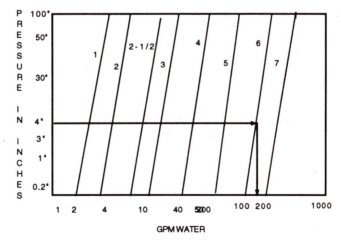

shows flow rate in gallons per minute versus measured pressure drop is provided with the flow meter.

Example 5.7: A 6 inch pipe has an annular flow meter installed. The measured pressure drop is 4 inches.

- Enter the curve (Fig. 5.7) on the left side at 4 inches.
- Draw a line to the right to intersect the correct pipe size (6 inches).
- From this intersection draw a line down to the gallons per minute scale at the bottom of the curve.
- The flow through the annular meter is 200 gpm.

How to Use Calibrated Balancing Valves

Calibrated balancing valves are a combination of a flow meter and a balancing valve. Calibrated balancing valves are similar to ordinary balancing valves except that they have a pressure tap in the inlet and outlet of the valve. Also, the flow meter has been calibrated by the manufacturer by measuring the resistance at various valve positions against known flow quantities. A capacity curve (Fig. 5.8) which shows flow rate in gallons per minute versus measured pressure drop is provided with the valve. Pressure drop is measured with any appropriate differential gauge. Calibrated balancing valves have a graduated scale or dial to show the degree the valve is open.

Example 5.8: A 1 inch pipe has a calibrated balancing valve installed. The measured pressure drop is 10 feet of water column. The valve setting is 12°.

- Enter on the left side in the lower section of the capacity curve (Fig. 5.8) at 10 feet pressure drop.
- Draw a line to the right.

FIGURE 5.8.   CALIBRATED BALANCING VALVE CAPACITY CURVE

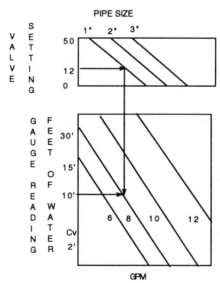

- Enter on the left side in the upper section of the capacity curve at 12°.
- Draw a line to the right to intersect the correct pipe size (1 inch).
- At the intersection of the pipe size and the degree setting, draw a line vertically down to intersect the pressure drop line.
- Read the gallons per minute curve.
- The flow through the calibrated balancing valve is 8 gpm.

How to Determine Water Flow Using the Valve Flow Coefficient

It is possible to determine the approximate flow through an automatic control valve if (1) the flow coefficient of the valve is known and (2) the valve has taps on the entering and leaving sides to take pressure drop across the valve. The valve flow coefficient is defined as the flow rate in gallons per minute that will cause a pressure drop of 1 pound per square inch across a wide open valve. Use Eq. 5.5 to determine flow rate. To adjust the flow rate, adjust the manual balancing valve. Use Eq. 5.6 to determine manual balancing valve position from pressure drop.

Equation 5.5

$$\text{gpm} = C_V\sqrt{\Delta P}$$

Equation 5.6

$$\Delta P = \left[\frac{\text{gpm}}{C_V}\right]^2$$

gpm = water flow rate in gallons per minute
C_V = valve flow coefficient
$\sqrt{\Delta P}$ = square root of the pressure drop across the valve in psi
ΔP = pressure drop across the valve in psi

Example 5.8: An automatic control valve (Fig. 5.9) has a flow coefficient rating of 10 (at 1 psi pressure drop through the wide open valve the flow rate is 10 gpm). The measured pressure drop is 1.5 psi. The calculated flow rate is 12.2 gpm.

$$\text{gpm} = C_V\sqrt{\Delta P}$$
$$\text{gpm} = 10\sqrt{1.5}$$
$$\text{gpm} = 12.2$$

Example 5.9: Set the flow through the automatic control valve for 14 gpm. Open the manual balancing valve until a 2 psi drop is read on the differential pressure gauge.

$$\Delta P = \left[\frac{\text{gpm}}{C_V}\right]^2$$

$$\Delta P = \left[\frac{14}{10}\right]^2$$

$$\Delta P = 1.96 \; or \; 2 \text{ psi}$$

FIGURE 5.9.   READING PRESSURE DROP ACROSS AN AUTOMATIC CONTROL VALVE TO OBTAIN GPM

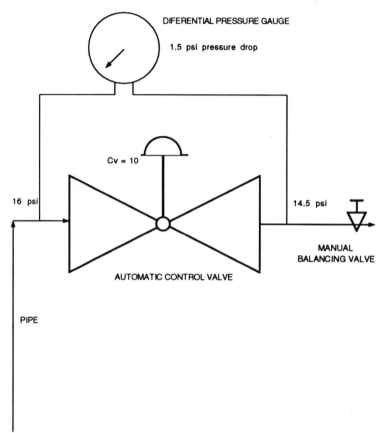

How to Determine Water Flow Using Rated Coils

It is sometimes possible to determine the approximate flow through a coil or heat exchanger if:

- The coil is new or in almost new condition. When this is the case, the coil is the same as any other flow meter with a known pressure drop.

- The coil has a manufacturer's actual tested and rated resistance to a tested and rated flow volume.

- The coil has properly located pressure taps for measuring differential pressure. Use Eq. 5.7 to determine flow rate.

Equation 5.7

$$\mathrm{gpm}_C = \mathrm{gpm}_R \sqrt{\frac{\Delta P_M}{\Delta P_R}}$$

gpm_C = calculated flow rate
gpm_R = rated flow rate
ΔP_M = measured pressure drop
ΔP_R = rated pressure drop

Example 5.10: A coil is rated at 35 gallons per minute with 6 feet of pressure drop. The measured pressure drop is 7 feet. The calculated flow rate is 37.8 gpm.

$$\text{gpm}_C = \text{gpm}_R \sqrt{\frac{\Delta P_M}{\Delta P_R}}$$

$$\text{gpm}_C = 35 \sqrt{\frac{7}{6}}$$

$$\text{gpm}_C = 37.8$$

Example 5.11: A coil is rated at 35 gallons per minute with 6 feet of pressure drop. The measured pressure drop is 5 feet. The calculated flow rate is 32 gpm.

$$\text{gpm}_C = \text{gpm}_R \sqrt{\frac{\Delta P_M}{\Delta P_R}}$$

$$\text{gpm}_C = 35 \sqrt{\frac{5}{6}}$$

$$\text{gpm}_C = 32$$

CHAPTER 6 Central Water System Components—Pumps

In this chapter you will learn about pumps and pump characteristics, including multiple pump arrangements. You will learn how to calculate pump power and pump efficiency. You will learn how to use the pump laws to determine system performance and how to use pump performance curves to predict pump performance. Finally, in this chapter, you will learn how to plot a system curve and determine the operating point of pumps in series and parallel. By using the pump laws, the pump performance curve and the pump system curve, you can calculate and graphically depict any change to the pump impeller diameter or the pipe system. By using the pump curve, you can predict the performance of a pump from available test data. Although pump curves can be used for troubleshooting, pump performance (as determined by field measurement), may be more or less than shown on the pump curve because of inaccuracies in measurement and/or improper locations of test points.

CHARACTERISTICS OF HVAC CENTRIFUGAL PUMPS

A typical HVAC centrifugal pump will have a volute (spiral) casing with one or more closed, backward curved radial flow impellers. If the pump has one impeller it is a single stage pump. If it has two or more impellers in series on a common shaft it is a multistage pump. The inlet to the pump may be on just one side. This is a single inlet pump. If the inlet to the pump is on both sides of the pump it is a double suction inlet pump. The suction inlet pipe may be the same size or larger than the discharge pipe. When the inlet is larger than the discharge there is a velocity head component in the calculation of total dynamic head. (See Chap. 5.)

Most HVAC water pumps are constant volume machines and are coupled directly (direct drive) to a constant speed motor. Some direct drive pumps are driven by a variable frequency-variable speed motor. Varying the speed of the motor changes the speed of the pump. Varying the pump speed makes the pump variable volume. Some pumps are belt driven. These pumps may be either constant speed and volume or variable speed and volume.

HOW A PUMP OPERATES

An HVAC centrifugal pump is a power-driven machine that is used to overcome system resistance and produce required water flow. As the pump impeller is rotated, centrifugal force throws the water outward from the impeller. The centrifugal force and other design characteristics reduce the pressure (a partial vacuum is created) at the inlet of the impeller and allow more water to be forced in through the pump suction opening by atmospheric or external pressure. This makes the pump's discharge pressure higher than the pump's suction pressure. After the water enters the pump's suction opening, there's a further reduction of pressure between this opening and the inlet of the impeller. The lowest pressure in the system is at the pump inlet.

The water leaves the impeller at a relatively high velocity. Then, in the pump casing, the velocity is reduced and converted into static pressure. The size of the pump impeller and its rotational speed determines the static head pressure developed by the pump. Equation 6.1 is used to find approximate head developed by the pump. Equation 6.2 can be used to determine approximate impeller diameter.

Equation 6.1: Approximate head developed by the pump.

$$h = \left[\frac{d \times \text{rpm}}{1,840} \right]^2$$

Equation 6.2: Approximate pump impeller diameter.

$$d = \frac{1,840 \sqrt{h}}{\text{rpm}}$$

h = pressure in feet of head developed by the pump
d = diameter of the pump impeller in inches
rpm = impeller speed
1,840 = constant

Example 6.1: A pump has an 11 inch impeller. The pump speed is 1,750 rpm. The approximate head developed by this pump is 109 feet.

$$h = \left[\frac{d \times \text{rpm}}{1,840} \right]^2$$

$$h = \left[\frac{11 \text{ in.} \times 1,750}{1,840} \right]^2$$

$$h = 109 \text{ ft.}$$

HOW NET POSITIVE SUCTION HEAD AFFECTS PERFORMANCE

One of the factors that can significantly affect the pump performance is not having enough "net positive suction head" at the pump inlet. The net positive suction head is the minimum suction pressure needed at the pump inlet to overcome design and installations considerations such as:

- The pump's internal losses.
- Piping and component friction and dynamic losses.
- Water or other fluid vapor pressure at the fluid temperature.
- Elevation of the suction supply source.
- Altitude of the system installation.

Net positive suction head is further categorized as net positive suction head required (*NPSHR*) and net positive suction head available (*NPSHA*).

Net positive suction head required (*NPSHR*) is a characteristic of the pump design and is the actual absolute pressure needed to overcome the pump's internal losses and allow the pump to operate properly. NPSHR is a fixed value for a given capacity. The NPSHR is based on the velocity and friction at the pump inlet. It varies directly with pump capacity and speed change, but does not vary with altitude or temperature. Pump NPSHR is published by the manufacturer and is found on the pump performance curve, submittal data and in catalogs. A pump curve will give the full range of NPSHR values for each impeller size and capacity.

Net positive suction head available (*NPSHA*) is a characteristic of the piping system. The available net positive suction head depends on the elevation of the suction supply in relation to the pump, friction loss in the pipe, water vapor pressure (Table 6.1) and pressure on the suction supply. The NPSHA at the pump inlet should exceed the NPSHR of the pump by a margin of at least two feet. The NPSHA varies inversely with the pump's capacity and speed because of the increased friction losses in the suction piping. To calculate NPSHA use Eq. 6.3.

TABLE 6.1.   WATER PROPERTY TABLE

Temp. (°F)	Den.	Wt	VP	SG
50	62.38	8.34	0.41	1.002
60	62.35	8.33	0.59	1.001
70	62.27	8.32	0.84	1.000
80	62.19	8.31	1.17	0.998
90	61.11	8.30	1.62	0.997
100	62.00	8.29	2.20	0.995
110	61.84	8.27	2.96	0.993
120	61.73	8.25	3.95	0.990
130	61.54	8.23	5.20	0.988
140	61.40	8.21	6.78	0.985
150	61.20	8.18	8.74	0.982
160	61.01	8.16	11.20	0.979
170	60.00	8.12	14.20	0.975
180	60.57	8.10	17.85	0.972
190	60.35	8.07	22.30	0.968
200	60.13	8.04	27.60	0.965
210	59.88	8.00	34.00	0.961

Temp. = temperature, °F
Den. = density, pounds per cubic foot
WT = weight, pounds per gallon
VP = vapor pressure, feet of water
SG = specific gravity

Equation 6.3: Net positive suction head available.

$$NPSHA = P \pm H_S + VH - VP_A$$

NPSHA = net positive suction head available expressed in feet
  P = atmospheric pressure, at the elevation of the installation, expressed in feet
  H_S = suction head (pressure or vacuum) corrected to pump centerline and expressed in feet. If H_S is above atmosphere it is added, and if it is below atmosphere it is subtracted.
  VH = velocity head of the water at the point of measurement of H_S, expressed in feet
  VP_A = absolute vapor pressure of the liquid at the pumping temperature, expressed in feet

Example 6.2: The suction pressure of a hot water pump located at sea level is 14 psig. The velocity of the water is 4 feet per second. The temperature of the water is 180°F. The NPSHA is 48.6 feet.

$$NPSHA = P \pm H_S + VH - VP_A$$

$$NPSHA = 33.9 \text{ feet} + 32.34 \text{ feet} + 0.25 \text{ feet} - 17.85 \text{ feet}$$

$$NPSHA = 48.6 \text{ feet}$$
  P = 33.9 feet (atmospheric pressure for sea level from altitude table)
  H_S = 32.34 feet (14 psig × 2.31 feet/psi)
  HV = 0.25 feet (from velocity head table or using the velocity head equation)

$$HV = \frac{V^2}{2g} \qquad HV = \frac{4^2}{2 \times 32.2}$$

  VP_A = 17.85 feet (from vapor pressure Table 6.1)

WATER HORSEPOWER

Water horsepower (whp) is the theoretical horsepower required to drive a pump (Eq. 6.4). It assumes 100% efficiency.

Equation 6.4: Water horsepower.

$$whp = \frac{gpm \times h \times SG}{3,960}$$

whp = water horsepower
gpm = gallons per minute
  h = pressure (head) against which the pump operates, in feet of water
  SG = specific gravity. For water temperatures between freezing (0°C, 32°F) and boiling (100°C, 212°F) a specific gravity of 1.0 is used and is therefore dropped from the equations for brake horsepower and efficiency.
3,960 = constant, 33,000 ft-lb/min divided by 8.33 lb/gal

However, since pumps are not 100% efficient, the actual power required to drive a pump is its brake horsepower (bhp). Use Eq. 6.5 to calculate brake horsepower. Use 0.70 as an efficiency to determine brake horsepower when pump efficiency is unknown.

Equation 6.5: Pump brake horsepower.

$$bhp = \frac{gpm \times TDH}{3,960 \times eff_p}$$

bhp = brake horsepower
TDH = total dynamic head, against which the pump operates, in feet of water
3,960 = constant, 33,000 ft-lb/min divided by 8.33 lb/gal
eff_P = pump efficiency, percent

Example 6.3: A pump is operating at 1,600 gpm, 280 feet TDH and 83% efficiency. The calculated brake horsepower is 136 bhp.

$$bhp = \frac{gpm \times TDH}{3,960 \times eff_P}$$

$$bhp = \frac{1,600 \times 280ft.}{3,960 \times .83}$$

$$bhp = 136$$

HOW TO DETERMINE PUMP EFFICIENCY

Efficiency is useful energy output divided by the power input. Pump efficiencies can be determined from Eq. 6.6.

Equation 6.6: Pump efficiency.

$$eff_P = \frac{gpm \times TDH}{3,960 \times bhp}$$

HOW TO USE PUMP LAWS TO PREDICT PERFORMANCE

The performance of a pump can be predicted from the pump laws. The pump laws predict:

- water volume in gallons per minute (gpm)
- total dynamic head (TDH)
- brake horsepower (bhp) at varying pump speeds (rpm)
- impeller diameters (d).

Simply stated, the pump laws are as follows:

1. **a.** Water volume (gpm) varies in direct proportion to pump speed (rpm).
   b. Water volume (gpm) varies in direct proportion to impeller diameter (d).

$$\frac{\text{gpm}_2}{\text{gpm}_1} = \frac{\text{rpm}_2}{\text{rpm}_1}$$

$$\frac{\text{gpm}_2}{\text{gpm}_1} = \frac{d_2}{d_1}$$

2. **a.** Pressure [for most calculations total dynamic head (TDH) is used] varies as the square of the pump speed (rpm).

   b. Total dynamic head (TDH) varies as the square of the impeller diameter (d).

   c. Total dynamic head (TDH) varies as the square of the water volume (gpm).

$$\frac{TDH_2}{TDH_1} = \left[\frac{\text{rpm}_2}{\text{rpm}_1}\right]^2$$

$$\frac{TDH_2}{TDH_1} = \left[\frac{d_2}{d_1}\right]^2$$

$$\frac{TDH_2}{TDH_1} = \left[\frac{\text{gpm}_2}{\text{gpm}_1}\right]^2$$

3. **a.** Brake horsepower (bhp) varies as the cube of the pump speed (rpm).

   b. Brake horsepower (bhp) varies as the cube of the impeller diameter (d).

   c. Brake horsepower (bhp) varies as the cube of the water volume (gpm).

   d. Brake horsepower (bhp) varies as the square root of the total dynamic head (TDH) cubed.

   e. Brake horsepower (bhp) varies as the total dynamic head (TDH) to the 1.5 power.

$$\frac{\text{bhp}_2}{\text{bhp}_1} = \left[\frac{\text{rpm}_2}{\text{rpm}_1}\right]^3$$

$$\frac{\text{bhp}_2}{\text{bhp}_1} = \left[\frac{d_2}{d_1}\right]^3$$

$$\frac{\text{bhp}_2}{\text{bhp}_1} = \left[\frac{\text{gpm}_2}{\text{gpm}_1}\right]^3$$

$$\frac{\text{bhp}_2}{\text{bhp}_1} = \sqrt{\left[\frac{TDH_2}{TDH_1}\right]^3}$$

gpm_1 = initial water volume
gpm_2 = final water volume
rpm_1 = initial pump speed
rpm_2 = final pump speed
d_1 = initial impeller diameter
d_2 = final impeller diameter
TDH_1 = initial total dynamic head
TDH_2 = final total dynamic head
bhp_1 = initial brake horsepower
bhp_2 = final brake horsepower

Example 6.4: A pump is operating at 1,150 rpm, 40 gpm of 100 feet TDH and 14 brake horsepower. The impeller is 15 inches in diameter. The pump capacity is reduced to 340 gpm by trimming the impeller. Calculate new impeller size, new total dynamic head and new brake horsepower requirements.

The new impeller size is 12.75 inches. The new TDH is 72 feet. The new brake horsepower is 8.6.

$$\frac{gpm_2}{gpm_1} = \frac{d_2}{d_1}$$

$$d_2 = d_1 \times \frac{gpm_2}{gpm_1}$$

$$d_2 = 15 \times \frac{340}{400}$$

$$d_2 = 12.75 \text{ in.}$$

$$TDH_2 = TDH_1 \times \left[\frac{gpm_2}{gpm_1}\right]^2$$

$$TDH_2 = 100 \times \left[\frac{340}{400}\right]^2$$

$$TDH_2 = 72.25 \text{ ft.}$$

$$bhp_2 = bhp_1 \times \left[\frac{gpm_2}{gpm_1}\right]^3$$

$$bhp_2 = 14 \times \left[\frac{340}{400}\right]^3$$

$$bhp_2 = 8.6$$

Example 6.5: A pump is operating with the motor at 40 brake horsepower. The diameter of the impeller is 10 inches. Calculate the new impeller size to bring the brake horsepower down to 35. The new impeller size is 9.56 inches.

$$d_2 = d_1 \times \sqrt[3]{\frac{bhp_2}{bhp_1}}$$

$$d_2 = 10 \times \sqrt[3]{\frac{35}{40}}$$

$$d_2 = 9.56 \text{ in.}$$

Example 6.6: A variable speed pump is designed for 700 gpm at 1,750 rpm. The measured speed is 1,200 rpm. Calculate the gpm. The pump is operating at 480 gpm.

$$gpm_2 = gpm_1 \times \frac{rpm_2}{rpm_1}$$

$$gpm_2 = 700 \times \frac{1,200}{1,750}$$

$$gpm_2 = 480$$

PUMP CURVE

A pump curve (Fig. 6.1) is developed by the pump manufacturer and is a graphic representation of the performance of a specific pump. The curve is for a constant speed. Most curves will show:

- pressure (TDH, feet of water)
- flow (gpm)
- pump efficiency (%)
- horsepower/brake horsepower (hp/bhp)
- pump inlet and discharge size (inches)
- speed (rpm)
- maximum and minimum impeller diameter (inches)
- net positive suction head required (NPSHR, feet of water).

Pressure is indicated on the vertical scale (*y*-axis) and flow (gpm) is on the horizontal scale (*x*-axis). Pump efficiency is indicated on the vertical curves. Selected horsepower (and operating brake horsepower) are the diagonal lines running top left to bottom right (NW to SE). Net positive suction head (NPSH) is along the top to the right. At the top of the graph are the pump inlet and discharge sizes (in this example; 8 inch inlet and 6 inch discharge), minimum and maximum impeller diameter, and motor/pump speed. Notice that pump curves for direct drive pumps are referenced to impeller size since the pump speed stays constant with motor speed. For HVAC pumps, the pressure in feet of head is normally plotted against flow rate in gallons per minute as shown in Fig. 6.1. This type of pressure-capacity curve is used because it gives a general description of pump operation without

FIGURE 6.1.   PUMP CURVES

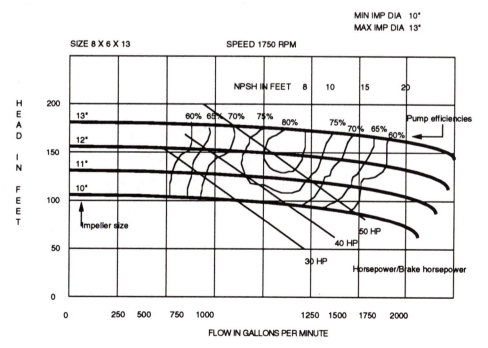

being affected by water temperature or density. Some pump curves however, are for a specific fluid, water temperature, or density. Pressure versus flow is shown on these curves as "psi versus gpm" or "psi versus flow in pounds per hour."

HOW TO USE A SYSTEM CURVE

In a fixed piping system the pressure (*TDH*) changes as the square of the flow (gpm). A system curve is a plot of the change in pressure versus the change in flow. The system curve can be used to analyze pump operation and to identify problem areas associated with friction and dynamic losses in the piping system. To construct the system curve, determine the gpm and total dynamic head from field measurements. Plot this point (the operating point) on the pump curve. Then, any other point on the system curve can be found by using the pump law which states that total dynamic head varies as the square of the gpm.

Example 6.7: In Fig. 6.2, the pump is operating at 1,000 gpm at 100 feet TDH.

Point A is 500 gpm

Point B is 750 gpm

Point C is 1,250 gpm

To plot the system curve use the following equation:

$$\frac{TDH_2}{TDH_1} = \left[\frac{\text{gpm}_2}{\text{gpm}_1}\right]^2$$

FIGURE 6.2.   THE SYSTEM CURVE

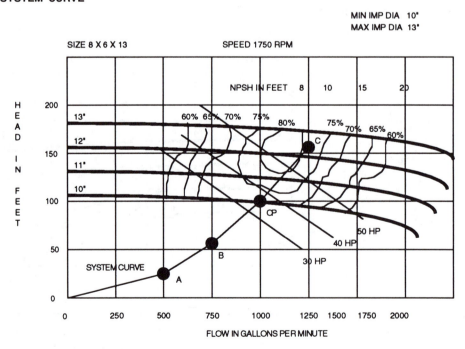

Point A

$$TDH_2 = TDH_1 \times \left[\frac{gpm_2}{gpm_1}\right]^2$$

$$TDH_2 = 100 \times \left[\frac{500}{1,000}\right]^2$$

$$TDH_2 = 25 \text{ ft.}$$

Point B

$$TDH_2 = 100 \times \left[\frac{750}{1,000}\right]^2$$

$$TDH_2 = 56 \text{ ft.}$$

Point C

$$TDH_2 = 100 \times \left[\frac{1,250}{1,000}\right]^2$$

$$TDH_2 = 156 \text{ ft.}$$

A system curve shows the pressure required to overcome the friction and dynamic losses in the water distribution system in order to deliver various water quantities. A water distribution system operates only along its system curve. If there are no changes in the system resistance (closing or opening of valves, changes in the condition of the coils, etc.) an increase or decrease in system resistance results only from an increase or decrease in gpm. This change in resistance will fall along the system curve. If however, valves are operated toward their closed positions, coils and pipes become scaled, etc., this system curve will no longer apply and a new system curve must be plotted based on the new operating point.

HOW TO CALCULATE THE PUMP/SYSTEM OPERATING POINT

As discussed in the last section, the operating point for the pump is the intersection of the system curve and the pump performance curve. Any change to the pump or the water distribution system can be calculated and graphically depicted using the pump curve, the system curve and the pump laws. For the pump, a decrease in system resistance will mean an increase in gpm, while an increase in system resistance results in a decrease in gpm. For the system curve, the system resistance increases or decreases as the square of the water volume increases or decreases.

If the operating total dynamic head exceeds the design total dynamic head, the water volume, horsepower and net positive suction head required will be less than design. If, on the other hand, the operating head is lower than total dynamic design, the water volume, horsepower and net positive suction head required will be greater than design. To increase or decrease the water volume, a physical change must be made to either the piping system (opening or closing a valve) or the pump impeller (or speed). Any change to the pump impeller, pump speed or the distribution system can be calculated and graphically depicted using the pump laws, pump curve and system curve.

Example 6.8: A pump impeller is increased to increase volume. The operating point (OP) for the pump operating on the 10 inch curve in piping system X is shown in Fig. 6.3. A new

FIGURE 6.3.   SYSTEM CURVE

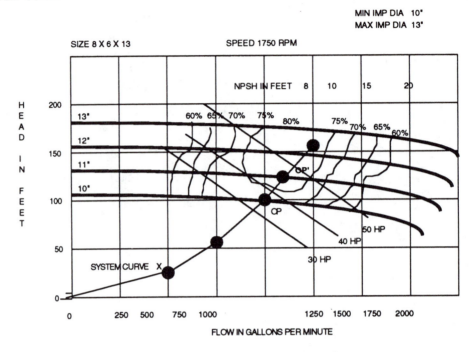

11 inch impeller is installed. The pump now operates on the new 11 inch pump curve. Since the piping system has remained unchanged, the system curve also remains unchanged. The pump is now operating at a higher gpm and higher head at OP prime.

Example 6.9: Water volume is increased by lowering the resistance in the piping system by opening a valve. The pump is now operating at a higher gpm and lower head at OP prime. A new system curve (system curve Y, Fig. 6.4) is established from the new operating point on the 10 inch pump curve.

Example 6.10: The Pump curve 1 is for a centrifugal pump. The pump is direct drive and the motor/pump speed is 1,750 rpm. Minimum impeller diameter is 11.00 inches. The maximum impeller diameter is 15.00 inches. Impeller sizes between 11 inches and 15 inches are in ⅛ inch increments. The pump outlet size is 5 inches and the inlet size is 6 inches.

- Pump Curve 1: Determine the impeller diameter. Hook up the test gauge. Turn the pump off. Close the discharge valve, turn the pump on and take the following readings.
- Discharge pressure = 98.5 psig
- Suction pressure = 20 psig
- The psi rise across the pump is 78.5 psi
- The feet of head rise across the pump is 181 ft. hd.
- The impeller size is verified to be 13 inches
- Pump Curve 1: Determine the operating point. Leave the test gauge in position. Turn the pump off. Open the discharge valve, turn the pump on and take the following readings.
- Discharge pressure = 84 psig

FIGURE 6.4.   SYSTEM CURVE Y

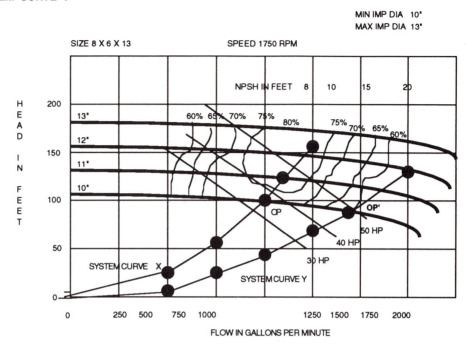

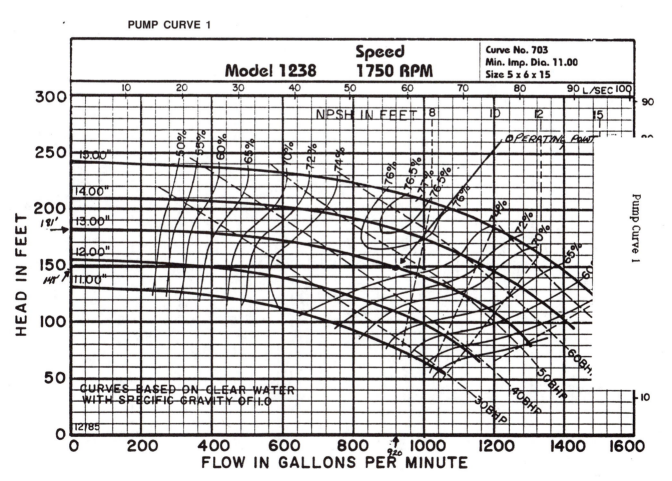

- Suction pressure = 20 psig

- The psi rise across the pump is 64 psi

- The total dynamic head (TDH) rise across the pump is 148 ft. hd.

- The pump efficiency is 0.75

- From the pump curve or by calculation the pump is operating at 45.8 bhp.

- Pump Curve 2: Plot the system curve. The arbitrarily selected points are

GPM
1,200
1,000
920
800
600
400
200
0

- Pump Curve 2: The head pressures are plotted and the system curve is drawn.

PUMP CURVE 2

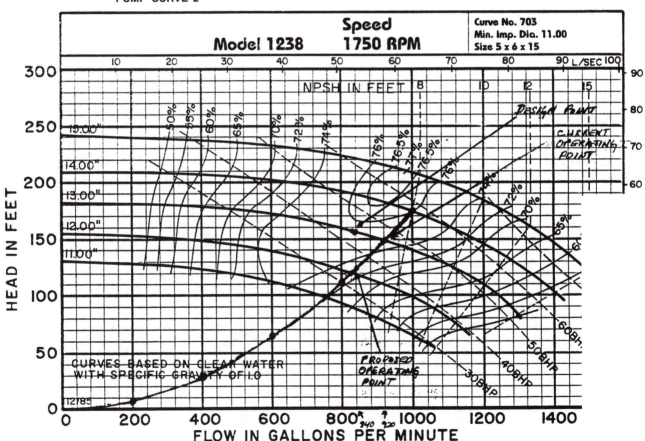

$$HD_2 = HD_1 \left[\frac{gpm_2}{gpm_1} \right]^2 \quad HD_2 = 148 \left[\frac{1,000}{920} \right]^2 \quad HD_2 = 175 \text{ ft.}$$

GPM	HD
1,000	175
920	148
840	123
800	112
600	63
400	28
200	7
0	0

- Pump Curve 2: The design point is 840 gpm at 156 feet TDH, 0.75 eff., 44.1 bhp, 13 inch impeller. The current operating point is 920 gpm at 148 feet TDH, 0.75 eff., 45.8 bhp, 13 inch impeller.

- Option 1: To reduce flow to design gpm, close the main discharge valve to impose 8 (156 − 148) feet of additional resistance. By imposing this additional resistance the pump is "backed up on its curve." Closing the main discharge valve will save approximately 1.7 bhp.

- Option 2: Reduce the impeller size to approximately 11 and seven eighths inches.*

$$D_2 = D_1 \left[\frac{gpm_2}{gpm_1} \right]$$

The calculated head is 123 feet. The calculated bhp is 34.9. Reducing impeller size saves approximately 10.9 bhp.

$$HD_2 = HD_1 \left[\frac{gpm_2}{gpm_1} \right]^2 \quad BHP_2 = BHP_1 \left[\frac{gpm_2}{gpm_1} \right]^3$$

MULTIPLE PUMP ARRANGEMENTS

Pumps may be selected to operate in series and/or in parallel. When they operate in series, the flow remains the same but the pressures and horsepowers are increased (Fig. 6.5). Compare this to pumps that operate in parallel (Fig. 6.6). With parallel pumping the pressure remains constant, while the capacities and horsepowers are increased. In addition, when pumps are operating in parallel and one pump is removed from operation, the single pump is able to deliver much of the total design flow. The gpm from the single pump actually increases from the gpm that it delivers when both pumps are operating. This can be a benefit

*Notice that when the head and bhp are plotted on the system curve the impeller size is 12 ⅛ inches (not 11 ⅞ inches) for 840 gpm, 123 feet, 34.9 bhp. This is because the curve widens out a little between 12 and 13 inches. When you have the pump curve, plot the impeller size from the curve. This is another reason for getting the pump curve. If you didn't have the curve, then you would go by the pump laws to select impeller size. The gpm produced would be approximately 825 at 118 feet and 33.7 bhp, CEFAPP. When using the curve, and depending on the difference in costs, select either a 12 inch or 12 ⅛ inch impeller to be sure that you have required gpm. The bhp savings are still approximately 10.9.

FIGURE 6.5.   PUMP OPERATING IN SERIES

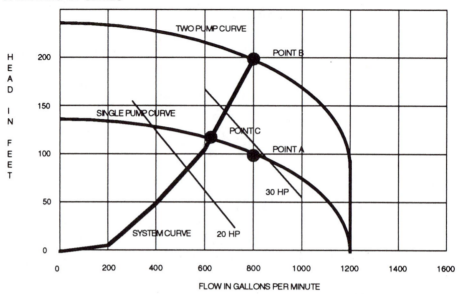

if standby protection is required, since the purpose of a standby pump is to continue the pumping operation when the normal pump goes out of service.

Use the pump and system curves to determine the flow rate of a single pump and the flow rate of both pumps in either a series arrangement or a parallel arrangement.

Example 6.11: Two pumps, each separately handling 800 gpm at 100 feet of head and 29 bhp would operate at 800 gpm at 200 feet of head and 58 bhp if placed in series (Fig. 6.5) and 1,600 gpm at 100 feet of head and 58 bhp if placed in parallel (Fig. 6.6).

Example 6.12 (Fig. 6.5.): The operating point of each pump in series (all pumps are on) is on the single pump curve (point A, 800 gpm, 100 feet TDH, 29 bhp). The design operating

FIGURE 6.6.   PUMP OPERATING IN PARALLEL

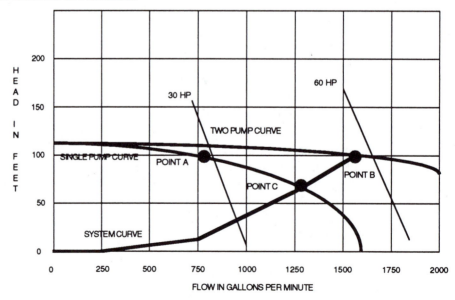

condition for two pumps in series is point B, 800 gpm, 200 feet TDH, 58 bhp. If only one pump is operating, the point of operation shifts to the intersection of the single pump curve with the system curve (point C, 625 gpm, 122 feet TDH, 27.5 bhp). Notice that the head (122 feet TDH vs. 100 feet TDH) increases while the volume (625 gpm vs. 800 gpm) and brake horsepower (29 bhp vs. 27.5 bhp) decrease. Each pump motor must be sized for its maximum horsepower. In series pumping, this will happen when both pumps are operating.

Example 6.13 (Fig. 6.6.): The operating point of each pump in parallel (all pumps are on) is on the single pump curve (point A, 800 gpm, 100 feet TDH, 29 bhp). The design operating condition for two pumps in parallel is point B, 1,600 gpm, 100 feet TDH, 58 bhp.

If only one pump is operating, the point of operation shifts to the intersection of the single pump curve with the system curve (point C, 1,300 gpm, 66 feet TDH, 31 bhp). Notice that the head (66 feet TDH vs. 100 feet TDH) decreases while the volume (1,300 gpm vs. 800 gpm) and brake horsepower (31 bhp vs. 29 bhp) increase. Each pump motor must be sized for its maximum horsepower. With parallel pumping, this will happen when only one pump is operating.

CHAPTER 7 Water Distribution Components

A typical HVAC water distribution system will generally consist of most of the following components:

- Piping system
- Water filtration
- Water flow control
- Flow meters
- Temperature measuring stations
- Pressure measuring stations
- Balancing stations
- Pressure control components
- Air control components
- Heat exchangers
- HVAC coils
- Centrifugal pumps (see Chap. 6)
- Heat conversion units (including boilers, covered in Chap. 9, and chillers, covered in Chap. 10).

PIPING SYSTEMS

Open and Closed Systems

An open pipe system is one in which there's a break in the piping circuit and the water is open to the atmosphere. An example of an open system is a water-cooled condenser and cooling tower (Fig. 7.1). A closed pipe system, such as the chilled water system in Fig. 7.2, is one in which there's no break in the piping circuit and the water is closed to the atmosphere.

FIGURE 7.1.   OPEN SYSTEM

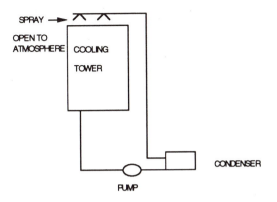

FIGURE 7.2.   CLOSED SYSTEM

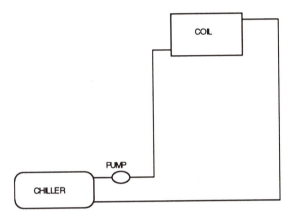

One-Pipe System

One-pipe systems are used in residences, small commercial buildings and industrial buildings. A one-pipe system (Fig. 7.3) uses a single loop main distribution pipe. Each terminal (coil) is connected by a supply and return branch pipe to the main. A diverting tee (Fig. 7.4) is installed in either the supply branch, return branch or sometimes both branches. Diverting tees are selected to create the proper amount of resistance in the main to direct water to the terminal. If the diverting device is not installed, the water circulating in the main will tend to flow through the straight run of a normal tee and not be directed into the coil. This is because the terminal has a higher pressure drop than the main. If this happens the coil is starved.

You can separately control and service the one-pipe main arrangement by installing control valves and service valves in the branches. If the system has control valves and there are too many terminals, the terminals farthest from the boiler may not receive water at a temperature that is high enough to maintain desired space temperatures. For larger hydronic systems, use two-pipe arrangements to try to maintain the water temperature to each coil equal to the boiler temperature.

Two-Pipe Direct and Reverse Return Systems

Two-pipe systems have a supply pipe and a return pipe to each coil. There are separate automatic control valves and manual service valves for each water coil. The two-pipe direct

FIGURE 7.3.   OPEN-PIPE SYSTEM

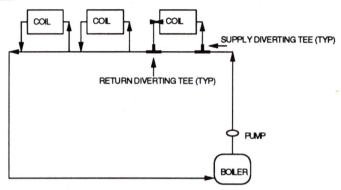

FIGURE 7.4.

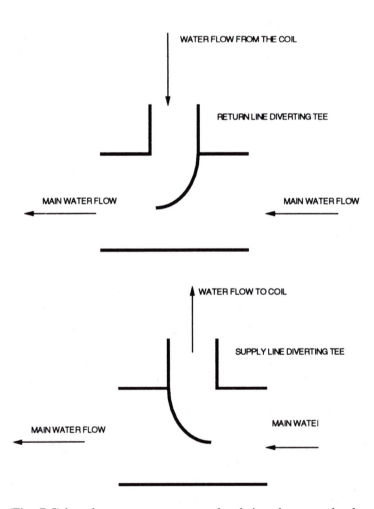

return system (Fig. 7.5) has the return water routed to bring the water back to the pump by the shortest possible route. The water coils are piped so that the first coil to be supplied is the first coil returned. With the direct return arrangement there is less main pipe needed and therefore a lower initial cost. If the system is to be water balanced, flow meters and balancing valves are required throughout the system. With two-pipe reverse return systems (Fig. 7.6), the water coils are piped so that the first coil to be supplied is the last coil returned. Reverse return systems require more piping than direct return systems and therefore have a higher initial piping cost. Since all the piping circuits are essentially the same length, the

FIGURE 7.5.   TWO-PIPE DIRECT RETURN SYSTEM

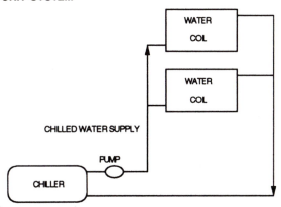

FIGURE 7.6.   TWO-PIPE REVERSE RETURN SYSTEM

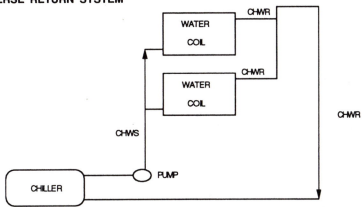

impression is that reverse return systems are "self-balancing." Reverse return systems are not self-balancing. Flow meters and balancing valves are required to water balance the system.

Three-Pipe Direct and Reverse Return Systems

Three-pipe systems (Fig. 7.7) have two supplies and one return. One supply main provides chilled water and the other supply main provides heated water. The chilled water and heated water supplies are not mixed. Each water coil has a three-way valve which switches to deliver either chilled water or heating water (but not both) to the coil. However, the return main receives water from every coil. This means that in some instances the return pipe may carry a mixture of chilled and heated water. This results in a waste of energy because the chiller receives warmer water and the boiler receives cooler water than design. Therefore, both heat conversion units must work harder to supply their proper discharge temperature water.

The return connections from the coils can be made either direct or reverse return.

Four-Pipe Direct and Reverse Return Systems

Four-pipe systems consist of two, two-pipe systems. One two-pipe system is for chilled water and the other two-pipe system is for heating water. The air handling unit (Fig. 7.8) normally has two separate water coils, one for heating water and one for chilled water.

FIGURE 7.7.   THREE-PIPE SYSTEM

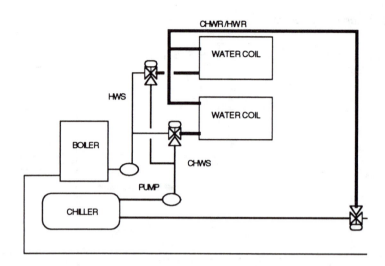

FIGURE 7.8.   FOUR-PIPE SYSTEM

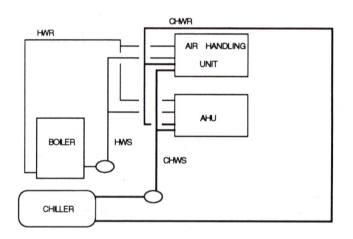

FIGURE 7.9.   FOUR-PIPE SYSTEM

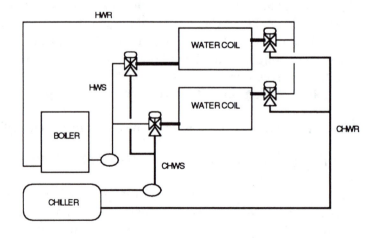

There is no mixing of the chilled and heated water. Each coil has its own automatic control valve, either two-way or three-way, and manual service valves.

Some air handling units have only one coil (Fig. 7.9). The coil is supplied with both heated and chilled water. A three-way valve is installed in the supply line to the coil. The chilled water and heated water supplies are not mixed. The three-way valve switches to deliver either chilled water or heating water (but not both) to the coil. In the return line from the coil a three-way, two-position valve diverts the leaving heated or chilled water to the correct return main. The return connections from the water coils can be made either direct or reverse return.

Primary-Secondary Piping Circuits

In primary only systems (Fig. 7.10A) only one pump is used to distribute the water. In primary-secondary systems (FIg.7.10B) the primary pump circulates the water through the primary loop and the secondary pump moves water through the secondary loop. A cross-over pipe, which may vary in length to a maximum of about two feet, is installed between the primary and secondary loop. The cross-over pipe has a negligible pressure drop which ensures the isolation of the two loops. In order to overcome the pressure loss in the secondary circuit and provide water flow to the coil, a secondary pump is installed. The water flow in the secondary loop may be less than, equal to, or greater than the flow in the primary loop (Fig. 7.11).

Example 7.1: Primary-secondary systems reduce pumping horsepower requirements and balance valve settings while increasing system control. In Fig. 7.10A, there is no primary-secondary pumping. The pump is operating at 1,825 gpm and 153 feet TDH. For a pump that is operating at 70% efficiency, this is 101 brake horsepower. The longest run is circuit B at 153 feet of pressure loss. If this system is retrofitted to a primary-secondary system (Fig. 7.10B), the longest run is changed. The longest run is now the piping out to and back from secondary circuit C. The longest run is now 83 feet. This means that the primary pump still must circulate 1,825 gpm but now at 83 feet TDH. The power needed to do this is only 55 brake horsepower. The secondary pump in circuit B is sized to circulate 450 gpm through a 100 foot pressure loss circuit. The secondary pump in circuit C is sized to circulate 575 gpm through a 45 foot pressure loss circuit. The total power requirement of the

FIGURE 7.10A.   TWO-PIPE SYSTEM

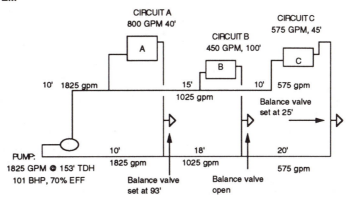

CIRCUIT "A"   10 + 40 + 10 = 60'

CIRCUIT "B"   10 + 15 + 100 + 18 + 10 = 153' (LONGEST RUN)

CIRCUIT "C"   10 + 15 + 10 + 45 + 20 + 18 + 10 = 128'

FIGURE 7.10B.   TWO-PIPE PRIMARY-SECONDARY SYSTEM

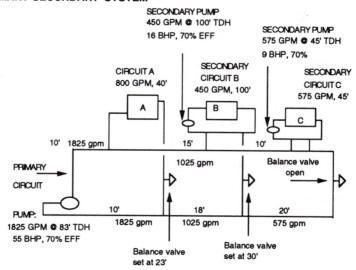

CIRCUIT "A"   10 + 40 + 10 = 60'

CIRCUIT "B"   10 + 15 + 18 + 10 = 53'

CIRCUIT "C"   10 + 15 + 10 + 20 + 18 + 10 = 83' (LONGEST RUN)

FIGURE 7.11.

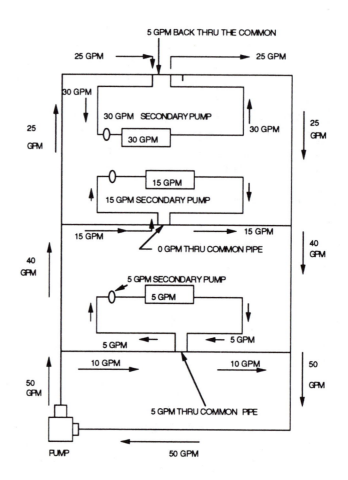

three pumps is 80 bhp (55 + 16 + 9), a savings of 21 bhp. If this retrofitted system operates 5,000 hours per year at $0.08 per kwh, the savings is over $7,000 per year versus the operating cost of the system before retrofit.

WATER FILTRATION

A strainer is a water filter. Inside the strainer body is a mesh screen. The screen is in the shape of a sleeve or basket and is designed to catch sediment or other foreign material in the water. If the system is open, the screen must be periodically removed and cleaned. If the system is closed, the screen must be removed and cleaned during startup. If the screen is not cleaned, a higher than normal pressure drop across the strainer and lower water flow will occur.

There will also be a higher than normal pressure drop when the screen has too fine a mesh. If the system is closed and the strainer has a fine mesh screen (construction screen), replace it with a larger mesh screen during startup. In addition to pump strainers, individual fine mesh strainers may also be installed before automatic control valves or spray nozzles (which operate with small clearances and require protection from materials that might pass through the pump strainer.)

To avoid cavitation, strainers placed in the suction side of the pump must be properly sized and kept clean. For instance, you may be able to remedy cavitating condenser pumps by removing the strainer altogether or moving the strainer to the pump discharge. Only use strainers where necessary to protect the components in the system. For example, in the case of cooling towers, the strainer in the tower basin may provide adequate protection and you may not need a pump strainer.

CONTROLLING WATER FLOW

Water flow is controlled through the use of various types of valves:

- Automatic and manual flow control valves and manual balancing valves are used to regulate flow rate.
- Service valves are used to isolate part or all of the system.
- Check valves are used to limit the direction of flow.

Manual Control Valves

There are three basic types of manual valves: flow control and balancing valves, service valves and check valves. Specific types are described in the following sections.

Flow Control and Balancing Valves

Ball Valves. Ball valves have a low pressure drop, good flow characteristics and are often used for water balancing.

Butterfly Valves. Butterfly valves have a low pressure drop and are sometimes used as balancing valves. However, they do not have the good throttling characteristics of ball or plug valves.

Globe Valves. Globe valves are normally used in water make-up lines. Although globe valves are sometimes used for throttling flow, they have a high pressure drop and therefore should not be used for balancing.

Combination Valves. Combination valves are also called multipurpose or triple-duty valves. These valves regulate flow and limit direction. They come in a straight or angle pattern and combine a check valve, calibrated balancing valve and shutoff valve into one casing. The valve acts as a check valve preventing backflow when the pump is off and can be closed for tight shutoff for servicing. Combination valves also have pressure taps for connecting flow gauges and reading pressure drop. A calibration chart is supplied with the valve for conversion of pressure drop to gpm for balancing. The valves generally have a memory stop.

Plug Valves. Plug valves are used primarily to balance water flow, but they are also used for shutoff. Plug valves have a low pressure drop and good throttling characteristics. Some plug valves have adjustable memory stops. The memory stop is set during the final balance. If the valve is closed for any reason it can later be reopened to the original setting.

Calibrated Balancing Valves. Calibrated balancing valves are basically plug valves with pressure taps in the valve casing at the inlet and outlet. They have also been calibrated by the manufacturer for flow versus pressure drop. A graduated scale or dial on the valve shows the degree that the valve is open. Calibration data which shows flow rate in gallons per minute (gpm) versus measured pressure drop is provided by the manufacturer (see Chap. 5).

Service Valves

Gate Valves. Gate valves are service valves used for tight shutoff to service or remove equipment. Gate valves regulate flow only to the extent that they are either fully open or fully closed. Even though gate valves have a low pressure drop, they cannot be used for throttling. The internal construction of the gate valve is such that when the plug is only partly opened, the resulting high velocity water stream will cause erosion of the valve plug and seat. The erosion of the plug and seat will allow water leakage when the valve is used for tight shutoff.

Check Valves

Check valves are installed on the discharge of the pump to prevent backflow. Check valves allow the water to flow in one direction only. The operation of check valves is such that when there is water pressure in the correct direction, the water forces the gate in the valve to open. The gate will close due to gravity (swing check valve) or spring action (spring-loaded check valve) when the system is off or when there is water pressure in the wrong direction.

Automatic Control Valves

Automatic control valves (*ACV*) can be classified according to type of construction and control:

- Construction:
  —Two-way valves—single seated and double seated.
  —Three-way valves—single-seated mixing valves and double-seated diverting valves
- Control:
  —Modulating
  —Two position

Two-way Valves

Two-way valves are used to regulate water flow that controls the heat transfer in the water terminal. They close off when heat transfer is not required and open up when heat transfer is needed. Single-seated, two-way control valves are the type most used in HVAC systems. Double-seated, two-way valves may be used when there is a high differential pressure and tight shutoff is not a requirement. The flow through double-seated valves tend to close one port while opening the other port. This design creates a balanced thrust condition which enables the valve to close off smoothly without water hammer, despite the high differential pressure.

Valves must be installed with the direction of flow opposing the closing action of the valve plug. The water pressure tends to push the valve plug open (Fig. 7.12). If the valve is installed the opposite way the valve may chatter. Chattering occurs when the valve plug (in an incorrectly installed valve) modulates to the almost full closed position (Fig. 7.13). The velocity of the water around the plug becomes very high because the area through which the water flows has been reduced. This high velocity (and resulting high velocity pressure) overcomes the spring resistance and forces the plug closed. When the plug seats flow is stopped the velocity and velocity pressure goes to zero. At this point, the spring force takes over and opens the plug. When the plug is opened (to the almost closed position) the cycle is repeated and chattering is the result.

Three-way Valves

Three-way control valves may be either single-seated (mixing valve) or double-seated (diverting valve). The single-seated, mixing valve is the most common. A mixing valve has two inlets and one outlet. A double-seated, diverting valve has one inlet and two outlets. The terms "mixing" or "diverting" do not indicate the valve application, but refer to the internal construction of the valve. The determination of which valve to use is based on

FIGURE 7.12.

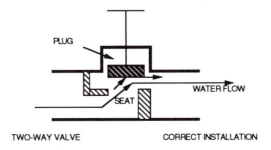

FIGURE 7.13.

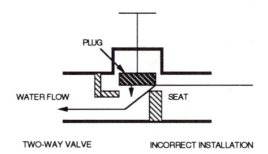

where the valve will be installed so that the plug will seat against flow. Substituting one type of valve for the other in a system (or installing either design incorrectly) will tend to cause chatter. Either valve may be installed for a flow control action (bypassing application, Fig. 7.14) or a temperature control action (mixing application, Fig. 7.15) depending on its location in the system.

Another type of modulating three-way valve is used in the supply line to coils in a three-pipe system. This valve has two inlets and one outlet. One inlet is supplied with heating water and the other is supplied with chilled water. The valve varies the quantity of heating and chilled water, but does not mix the two streams. Depending on the thermostatic controls in the occupied space, the valve opens to allow either heating water only or chilled water only into the coil.

This same type of three-way modulating valve is used in the supply line to a four-pipe, one coil system. The return line also has a three-way valve, but it is two position only. The return valve has one inlet and two outlets. Depending on temperature of the water entering the coil, the water leaving the coil enters the valve and is diverted to either the heating water return main or the chilled water return main.

FIGURE 7.14.   THREE-WAY VALVE IN A BYPASS APPLICATION

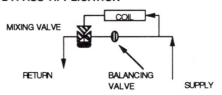

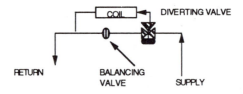

FIGURE 7.15.   THREE-WAY VALVE IN A MIXING APPLICATION

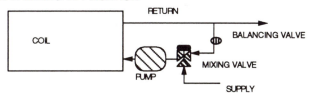

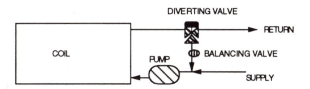

FLOW METERS

Flow metering devices (covered in Chap. 5) such as annular flow meters, orifice plate flow meters, Venturi flow meters and calibrated balancing valve flow meters are permanently installed devices used to measure flow through pumps, heat exchangers, pipes and coils. Flow meters must be installed away from any source of flow disturbance such as elbows, reducers, valves, etc., to allow the turbulence to subside and the water flow to regain uniformity. The manufacturer will specify the lengths of straight pipe upstream and downstream of the meter needed to get accurate, reliable readings. Straight pipe lengths vary with the type and size of flow meters. Typical specifications are between 5 to 25 pipe diameters upstream and 2 to 5 pipe diameters downstream of the flow meter.

Annular Flow Meters

Annular flow meters have a multiported flow sensor installed in the pipe. The holes in the sensor are spaced to represent equal annular areas of the pipe. The flow meter is designed to sense the velocity of the water as it passes the sensor. The upstream ports sense high pressure and the downstream port senses low pressure. The resulting differential pressure is measured with an appropriate differential pressure gauge. Calibration data which shows flow rate in gallons per minute (gpm) versus measured pressure drop is furnished with the flow meter.

Orifice Plate Flow Meters

Orifice plates are fixed circular openings in the pipe. The orifice is smaller than the pipe's inside diameter. A measurable "permanent" pressure loss is created as the water passes through the orifice. An abrupt change in velocity occurs causing turbulence and friction which results in a pressure drop across the orifice. Calibration data which shows flow rate in gallons per minute (gpm) versus measured pressure drop is furnished with the orifice plate. A differential pressure gauge is connected to the pressure taps and flow is read.

Venturi Flow Meters

Venturi flow meters operate on the same principle as orifice plate flow meters, but the shape of the venturi allows gradual changes in velocity. The "permanent" pressure loss is less than the loss created by an orifice plate. Calibration data which shows flow rate in gallons per minute (gpm) versus measured pressure drop is furnished with the venturi. The pressure drop is measured with a differential gauge.

Calibrated Balancing Valves

A balancing valve is needed with the venturi, orifice plate, annular flow meter and other types of flow meters. Calibrated balancing valves, however, are designed to do both duties of a flow meter and a balancing valve. The manufacturer calibrated the valve by measuring pressure drop through the valve at various positions against known flow quantities. Calibration data which shows flow rate in gallons per minute (gpm) versus measured pressure drop is provided with the valve. Pressure drop is measured with a differential gauge.

TEMPERATURE MEASURING STATIONS

Temperature measuring stations are installed at various points in the pipe. They are on the entering and leaving sides of chillers, condensers, boilers and coils. Temperature measuring stations called **thermometer wells** are sometimes installed in the pipe so that a test thermometer can be inserted into the pipe to indirectly measure the temperature of the water. The well holds a heat conducting oil. Heat from the water is conducted through the well wall into the oil and onto the thermometer. In order to make good contact with the water, temperature wells must extend far down into the pipe. Thermometer wells should be installed vertically. If this is not possible, they should be installed not more than 45° from vertical. If the well is installed below 45°, the well will be dry, air will act as an insulator and the temperature reading will be incorrect.

PRESSURE MEASURING STATIONS

Pressure test wells can also be installed in the piping on entering and leaving sides of chillers, condensers, boilers and coils. A hand held pressure probe, which is connected to a differential gauge or test gauge, is inserted into the well. In many cases the measuring station can be used for either temperatures or pressures.

BALANCING STATIONS

A balancing station is an assembly to both measure and control water flow. It has a measuring device and a volume control device. Balancing stations must be installed with the recommended lengths of straight pipe entering and leaving the station. A calibrated balancing valve is one type of balancing station.

WATER SYSTEM PRESSURE CONTROL COMPONENTS

Pressure Control Valves

Pressure Reducing Valve

Pressure reducing valves (Fig. 7.16) are installed in the make-up water pipe to the system. They reduce the pressure of city water down to the pressure needed to completely fill the system. They generally come set at 12 psi (about 28 feet, 12×2.31). This is adequate pressure for one or two story buildings. For three story or higher buildings, pressure reducing valves (*PRV*) are adjusted, causing a minimum of an additional 5 psi pressure at the highest terminal.

Example 7.2: The pump, the boiler and compression tank for an HVAC system are located on the ground floor of a building. The fan-coil unit is located on the top floor of this three story building. The coil elevation is 30 feet. The PRV should be set for 18 psi (30 feet is equal to 13 psi + 5 psi safety). To change the adjustment on the PRV, remove the cap from the top of the valve, loosen the jam nut, and turn the adjusting screw in a clockwise direction. This increases the pressure on the system. Turning the adjusting screw in a counterclockwise direction lowers the pressure.

Pressure Relief Valves

Pressure relief valves (Fig. 7.16) are safety devices installed on boilers or other equipment to protect the system and human life. Pressure relief valves come preset to open at a pressure less than the maximum pressure rating of the system.

Pressure Control Tanks

After the hydronic system is constructed it is filled with water through the city supply main (Fig. 7.16). The PRV is adjusted and the system is tested. Now the system is ready for

FIGURE 7.16.

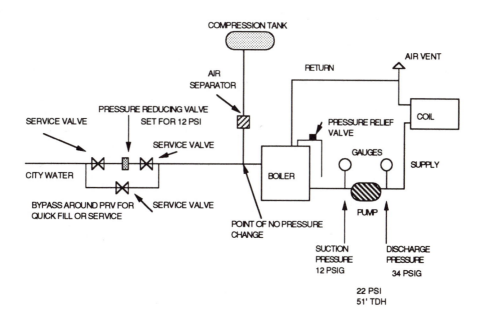

normal operation. As you probably know, water expands when heated and contracts when cooled. Let's say that the boiler is started. The water is heated and begins to expand. If this expanding water has nowhere to go, the increased pressure in the system could break a pipe or otherwise damage other equipment. Water expansion tanks are used to keep this from happening. These tanks maintain the proper pressure on the system and accommodate the fluctuations in water expansion and contraction while controlling pressure changes in the system. Expansion tanks are used in open systems. Compression tanks (Fig. 7.16) are used in closed systems.

Expansion Tank

An expansion tank is simply an open tank used in an open hydronic system to compensate for the normal expansion and contraction of the water. As the water temperature increases, the water volume in the system increases and the water in the expansion tank rises. Corrosion problems are associated with open expansion tanks as a result of the exposure to the air and evaporation and/or boiling of the water. Because of this, expansion tanks are limited to installations having operating water temperatures of 180°F or less.

Compression Tank

A compression tank is a closed vessel containing water and air or an air bladder. The tank is generally filled with water to about two-thirds full. The air in the compression tank or the bladder acts as a cushion to keep the proper pressure on the system. It accommodates the fluctuations in water volume and controls pressure changes in the system. Pressures in the hydronic system will vary from the minimum pressure required to fill the system (as described in the section on pressure reducing valves in this chapter) to the maximum allowable working pressure created by the boiler.

If the air in the compression tank leaks out, water will begin to fill the tank. This condition is called a "waterlogged" tank. Waterlogging can happen when the air leaks out of the compression tank and the pressure on the system is reduced below the setpoint on the pressure reducing valve. The PRV will then open to allow in more water to fill the tank until the setpoint on the pressure reducing valve is reached.

When the tank becomes waterlogged the fluctuations in water volume and the proper system pressures cannot be maintained. A waterlogged tank must be drained and the air leaks found and sealed. If the tank remains waterlogged when the water in the system is heated the water will expand to completely fill the tank. Since there is no longer an air cushion, and nowhere else for the water to go, every time the boiler fires the pressure relief valve on the boiler will open to spill water in order to relieve the pressure in the system. When the pressure relief valve opens and reduces the pressure in the system, the pressure reducing valve opens to bring fresh water into the system. This cycle continues. Each time fresh water comes into the system it also brings in air (see the section on air separators in this chapter).

The point where the compression tank connects to the system is called "the point of no pressure change" (Fig. 7.16). Compression tanks must be installed on the suction side of the pump. If the compression tank is installed on the discharge of the pump, the pump must operate at the pressure set by the pressure reducing valve. This may create a pressure condition at the pump inlet which will be lower than the vapor pressure of the water. If the inlet pressure is lower than the water vapor pressure, the pump will cavitate. In addition, the

FIGURE 7.17.

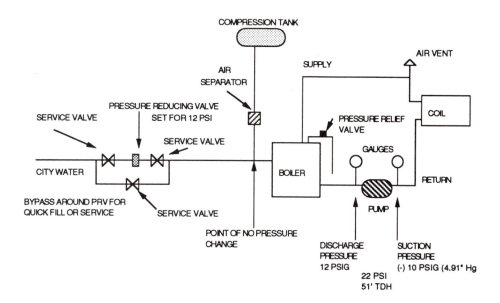

water pressure at the air vent may be negative instead of positive. This will bring air into the system and possibly create air and corrosion problems.

Example 7.3: In Fig. 7.16 the service valve is set for 12 psig. The compression tank is on the suction side of the pump. The suction gauge reads 12 psig static head. The pressure loss through the system is 51 feet TDH (22 psi). When the pump is operating the discharge pressure will be 34 psig (12 + 22). If the compression tank is installed on the discharge of the pump (Fig. 7.17), the pressure at the discharge gauge is 12 psig static head. The pressure loss through the system is still 51 feet TDH (22 psi). Therefore the suction pressure will be − 10 psig (12 − 22), or 4.9 in Hg vacuum.

AIR CONTROL COMPONENTS

The system is generally filled with water from the city main. City water also replaces water lost through leakage, shut down for repairs and evaporation. To prevent air problems, such as air locks in the bends of the piping or coils or corrosion, the city water should be introduced into the system at some point either in the air line to the compression tank or at the bottom of the compression tank.

In a closed hydronic system that has been correctly designed, installed, and operated, air in the system travels through the pipes and is vented out at the high points or collected in the compression tank. In addition to the air that is already in the system, when water is heated, air entrained in the water is released.

A compression tank can become waterlogged if there is inadequate air control in the system. For example, when the boiler is off, the water in the compression tank cools. The cool water absorbs air from the compression tank. The water flows back into the boiler by gravity. When the water is heated the air is again released and vented out. After several cycles, all the air is removed from the tank and the tank fills with water. Now the water has nowhere to go except the pressure relief valve on the boiler, which will begin opening on every boiler firing cycle to relieve the pressure caused by the water expansion.

Air control devices are designed to free the air entrained in water in a hydronic system. Air control is accomplished with the following components:

- Air separators
- Manual air vents
- Automatic air vents

Air Separators

There are several types of air separators: centrifugal, dip tube and inline. The centrifugal air separator (Fig. 7.16) uses centrifugal force and low water velocities for air separation. As water circulates through the air separator, centrifugal motion creates a vortex or whirlpool in the center of the tank and sends heavier, air free water to the outer part of the tank. The lighter air-water mixture moves to a low velocity air separation and collecting screen located in the vortex. The entrained air collects and rises into the compression tank.

The boiler dip tube air separator is a tube in the top or top side of the boiler. When the water is heated, air is released and collects at a high point in the boiler. The dip tube allows this collected air to rise into the compression tank. The in-line, low velocity, air separating tank with dip tube type of air separator is used when a boiler isn't available or useable as the point of air separation.

Air Vents

Air vents (Fig. 7.16) may be automatic or manual. One type of automatic air vent is the hydroscopic air vent, which contains a material that expands when wet and holds the air vent valve closed. When there is air in the system, the hydroscopic material dries out, causing it to shrink and open the air vent valve. The float type of automatic air vent has a float valve that keeps the air vent closed when there's water in the system and vent. When there is air in the system it rises into the air vent replacing the water. The float drops and opens the air vent valve.

Manual valves are installed at the high points and bends in the system. The vent is manually opened periodically to allow entrained air to escape.

HEAT EXCHANGERS

The basic types of heat exchangers are shell and tube, shell and coil, helical and plate. Typical HVAC heat exchangers are designed for a number of fluid combinations such as:

- steam to water (converter, steam coil)
- water to steam (generator, boiler)
- refrigerant to water (condenser)
- water to refrigerant (evaporator, cooler, chiller)
- water to water (heat exchanger)
- air to water (water cooling coil)

- air to refrigerant (evaporator cooling coil)
- water to air (heating coil)
- refrigerant to air (condenser)

HVAC COILS

HVAC coils come in a variety of types, materials and sizes. The coils are designed for various fluids such as water, brines, refrigerants or steam. The fluid circulating through the coil heats, cools or dehumidifies air passing over the coil.

The typical HVAC coil has a frame and a core. The core has tubing, return bends and fins. The coil frame is designed to support the core and prevent strains caused by expansion and contraction of the core. The core has ⅝ inch diameter copper tubing. Other tubing materials include carbon steel, stainless steel, brass and cupro-nickel. The number of tubes in the coil varies in both depth and height. Usually there are 1 to 12 rows in the direction of airflow (depth) and 4 to 36 tubes per row (height). The more rows and tubes per row, the more heat transfer (and for chilled water coils, the more dehumidification). With more tubes and rows there is more resistance to airflow (more horsepower is required) and initial cost of the coil.

Return bends are the U-shaped pieces of tubing at the end of the coil. The number and arrangement of the return bends will depend on the number of rows and the circuiting arrangement. The circuiting arrangement is the order in which the tubes are connected together in series and in parallel to produce the best heat transfer, capacity, flow and pressure drop combination.

Tube fins are normally either aluminum or copper. Fins on a coil increase the area of heat transfer surface to improve the efficiency and rate of transfer. They are generally spaced from 4 to 14 fins per inch (fpi). As with coil tubes—the more fins, the more heat transfer—but also the more resistance to airflow. Aluminum is usually picked over copper for fin material to trim costs. However, when cooling coils are sprayed with water, copper fins are needed to prevent electrolysis between dissimilar metals (copper tubes and aluminum fins). Coils wetted only by condensation are seldom affected by electrolysis and are usually copper headers, copper tubes, aluminum fins and steel frames. For applications where the air stream may contain corrosives, various protective coatings can be applied to the fins and tubes.

HVAC Water Coil Piping

HVAC water coils can be piped either counter flow or parallel flow (Fig. 7.18). However, for the greatest heat transfer for a given set of conditions, water coils should be piped counter flow. Counter flow means that the flow of air and water are in opposite directions to each other. In other words, the water enters on the same side of the coil that the air leaves. For cooling coils, this would mean that the coldest water is entering the coil on the same side that the coldest air is leaving the coil.

A coil that is piped parallel flow means that the flow of air and water are in the same direction to each other. The water and air enter on the same side. For parallel flow cooling coils, the coldest water enters the coil on the same side that the warmest air enters the coil. This also means less heat transfer.

FIGURE 7.18.

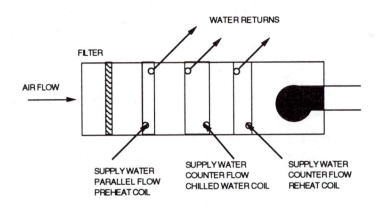

In some applications such as preheat coils, the coils are intentionally designed for parallel flow. For example, a preheat coil may be used to heat the outside air in cold climates to prevent the freeze-up of other downstream coils. Therefore, the coil is piped parallel flow so the hot water or steam enters the coil on the same side that the air enters. Heat transfer, in this example, is critical, and getting the most heat to the coil as quickly as possible is what is important.

In addition to being piped counter flow, water coils should also be piped so that the inlet is at the bottom and water flow is up through the coil and out the top. This will enable the air entrained in the water that's inside the coil to be pushed ahead of the water and accumulate in the top portion of the system where it can be easily vented. Steam coils are piped stream at the top and condensate at the bottom.

Calculating the Log Mean Temperature Difference in Heat Transfer

When two fluids are used in a heat transfer process, the temperature difference at the end of the process will be less than at the beginning. The exchange of heat will follow a logarithmic curve. It is this logarithmic average of the temperature differences called Log Mean Temperature Difference (*LMTD*) that establishes which heat exchanger is best. The higher the LMTD number, the greater the heat transfer.

Equation 7.1: Log mean temperature difference (*LMTD*):

$$LMTD = \frac{\Delta T_L - \Delta T_S}{L_n \left[\dfrac{\Delta T_L}{\Delta T_S} \right]}$$

$LMTD$ = log mean temperature difference
ΔT_L = the larger temperature difference
ΔT_S = the smaller temperature difference
L_n = natural log

Example 7.4: A parallel flow water cooling coil has an entering air temperature of 75°. The entering water temperature is 45°. The leaving air temperature is 60°. The leaving water temperature is 55°. The larger temperature difference is 30 (75 − 45). The smaller temperature difference is 5 (60 − 55). The LMTD is 13.9.

A counter flow water cooling coil has an entering air temperature of 75°. The entering water temperature is 45°. The leaving air temperature is 60°. The leaving water temperature is 55°. The larger temperature difference is 20 (75 − 55). The smaller temperature difference is 15 (60 − 45). The LMTD is 17.4.

A parallel flow water heating coil has an entering air temperature of 75°. The entering water temperature is 200°. The leaving air temperature is 110°. The leaving water temperature is 180°. The larger temperature difference is 125 (200 − 75). The smaller temperature difference is 70 (180 − 110). The LMTD is 94.9.

A counter flow water heating coil has an entering air temperature of 75°. The entering water temperature is 200°. The leaving air temperature is 110°. The leaving water temperature is 180°. The larger temperature difference is 105 (180 − 75). The smaller temperature difference is 90 (200 − 110). The LMTD is 97.3.

CHAPTER 8 Verifying the Performance of Electrical, Heating and Refrigeration Subsystems

ELECTRICAL SUBSYSTEM

Before beginning actual field measurements, try to obtain motor specifications and equipment catalogs. Study these documents to become familiar with the electrical system and its design intent. For clarity, label the motors and circuits on the drawings. While reviewing the documents, make note of any motor system component (or any condition) that specifically should be investigated during the field verification of performance.

Not all the documentation listed below will be available for every system, but the more information you get, the better your understanding of the system will be. You will be able to make better recommendations for system optimization. Try to get:

- Electrical plans
  Engineering drawings
  Shop drawings
  "As-built" drawings
  Schematics
- Equipment catalogs
  Motor description and rating
  Recommendations for testing
  Operation and maintenance instructions
- Motor performance curves
- Previous equipment surveys or balance reports

Reporting Forms

Prepare the following reporting forms for each motor as applicable:

- Motor information and test sheet

- Drive information sheet
- Summary sheet
- Instruments used and calibration information sheet
- Schematic

Verifying Electrical System Performance

Checking the Motor Service Factor

The motor service factor (*SF*) is the number by which the horsepower rating is multiplied to determine the maximum safe load that a motor may be expected to carry continuously at its rated voltage and frequency. For example, a service factor of 1.15 would, by manufacturer's warranty, allow the motor to operate safely at 115% of horsepower at its rated voltage and frequency. However, it is not acceptable to have a motor operating over rated nameplate horsepower (i.e., over 100%). Under some circumstances, such as a voltage drop, if the motor is operating over 100% (but is still within the service factor limit), damage to the motor windings could occur and shorten the life of the motor (Eq. 8.1).

Equation 8.1: Watts equals voltage times amperage ($w = va$). Therefore, amperage is inversely proportional to voltage.

$$w = va$$
$$w = \text{watts}$$
$$v = \text{volts}$$
$$a = \text{amps}$$

Equation 8.1: Power equals current times potential ($p = ie$).

$$p = ie$$
$$p = \text{power or watts}$$
$$i = \text{current or amperage}$$
$$e = \text{electromotive force, or potential or voltage}$$

Example 8.1: A motor is operating at 480 volts and 25 amps. If the voltage dropped to 460 volts the amperage draw would rise to 26 amps.

$$w = va$$
$$w = 480 \times 25$$
$$w = 12,000$$
$$a = \frac{w}{v}$$
$$a = \frac{12,000}{460}$$
$$a = 26$$

Checking Motor Speed

The motor is a constant speed machine unless connected to a variable frequency drive converter. The nameplate revolutions per minute (rpm) is recorded as the motor maximum operating speed. Actual motor operating speed is generally not measured. The nameplate speed is the rated speed that the motor will turn when it's operating at rated horsepower. In

**TABLE 8.1.   THERMAL OVERLOAD PROTECTORS FOR GENERAL PURPOSE
AND 40°C RISE MOTORS SERVICE FACTOR 1.15 CLASS A INSULATION**

| | Starter Size | | | |
	Size 00	Size 0	Size 1	Size 2
Heater No.	Full Load Amps			
AA4	0.19	0.19	0.19	
AA3	0.21	0.21	0.21	
AA2	0.23	0.23	0.23	
A10	1.40	1.40	1.40	
A11	1.55	1.55	1.55	
A12	1.70	1.70	1.70	
A13	1.80	1.80	1.80	
A14	2.00	2.00	2.00	
A25	5.30	5.30	5.30	
A26	6.30	6.30	6.30	
A27	7.00	7.00	7.00	
A28	7.70	7.70	7.70	
A29	8.50	8.50	8.50	
A40		15.6	15.6	16.0
A41		17.5	17.5	17.9
A42		18.9	18.9	19.4
A43		21.3	21.3	22.2
A44		23.4	23.4	24.7

some cases the speed may vary a small amount, but not enough to affect fan, pump or compressor performance or engineering calculations.

Verifying Motor Overload Protection

Motor overload protection, also called "heaters" or "thermals," is used to protect a motor from overheating. When a motor overload occurs, there is an increase in current causing an overheated condition in the motor. The overload protection devices sense the increased current and stop the motor before damage to the windings occur. Before the motor can be restarted, the overload protection devices must be manually reset.

Thermals must be matched to the motor starter and the rated full load current of the motor. Generally, overload protection is installed to protect the motor against current 125% greater than rated current. If the selected thermals are too large, the motor may not be adequately protected. On the other hand, if the thermals are too small, the motor may stop repeatedly. A chart listing thermals and their current ratings for a specific starter size (Table 8.1) is usually found inside the motor disconnect box cover. Thermals generally have a letter and/or number on them. The number/letter will correspond to amperage for proper sizing.

If the motor is new, or if changes have been recently made to the electrical system, verify that the proper motor overload protection devices are installed. To select the proper sized overload protection get the following information:

Motor starter size

Full load rated current

Service factor

The electrical supplier may also need to know the motor's insulation class, allowable temperature rise and classification. Also important is the average ambient temperature at the starter as compared to the ambient temperature at the motor. Sometimes, these temperatures may vary greatly and a temperature compensating overload device may be needed.

Taking Electrical Measurements

Taking Voltage Readings

Voltage is generally read at the motor control center or at the disconnect box using a digital or analog voltmeter. A voltmeter measures the difference in potential between each phase on three-phase systems, or between phase and neutral for a single-phase system. The measured voltage will normally be between plus or minus 10% of the motor nameplate voltage. For example, if the motor nameplate voltage is 230 volts, the voltage reading may be anywhere between 207 volts and 253 volts. If the voltage is out of this range, the system should be inspected. The voltage will usually not be identical from phase to phase.

Example 8.2: Phase 1 to Phase 2 = 233 V, Phase 2 to Phase 3 = 235 V, Phase 3 to Phase 1 = 238 V.

When there is a voltage imbalance there is also a current imbalance. The current imbalance can be as much as 10 times the percent of voltage imbalance. This means that the motor runs hotter than design. If the imbalance is large enough, it can reduce the life of the motor. Therefore, the maximum allowable phase voltage imbalance for a three-phase motor is 2%.

Equation 8.2: Phase voltage imbalance.

$$\%V = \frac{\Delta D_{max}}{V_{avg}} \times 100$$

$$\%V \ = \ \% \text{ voltage imbalance}$$
$$\Delta D_{max} = \ \text{maximum deviation from average voltage}$$
$$V_{avg} = \ \text{average voltage}$$

Example 8.3: Phase 1 to Phase 2 = 233 V, Phase 2 to Phase 3 = 235 V, Phase 3 to Phase 1 = 238 V. For this example the voltage imbalance does not exceed 2%.

$$233 + 235 + 238 \ = \ 706$$
$$706/3 \ = \ 235 \text{ average voltage}$$
$$235 - 233 \ = \ 2 \text{ volt difference}$$
$$235 - 235 \ = \ 0 \text{ volt difference}$$
$$235 - 238 \ = \ 3 \text{ volt difference}$$

$$\%V = \frac{\Delta D_{max}}{V_{avg}} \times 100$$

$$\%V = \frac{3}{235} \times 100$$

$$\%V = 1.3$$

Checking Current Readings

Current is generally read at the motor control center or at the disconnect box using a portable clamp-on digital or analog ammeter. Current is measured on only one phase of a single-phase system and on each phase of a three-phase system. Any phase reading should not exceed the motor nameplate current. When the operating current is over the nameplate current, but within the service factor and voltage limits, close the main discharge valve (water system) or main damper (centrifugal fan air system) until the current reading is nameplate or below.

When the operating current is over the nameplate current and above the service factor limit, turn the motor off. Determine and correct the condition causing the overload before operating the motor again. If the overloaded motor is serving a critical area do not stop the motor, but immediately notify the person responsible for the system that the motor is operating in an overloaded condition.

Checking the Power Factor

The power factor is generally read at the motor control center or at the disconnect box using a portable clamp-on digital or analog power factor meter. The power factor is measured on only one phase of a single-phase system and on each phase of a three-phase induction system. The power factor should be 0.85 or greater. If the power factor reading is less than 0.90, check with the local utility to determine if that utility charges a fee for having a low power factor. If the utility has a power factor charge it may be necessary to install capacitors on the electrical induction system to maintain a higher power factor.

HEATING SUBSYSTEM

Before beginning actual field measurements, try to obtain the specifications and catalogs for the heating equipment. Study these documents to become familiar with the heating system and its design intent. For clarity, label the equipment and air, water or steam distribution pipes on the drawings. Also, while reviewing the documents, make note of any piece of equipment, system component or other condition that specifically should be investigated during the field verification of performance.

Not all the documentation listed below will be available for every system, but the more information you get, the better your understanding of the system will be. You will be able to make better recommendations for system of optimization. Try to get:

- Mechanical plans

  Engineering drawings

  Shop drawings

  "As-built" drawings

  Schematics

- Equipment catalogs

  Heating equipment description and capacities

  Recommendations for testing equipment

  Operation and maintenance instructions

- Heating equipment performance curves or tables

- Previous heating equipment inspection reports or balance reports
- Daily log

Reporting Forms

Prepare the following reporting forms for each heating system as applicable:

- Heating equipment information and test sheet
- Summary sheet
- Instruments used and calibration information sheet
- Schematic

Verifying Heating System Performance

Verifying Boiler Performance

If the boiler is new, verify that it has been started and tested for proper and safe operation in accordance with the manufacturer's instructions and local codes. A daily log should be kept of the boiler's operating pressures and temperatures. Any major variation in the recorded pressures or temperatures may indicate that a problem exists. The cause of the variations should be investigated. If problems which hamper system operation, safety of equipment or personnel are found they should be corrected before continuing the performance survey. Where applicable, the following items need to be checked and noted during the verification of performance:

- Nameplate data

   Compare the nameplate data with the manufacturer's submittal data and design requirements.
- Controls

   Combustion controls—Verify that temperature and pressure settings are correct.

   Safety controls—Verify that temperature and pressure settings are correct.

   Water level controls—Verify that water levels are correct and low water cutoffs are operating properly.
- Accessory equipment—Verify operation and general condition.
- Boiler tubes—Note the general condition.
- Combustion air openings, Barometric dampers, Draft control dampers—Verify that they are correctly sized and working properly.

Inspecting the Pressure Relief Valve

Visually inspect the pressure relief valve to see that it is in place and not capped or wired closed. Do not open it. The pressure relief valve is a safety device. It will open at a preset value so the system pressure cannot exceed the maximum allowable working pressure (*MAWP*). Maximum allowable working pressure must not exceed the construction limits of the boiler. On a low temperature boiler, for example, the pressure relief valve might be set for 30 psig.

Checking Water Temperatures and Water Level Controls

On water boilers, minimum boiler water temperature of 170° should be maintained. This will mean a stack temperature of about 320°. Check the low water cutoff by manually tripping the control.

On steam boilers, the low water cutoff and water column should be blown daily to remove the solids. Additionally, the low water cutoff should be checked under operating conditions at least once a week. To do this, turn off the feedwater pump and let the system operate as normal. Watch the gauge glass and mark the glass at the precise level where the low water cutoff shuts off the boiler. This is now a reference point. The cutoff control should shut down the boiler at the same water level each time its checked. If it doesn't, the controls may need to be replaced.

Determining Boiler Combustion Efficiency

To determine the combustion efficiency of the boiler, take a flue gas analysis using an electronic flue gas analyzer or a chemical test kit. The test will include taking percent of carbon dioxide (CO_2), percent of excess air (O_2), flue gas stack temperature and boiler room air temperature. These readings will determine the percent of stack loss and the combustion efficiency. The instrument may read out directly in percent of carbon dioxide, oxygen and efficiency, or a chart supplied with the instrument may be needed to determine these percentages. The maximum efficiency attainable for both natural gas- and oil-fired boilers will be about 80%.

Here's how to take a flue gas analysis:

- Drill a hole in the flue stack between the boiler and the stack damper. The hole should be at least 6 inches from the stack damper. The hole must be large enough to accommodate the instrument probe.
- Record the boiler firing rate.
- Use a digital or analog thermometer to take the ambient temperature in the boiler room.
- Use a thermometer that is graduated to 1,000° to read the boiler stack temperature.
- Calculate net stack temperature (stack temperature minus room temperature).
- Insert the instrument probe in the test hole in the stack. Take carbon dioxide and oxygen readings.
- If the instrument is digital, it will calculate and read out the boiler efficiency on the meter. If the meter is an analog instrument, go to a table supplied by the instrument manufacturer or a nomograph such as shown in Fig. 8.1 to determine boiler efficiency.

Taking the Temperature of the Boiler Stack

The boiler stack temperature should be at least 320°, but not more than 150 degrees above the steam or water temperature. If the stack temperature is too low, the water vapor in the

FIGURE 8.1.　HEATING-EFFECT OF FLUE GAS COMPOSITION AND TEMPERATURE ON BOILER EFFICIENCY

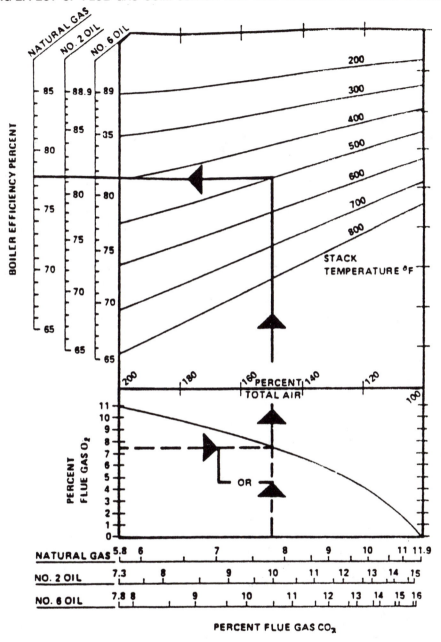

flue gas will start to condense in the stack. This water mixes with the sulfur in the flue gas and creates sulfuric acid which will corrode the stack and the tubes. If the stack temperature is too high the boiler is not working efficiently. The rule of thumb for stack temperature is that for each 100° the stack temperature can be lowered, there is a 2.5% increase in efficiency. A high stack temperature means that either there's poor combustion, the tubes are fouled or there's too much air being brought in and it's pushing the gases through the boiler without the proper heat exchange taking place.

Stack temperatures can vary 100° within a few minutes during load changes, so always note the firing rate when logging temperatures and pressures.

Determining the Amount of Excess Air and the Oxygen Level

The amount of excess air, that is, the air needed for complete combustion plus some extra for a safety factor, should not exceed 10%. The oxygen level in the flue gas should be at least 1%, but should not exceed 2%. A rule of thumb says that there is approximately 5% excess air for each 1% of oxygen in the flue gas.

Determining the Amount of Carbon Dioxide and Carbon Monoxide

The amount of carbon dioxide in the flue gas should be as high as possible. For maximum efficiency in natural gas boilers this will be about 10%, while oil-fired boilers should have about 13 to 14%. A test should be made for the presence of carbon monoxide. No carbon monoxide should be present. Carbon monoxide is a deadly gas. The existence of carbon monoxide indicates incomplete combustion. If carbon monoxide is found, either there is not enough air being brought into the boiler, or there is a problem with the burner. The boiler should be shut down and the problem corrected before the boiler is operated.

Checking Fuel Pressure

On a gas-fired boiler, a drop in gas pressure may mean that there's a drop in city supply pressure or that there's a malfunction in the gas regulator. For an oil-fired boiler, a drop in the oil pressure may mean that the strainer is clogged, the regulating valve is malfunctioning or there's a leak in the line.

Testing for Evidence of Smoke

In addition to the flue gas test, a smoke test should be done on oil-fired boilers. Excessive smoke is evidence of incomplete combustion of the oil. Incomplete combustion means that fuel is being wasted and can also result in soot being deposited on the heat transfer surfaces. This results in lower operating efficiencies. For instance, a 1/8 inch thick soot deposit increases fuel consumption by approximately 10%.

Testing Steam Traps for Leakage

Testing steam traps for leakage can be done using sound detection devices ranging from a simple stethoscope with a steel probe to sophisticated electronic sonic testers. On some systems it may be possible to detect trap leakage thermally using thermography, contact pyrometers or temperature sensing crayons or other materials.

REFRIGERATION SUBSYSTEM

Before beginning actual field measurements, try to obtain refrigeration equipment specifications and catalogs. Study these documents to become familiar with the refrigeration system and its design intent. For clarity, label the equipment on the drawings. While reviewing the documents, make note of any piece of equipment, system component, or any other condition that specifically should be investigated during the field verification of performance.

Not all the documentation listed below will be available for every system, but the more information you get, the better your understanding of the system will be. You will be able to make better recommendations for system optimization. Try to get:

- Mechanical plans

    Engineering drawings

    Shop drawings

    "As-built" drawings

    Schematics
- Equipment catalogs

    Refrigeration equipment description and capacities

    Recommendations for testing equipment

    Operation and maintenance instructions
- Refrigeration equipment performance curves and tables, and refrigerant pressure-enthalpy (*P-E*) diagrams and pressure-temperature (*PT*) tables
- Previous refrigeration equipment inspection reports or balance reports
- Daily log

Reporting Forms

Prepare the following reporting forms for each refrigeration system as applicable:

- Refrigeration equipment information and test sheet
- Summary sheet
- Instruments used and calibration information sheet
- Schematic

Verifying Refrigeration Performance

Measuring Compression System Pressure and Temperature

Depending on the type and size of the system, you can take pressure and temperature measurements at the compressor using a manifold, called a service or gauge manifold. The manifold includes high and low side gauges, hand valves and two or three hoses. They allow access to the refrigeration system to take operating pressures, evacuate the system, add or remove refrigerant, etc. The hoses are generally color coded blue for the low side hose, red for the high side hose, and white or yellow for the middle hose. The manifold also has a hook for hanging it out of the user's way. The right-hand gauge is the high pressure gauge, which measures condensing or high-side pressures (psig). The typical gauge is usually graduated in 5-pound increments from 0 to 500 psi. The gauge on the left measures suction or low-side pressures. It is a compound gauge that is usually graduated in inches of mercury increments from 30 inches of vacuum to zero and one pound increments from zero to 120 or 200 psi. While other pressure ranges are available, these are the most common. The gauges are marked for various refrigerants such as R-12, R-22, and R-502. Temperature scales which correspond to the pressure scale for the refrigerant being measured are on the dial face inside the pressure scale. For instance, if the pressure read was 85 psig, the corresponding temperature on the gauge would be approximately 80°F for R-12 or 50°F for R-22. Gauges for other refrigerants and metric measurements are also available. On larger systems, pressures and temperatures are recorded directly from installed gauges.

Measuring Air-cooled Condenser Temperature and Air Velocity

Measure the entering and leaving dry bulb temperatures across the condenser coil. Normal condenser temperature TD ranges from 15° to 30°F. Measure the air velocity at the face of the condenser coil as applicable. Normal air velocities across air-cooled condensers are between 500 and 1000 feet per minute. To determine the total quantity of air movement (cfm) across the coil, multiply the face velocity (fpm) times the face area (square feet) of the coil.

Measuring Water-cooled Condenser Temperature and Water Flow

Measure the condenser's entering and leaving water temperature. This should be done in temperature wells as close to the inlet and outlet of the condenser as possible. Normal condenser water temperature TD range is 10°. If possible, measure the water flow (gpm) at the condenser coil. If this is not possible, use the condenser pump as a flow meter (see Chap. 5).

Assessing the Thermal Performance of the Cooling Tower

Cooling towers play a critical role in the proper conditioning of buildings. If the tower fails to perform as specified, it may not be able to maintain design temperatures. To accurately assess the thermal performance of a cooling tower, a qualified testing agency should be used. The following lists the cooling tower tests that should be performed and the normal instrumentation required for each test:

- Air Temperatures: Precision thermometers with 0.2°F or smaller divisions
- Wind speed: Anemometer
- Water Temperatures: Precision thermometers with 0.2°F or smaller divisions
- Water flow: Pitot tube, manometer and accessories
- Pump discharge pressure: Pressure gauge
- Fan horsepower: Wattmeter

Gather the following design data:

- Flow rate of the water recirculating through the cooling tower.
- Motor brake horsepower.
- Density of the air entering the tower.

- Barometric pressure.
- Temperature of water leaving the cooling tower.
- Temperature of water entering the cooling tower.
- Wet-bulb temperature of the air entering the cooling tower.
- Dry-bulb temperature of the air entering the cooling tower.
- Liquid-to-gas ratio.
- Pump efficiency.
- Motor efficiency.

Gather the following test data:

- Temperature of cold water leaving the cooling tower. Minimum number of readings is 12 per hour per station.
- Temperature of hot water entering the cooling tower. Minimum number of readings is 12 per hour per station.
- Wet-bulb temperature of the air entering the cooling tower. The entering wet-bulb temperature must be within +/− 10% of design. Minimum number of readings is 12 per hour per station.
- Dry-bulb temperature of the air entering the cooling tower.
- Temperature of the make-up water entering the tower (if not shut off during the duration of the test). Minimum number of readings is 2 per hour per station.
- Flow rate of the make-up water entering the tower (if not shut off during the duration of the test). Minimum number of readings is 2 per hour per station.
- Flow rate of the water recirculating through the cooling tower. The flow rate must be within +/− 10% of design. Minimum number of readings is 3 per hour per station.
- Flow rate of the blowdown (bleed) discharged from the tower (if not shut off during the duration of the test). Minimum number of readings is 2 per hour per station.
- Power input to the fan motor(s). The power input must be within +/− 10% of design. Minimum number of readings is 1 per hour per station.
- Pump discharge pressure.
- Density of the entering air.
- Barometric pressure.
- Tower range. The range must be within +/− 20% of design.
- Wind velocity: The average wind velocity must be 10 mph or less and cannot exceed 15 mph for more than one minute.

CHAPTER 9 Heating Subsystem Components

For the purposes of this book, combustion is defined as a chemical reaction between a fossil fuel such as coal, natural gas, liquid petroleum gas or fuel oil and oxygen. Fossil fuels consist mainly of hydrogen and carbon molecules. These fuels also contain minute quantities of other substances (such as sulphur) which are considered impurities. When combustion takes place, the hydrogen and the carbon in the fuel combine with the oxygen in the air to form water vapor (H_2O) and carbon dioxide (CO_2). If the conditions are ideal, the fuel-to-air ratio is controlled at an optimum level, and the heat energy released is captured and utilized to the greatest practical extent. Complete combustion (a condition in which all the carbon and hydrogen in the fuel would be combined with all the oxygen in the air) is a theoretical concept and cannot be attained in HVAC equipment. Therefore, what is attainable is called incomplete combustion. The products of incomplete combustion may include unburned carbon in the form of smoke and soot, carbon monoxide (a poisonous gas) as well as carbon dioxide and water.

STEAM HEATING SYSTEMS

Steam has some design and operating advantages for heating systems. For instance, one pound of steam at 212°F when condensed gives up approximately 1000 Btu per pound. On the other hand, a hot water heating system with supply water temperatures at 200°F and return water temperatures at 180°F only gives up 20 Btu per pound. Steam, based on its operating pressure, flows throughout the system on its own. Therefore, a pump and motor is not needed to circulate the steam.

In an open vessel, at standard atmospheric pressure, water vaporizes or boils into steam at a temperature of 212°F. But the boiling temperature of water, or any liquid, is not constant. The boiling temperature can be changed by changing the pressure on the liquid. If the pressure is to be changed, the liquid must be in a closed vessel. In the case of water in a heating system, the vessel is the boiler. Once the water is in the boiler it can be boiled at a temperature of 100°F or 300°F as easily as at 212°F. The only requirement is that the pressure in the boiler be changed to the one corresponding to the desired boiling point. For instance, if the pressure in the boiler is 0.95 psia, the boiling temperature of the water will be 100°F. If the pressure is raised to 14.7 psia, the boiling temperature is raised to 212°F. If the pressure is raised again to 67 psia, the temperature is correspondingly raised to 300°F (Table 9.1). A common low pressure HVAC steam heating system will operate around 15 psig (30 psia) or about 250°F.

The amount of heat required to bring the water to its boiling temperature is its sensible heat. Additional heat is then required for the change of state from water to steam. This addition of heat is steam's latent heat content or "latent heat of vaporization." To vaporize

TABLE 9.1. DRY SATURATED STEAM—TEMPERATURE TABLE

TEMP	ABS PRESSURE		SPECIFIC VOLUME		ENTHALPY		
F	PSI	IN. HG	SAT. LIQUID	SAT. VAPOR	SAT LIQUID	LATENT HEAT	SAT. VAPOR
t	p	p	V_t	V_g	H_f	H_{fg}	H_g
COL 1	COL 2	COL 3	COL 4	COL 5	COL 6	COL 7	COL 8
32	0.0885	0.1803	0.01602	3306.	0.00	1075.8	1075.8
34	0.0960	0.1955	0.01602	3061.	2.02	1074.7	1076.7
36	0.1040	0.2118	0.01602	2837.	4.03	1073.6	1077.6
38	0.1126	0.2292	0.01602	2632.	6.04	1072.4	1078.4
40	0.1217	0.2478	0.01602	2444.	8.05	1071.3	1079.3
45	0.1475	0.3004	0.01602	2036.4	13.06	1068.4	1081.5
50	0.1781	0.3626	0.01603	1703.2	18.07	1065.6	1083.7
55	0.2141	0.4359	0.01603	1430.7	23.07	1062.7	1085.8
60	0.2563	0.5218	0.01604	1206.7	28.06	1059.9	1088.0
65	0.3056	0.6222	0.01605	1021.4	33.05	1057.1	1090.2
70	0.3631	0.7392	0.01606	867.9	38.04	1054.3	1092.3
75	0.4298	0.8750	0.01607	740.0	43.03	1051.5	1094.5
80	0.5069	1.0321	0.01608	633.1	48.02	1048.6	1096.6
85	0.5959	1.2133	0.01609	543.5	53.00	1045.8	1098.8
90	0.6982	1.4215	0.01610	468.0	57.99	1042.9	1100.9
95	0.8153	1.6600	0.01612	404.3	62.98	1040.1	1103.1
100	0.9492	1.9325	0.01613	350.4	67.97	1037.2	1105.2
110	1.2748	2.5955	0.01617	265.4	77.94	1031.6	1109.5
120	1.6924	3.4458	0.01620	203.27	87.92	1025.8	1113.7
130	2.2225	4.5251	0.01625	157.34	97.90	1020.0	1117.9
140	2.8886	5.8812	0.01629	123.01	107.89	1014.1	1122.0
150	3.718	7.569	0.01634	97.07	117.89	1008.2	1126.1
160	4.741	9.652	0.01639	77.29	127.89	1002.3	1130.2
170	5.992	12.199	0.01645	62.06	137.90	996.3	1134.2
180	7.510	15.291	0.01651	50.23	147.92	990.2	1138.1
190	9.339	19.014	0.01657	40.96	157.95	984.1	1142.0
200	11.526	23.467	0.01663	33.64	167.99	977.9	1145.9
212	14.696	29.922	0.01672	26.80	180.07	970.3	1150.4
250	29.825	60.725	0.01700	13.821	218.48	945.5	1164.0
300	67.013	136.44	0.01745	6.466	269.59	910.1	1179.7
350	134.63	274.11	0.01799	3.342	321.63	870.7	1192.3
400	247.31	503.52	0.01864	1.8633	374.97	826.0	1201.0
450	422.6	860.41	0.0194	1.0993	430.1	774.5	1204.6
500	680.8	1386.1	0.0204	0.6749	487.8	713.9	1201.7
600	1542.9	3141.3	0.0236	0.2668	617.0	548.5	1165.5
700	3093.7	6298.7	0.0369	0.761	823.3	172.1	995.4
705.4	3206.2	6527.8	0.0503	0.0503	902.7	0.	902.7

one pound of water at 212°F to one pound of steam at 212°F requires 970 Btu (Table 9.1). The amount of heat required to bring water from any temperature to steam is called "total heat." It is the sum of the sensible heat and latent heat.

Example 9.1: One pound of water is in an open container at standard atmospheric pressure. The temperature of the water is 32°F. It will take 180 Btu to raise the one pound of water to 212°F (212–32). To convert that one pound of water to one pound of steam at 212°F will require an additional 970 Btu. Therefore, the total heat required to change one pound of water at 32°F to one pound of steam at 212°F is 1,150 Btu. From Table 9.1 the enthalpy (or heat content) of the saturated liquid (H_f) at 212°F is 180.07 Btu per pound. The enthalpy of the mixture of liquid and vapor (H_{fg}) is 970.3 Btu per pound. The enthalpy of the saturated vapor (H_g) or total heat content is 1,150.4 Btu per pound (180.07 + 970.3).

Pressure Classes of Steam Heating Systems

Steam systems may be classified as low, medium, intermediate or high pressure. It is important to note that low pressure steam contains more latent heat per pound than high pressure steam. Compare low pressure steam at 250°F and 30 psia (946 Btu per pound) to high pressure steam at 700°F and 3,094 psia (172 Btu per pound). This indicates that while high pressure steam may be required to provide very high temperatures and pressures for process functions, low pressure steam provides more economical operation.

Steam Traps

The purpose of a steam trap is to separate the steam side of the heating system from the condensate side. A steam trap collects condensate and allows the trapped condensate to be drained from the system, while still limiting the escape of steam. Condensate must be trapped and then drained immediately from the system. If it isn't, the operating efficiency of the system is reduced because the heat transfer rate is slowed. In addition, the build up of condensate can cause physical damage to the system from "water hammer." Water hammer can occur in a steam distribution system when the condensate is allowed to accumulate on the bottom of horizontal pipes and is pushed along by the velocity of the steam passing over it. As the velocity increases, the condensate can form into a noncompressible slug of water. If this slug of water is suddenly stopped by a pipe fitting, bend or valve the result is a shock wave which can, and often does, cause damage to the system (such as blowing strainers and valves apart). The condensate may be returned to the boiler by a gravity return system or by a mechanical return system using a vacuum pump (closed system) or condensate pump (open system).

In a steam heating system, water enters a heat conversion unit (the boiler) and is changed into steam. When the water is boiled, some air in the water is also released into the steam and is piped along with the steam to the heat exchanger. As the heat is released at the heat exchangers (and through pipe radiation losses), the steam is changed into condensate water. Some of the air in the piping system is absorbed back into the water. However, much of the air collects in the heat exchanger. Steam traps are used to allow the air to escape, preventing the build up of air (which reduces the heat transfer efficiency of the system and may cause air binding in the heat exchanger).

Steam traps are classified as thermostatic, mechanical or thermodynamic. Thermostatic traps sense the temperature difference between the steam and the condensate using an expanding bellows or bimetal strip to operate a valve mechanism. Mechanical traps use a float to determine the condensate level in the trap and then operate a discharge valve to release the accumulated condensate. Some thermodynamic traps use a disc which closes to the high velocity steam and opens to the low velocity condensate. Other types will use an orifice which flashes the hot condensate into steam as the condensate passes through the orifice.

Steam traps are installed in locations where condensate is formed and collects, such as all low points, below heat exchangers and coils, at risers and expansion loops, at intervals along horizontal pipe runs, ahead of valves, at ends of mains, ahead of pumps. etc.

HOT WATER SYSTEMS

The heating systems most often encountered in HVAC work will be low temperature hot water systems with boiler water temperatures generally in the range of 170 to 200°F.

BOILERS

The basic construction of a fossil fuel boiler is the same whether it is a water or a steam boiler. The internal construction of the boiler may be either fire tube or water tube. A fire tube boiler has the hot flue gases from the combustion chamber, the chamber in which combustion takes place, passing through tubes and out the boiler stack. These tubes are surrounded by water. The heat from the hot gases transfers through the walls of the tubes and heats the water. Fire tube boilers may be further classified as externally fired, meaning that the fire is entirely external to the boiler, or they may be classified as internally fired, in which case, the fire is enclosed entirely within the steel shell of the boiler. Two other classifications of fire tube boilers are wet-back or dry-back. This refers to the compartment at the end of the combustion chamber. This compartment is used as an insulating chamber so that the heat from the combustion chamber, which can be several thousand degrees, does not reach the boiler's steel jacket. If the compartment is filled with water it is known as a wet-back boiler and conversely, if the chamber contains only air it is called a dry-back boiler.

Fire tube boilers are also identified by the number of passes that the flue gas takes through the tubes. Boilers are classified as two-, three- or four-pass. The combustion chamber is considered the first pass. Therefore, a two-pass boiler would have one-pass down the combustion chamber looping around and the second pass coming back to the front of the boiler and out the stack. A three-pass boiler would have an additional row of tubes for the gas to pass through going to the back of the boiler and out the stack. An easy way to recognize a two-, three- or four-pass boiler is by the location of the stack. A two- or four-pass boiler will have the stack at the front, while a three-pass boiler will have the stack at the back.

Still another grouping of fire tube boilers is by appearance or usage. The two common types used today in HVAC heating systems are the marine (or Scotch marine boiler) and the firebox boiler. The marine boiler was originally used on steam ships and is long and cylindrical in shape. The firebox boiler has a rectangular shape.

In a water tube boiler, the water is in the tubes while the fire is under the tubes. The hot flue gases pass around and between the tubes, heating the water and then out the boiler stack. Most of the water tube boilers used in heating systems today are rectangular in shape with the stack coming off the top, in the middle of the shell.

Example 9.2: A boiler is identified as a three-pass, internally fired, fire tube, natural gas fired, forced draft, marine, wet back boiler. The boiler has a cylindrical steel shell which is called the pressure vessel. The vessel is covered with several inches of insulation to reduce heat loss. The insulation is then covered with an outer jacket to prevent damage to the insulation. The other components of the boiler are a natural gas power burner, a forced draft fan and various controls. When the boiler is started it will go through a purge cycle in which the draft fan at the front of the boiler will force air through the combustion chamber and out the stack. This purges any combustibles that might be in the combustion chamber. An electrical signal from a control circuit will open the pilot valve allowing natural gas to flow to the burner pilot light. A flame detector will verify that the pilot is lit and gas will then be supplied to the main burner. The draft fan forces air into the combustion chamber and combustion takes place. The hot combustion gases flow down the chamber and into the tubes for the second pass back to the front of the boiler. As the gases pass through the tubes they give up heat into the water. The gases enter into the front chamber of the boiler (called the header) and make another loop to the back of the boiler and out the stack for the third pass. The temperature in the combustion chamber is several thousand degrees. The temperature of the gases exiting the stack is about 320°F.

CHAPTER 10 Refrigeration Subsystem Components

A refrigeration system removes heat from a place where it is not wanted and rejects it into another place where the heat is unobjectionable. Refrigeration is based on the physical principle of boiling liquids. When water is put into a container at sea level and heated to 212°F it begins to boil. As the water boils, it changes to steam. This change of state from water (a liquid) to steam (a vapor) is called vaporization. Steam is a colorless vapor. As the steam rises above the container it starts to cool, gives up its heat, and changes back to a liquid. This change of state, from a vapor to a liquid, is called condensation.

The amount of sensible heat needed to raise one pound of water from any given temperature to its standard boiling point at 212°F is equal to the difference in temperatures times the specific heat of water, which is 1 Btu per pound. For example, the sensible heat needed to raise one pound of water at 70°F to boiling is 142° (212 − 70). The amount of heat required to change one pound of water at 212°F to one pound of steam at 212°F is called water's "latent heat of vaporization," and is equal to 970 Btu. Therefore, the total heat (sensible + latent) needed to change one pound of water at 70°F to one pound of steam at 212°F is 1,112 Btu (142 + 970). "Latent heat of condensation" is the reverse of latent heat of vaporization. It is the amount of heat that one pound of any vapor would release when it changes back to a liquid. For steam at 212°F, latent heat of condensation is 970 Btu per pound.

Refrigerants

A refrigerant is a fluid that absorbs heat through the process of vaporization and releases heat through condensation. Water fits this definition and is listed as refrigerant R-718. However, water is not used as a refrigerant in mechanical vapor-compression systems because its boiling point is too high (212°F) at atmospheric pressures. In other words, at standard atmospheric conditions, it is impossible to vaporize water with room temperature air. Therefore, if water were to be used as a refrigerant, the system would have to be in a vacuum. To boil water at 80°F the system would have to be maintained at 0.5 psia (Table 9.1, Chap. 9). This is equal to approximately 1 inch of mercury ("Hg). Stated in terms of a vacuum this would be equal to approximately 29" Hg. A perfect vacuum is 30" Hg. Even though water is not used in a vapor-compression system, it is used as a refrigerant in some absorption systems.

Most HVAC systems use refrigerants which have low boiling points, such as monochlorodifluoromethane (R-22). The boiling point of a liquid changes in direct proportion to the pressure on the liquid. Increasing the pressure on a liquid refrigerant increases its boiling point while decreasing the pressure decreases the boiling point.

Example 10.1: At sea level, 14.7 pounds per square inch absolute, liquid refrigerant R-22 boils at − 40°F. Water at sea level boils at 212°F. Water subjected to 25 pounds per square inch of pressure absolute boils at approximately 240°F. Liquid refrigerant R-22 subjected to 83 pounds per square inch of pressure absolute boils at + 40°F.

Pressure-Temperature Relationship

Each refrigerant has a given pressure (saturation pressure) which corresponds to a given temperature (saturation temperature).

Refrigerant	Temp (°F)	Gauge Pressure (psig)
R-11	86	3.5
R-11	104	10.5
R-22	86	158.2
R-22	104	207.7
R-502	86	176.6
R-502	104	228.5

Vapor-Compression Mechanical Refrigeration System

A vapor-compression mechanical refrigeration system is a closed system in order to:

- Maintain the pressure-temperature relationships required to vaporize and condense the refrigerant.

- Reuse the refrigerant. Each time the refrigerant passes through the cycle it removes heat from a place where it is not wanted and rejects it into another place where the heat is unobjectionable.

- Keep the refrigerant from becoming contaminated.

- Keep air and other noncondensable gases out of the system.

- Keep moisture out of the system. Water can be very harmful to the system when it combines with the refrigerant to form an acid.

- Control the flow of the refrigerant.

Vapor-Compression Refrigeration Cycle

The vapor-compression mechanical refrigeration cycle has four stages. Vaporization, compression, condensation and expansion.

Example 10.2: To explain the basic refrigeration cycle and heat transfer, a water-to-water air-conditioning system (Fig. 10.1) will be used. The refrigerant is R-22 and the temperatures and pressures used are approximate. Pressure-enthalpy diagram (Fig. 10.2) shows the refrigeration cycle.

FIGURE 10.1.   WATER-TO-WATER AIR-CONDITIONING SYSTEM

FIGURE 10.2.   PRESSURE-ENTHALPY DIAGRAM

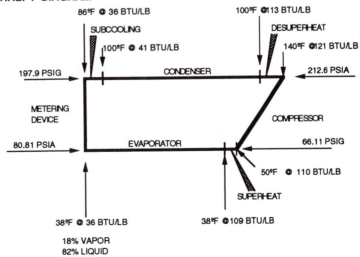

TABLE 10.1.   PROPERTIES OF MIXTURES OF AIR AND SATURATED WATER VAPOR TABLE BASED ON BAROMETRIC PRESSURE OF 29.92 INCHES.

Temp. °F	Humidity Ratio Weight of Saturated Vapor Per Pound of Dry Air		Enthalpy of 1 Lb. of Dry Air Above 0° F in Btu	Enthalpy of (Saturated) Vapor, Btu	Enthalpy of Mixture of 1 Lb. of Dry Air with Vapor to Saturate It In Btu
	Pounds	Grains			
0	0.000787	5.51	0.0	0.835	0.835
2	.000874	6.12	0.480	0.928	1.408
4	.000969	6.78	0.961	1.030	1.991
6	.001074	7.52	1.441	1.142	2.583
8	.001189	8.32	1.922	1.266	3.188
10	.001315	9.21	2.402	1.401	3.803
12	.001454	10.18	2.882	1.550	4.432
14	.001606	11.24	3.363	1.713	5.076
16	.001772	12.40	3.843	1.892	5.735
18	.001953	13.67	4.324	2.088	6.412
20	.002152	15.06	4.804	2.302	7.106
22	.002369	16.58	5.284	2.536	7.820
24	.002606	18.24	5.765	2.792	8.557
26	.002865	20.06	6.245	3.072	9.317
28	.003147	22.03	6.726	3.377	10.103
30	.003454	24.18	7.206	3.709	10.915
32	,003788	26.52	7.686	4.072	11.758
33	.003944	27.61	7.927	4.242	12.169
34	.004107	28.75	8.167	4.418	12.585
35	.004275	29.93	8.407	4.601	13.008
36	.004450	31.15	8.647	4.791	13.438
37	.004631	32.42	8.887	4.987	13.874
38	.004818	33.73	9.128	5.191	14.319
39	.005012	35.08	9.368	5.403	14.771
40	.005213	36.49	9.608	5.662	15.230
41	.005421	37.95	9.848	5.849	15.697
42	.005638	39.47	10.088	6.084	16.172
43	.005860	41.02	10.329	6.328	16.657
44	.006091	42.64	10.569	6.580	17.149
45	.00633	44.31	10.809	6.841	17.650
46	.00658	46.06	11.049	7.112	18.161
47	.00684	47.88	11.289	7.391	18.680
48	.00710	49.70	11.530	7.681	19.211
49	.00737	51.59	11.770	7.981	19.751
50	.00766	53.62	12.010	8.291	20.301
51	.00795	55.65	12.250	8.612	20.862
52	.00826	57.82	12.491	8.945	21.436
53	.00857	59.99	12.731	9.289	22.020
54	.00889	62.23	12.971	9.644	22.615

TABLE 10.1. (*Continued*)

Temp. °F	Humidity Ratio Weight of Saturated Vapor Per Pound of Dry Air		Enthalpy of 1 Lb. of Dry Air Above 0° F in Btu	Enthalpy of (Saturated) Vapor, Btu	Enthalpy of Mixture of 1 Lb. of Dry Air with Vapor to Saturate It In Btu
	Pounds	Grains			
55	.00923	64.61	13.211	10.01	23.22
56	.00958	67.06	13.452	10.39	23.84
57	.00993	69.51	13.692	10.79	24.48
58	.01030	72.10	13.932	11.19	25.12 ←
59	.01069	74.83	14.172	11.61	25.78
60	.01108	77.56	14.413	12.05	26.46
61	.01149	80.43	14.653	12.50	27.15
62	.01191	83.37	14.893	12.96	27.85
63	.01235	86.45	15.134	13.44	28.57
64	.01280	89.60	15.374	13.94	29.31
65	.01326	92.82	15.614	14.45	30.06
66	.01374	96.18	15.855	14.98	30.83 ←
67	.01424	99.68	16.095	15.53	31.62
68	.01475	103.3	16.335	16.09	32.42
69	.01528	107.0	16.576	16.67	33.25
70	.01582	110.7	16.816	17.27	34.09
71	.01639	114.7	17.056	17.89	34.95
72	.01697	118.8	17.297	18.53	35.83
73	.01757	123.0	17.537	19.20	36.74
74	0.1819	127.3	17.778	19.88	37.66
75	.01882	131.7	18.018	20.59	38.61
76	.01948	136.4	18.259	21.31	39.57
77	.02016	141.1	18.499	22.07	40.57
78	.02086	146.0	18.740	22.84	41.58
79	.02158	151.1	18.980	23.64	42.62
80	.02233	156.3	19.221	24.47	43.69
81	.02310	161.7	19.461	25.32	44.78
82	.02389	167.2	19.702	26.20	45.90
83	.02471	173.0	19.942	27.10	47.04
84	.02555	178.9	20.183	28.04	48.22
85	.02642	184.9	20.423	29.01	49.43
86	.02731	191.2	20.663	30.00	50.66
87	.02824	197.7	20.904	31.03	51.93
88	.02919	204.3	21.144	32.09	53.23
89	.03017	211.2	21.385	33.18	54.56
90	.03118	218.3	21.625	34.31	55.93
91	.03223	225.6	21.865	35.47	57.33
92	.03330	233.1	22.106	36.67	58.78
93	.03441	240.9	22.346	37.90	60.25
94	.03556	248.9	22.587	39.18	61.77

TABLE 10.1. (*Continued*)

Temp. °F	Humidity Ratio Weight of Saturated Vapor Per Pound of Dry Air		Enthalpy of 1 Lb. of Dry Air Above 0° F in Btu	Enthalpy of (Saturated) Vapor, Btu	Enthalpy of Mixture of 1 Lb. of Dry Air with Vapor to Saturate It In Btu
	Pounds	*Grains*			
95	.03673	257.1	22.827	40.49	63.32
96	.03795	265.7	23.068	41.85	64.92
97	.03920	274.4	23.308	43.24	66.55
98	.04049	283.4	23.548	44.68	68.23
99	.04182	292.7	23.789	46.17	69.96
100	.04319	302.3	24.029	47.70	71.73
101	.04460	312.2	24.270	49.28	73.55
102	.04606	322.4	24.510	50.91	75.42
103	.04756	332.9	24.751	52.59	77.34
104	.04911	343.8	24.991	54.32	79.31
105	.0507	355.	25.232	56.11	81.34
106	.0523	366.	25.472	57.95	83.42
107	.0540	378.	25.713	59.85	85.56
108	.0558	391.	25.953	61.80	87.76
109	.0576	403.	26.194	63.82	90.03
110	.0594	416.	26.434	65.91	92.34
111	.0614	430.	26.675	68.05	94.72
112	.0633	443.	26.915	70.27	97.18
113	.0654	458.	27.156	72.55	99.71
114	.0675	473.	27.397	74.91	102.31
115	.0696	487.	27.637	77.34	104.98
116	.0719	503.	27.878	79.85	107.73
117	.0742	519.	28.119	82.43	110.55
118	.0765	536.	28.359	85.10	113.46
119	.0790	553.	28.600	87.86	116.46
120	.0815	570.	28.841	90.70	119.54
125	.0954	668.	30.044	106.4	136.44
130	.1116	781.	31.248	124.7	155.9
135	.1308	916.	32.452	146.4	178.9
140	.1534	1074.	33.655	172.0	205.7
145	.1803	1262.	34.859	202.5	237.4
150	.2125	1488.	36.063	239.2	275.3
155	.2514	1760.	37.267	283.5	320.8
160	.2990	2093.	38.472	337.8	376.3
165	.3581	2507.	39.677	405.3	445.0
170	.4327	3028.9	40.882	490.6	531.5
175	.5292	3704.4	42.087	601.1	643.2
180	.6578	4604.6	43.292	748.5	791.8
185	.8363	5854.1	44.498	953.2	997.7
190	1.099	7693.	45.704	1255.0	1301.0
200	2.295	16065.	48.119	2629.0	2677.0

We'll begin the cycle at the chilled water coil. The temperature of the air entering the water coil is 76°F dry bulb and 66°F wet bulb. The temperature of the air leaving the water coil is 60°F dry bulb and 58°F wet bulb. From a psychrometric chart or table (Table 10.1), the enthalpy of the entering air is 30.83 Btu per pound and the enthalpy of the leaving air is 25.12 Btu per pound. The enthalpy difference (Δh) is 5.71 Btu. The air flow across the coil is 25,000 cfm. Therefore, the heat removed from the air by the water is 642,375 Btuh (53.5 tons).

$$\text{Btuh} = \text{cfm} \times 4.5 \times \Delta\text{h}$$
$$\text{Btuh} = 25,000 \times 4.5 \times 5.71$$
$$\text{Btuh} = 642,375$$

$$\text{Tons of refrigeration} = \frac{\text{Btuh}}{12,000 \text{ Btuh per ton}}$$

$$\text{Tons of refrigeration} = \frac{642,375}{12,000 \text{ Btuh per ton}}$$

$$\text{Tons of refrigeration} = 53.5$$

The coil is counter flow. The temperature of the entering water is 45°F. The temperature of the leaving water is 55°F. Therefore, the water flow through the coil is 128.5 gpm.

$$\text{gpm} = \frac{\text{Btuh}}{500 \times \Delta\text{t}} \qquad \text{gpm} = \frac{642,375}{500 \times 10} \qquad \text{gpm} = 128.5$$

The water leaves the coil and is circulated back to the chiller by the chiller water pump. The chiller has an evaporator called a "water cooler," a compressor, a water-cooled condenser, a metering device, piping and accessories and controls.

Stage 1: Vaporization

The vaporization stage is where unwanted heat is removed and absorbed into the refrigerant. The vaporization stage components include a heat exchanger (called the evaporator), a warm fluid (either air or water passing through the evaporator) and a cool refrigerant (in a liquid-vapor mixture) flowing through the evaporator. In order for heat to flow, there must be a difference in temperature. Heat always flows from a higher level or temperature to a lower level or temperature. In this example, the temperature of the refrigerant in the chiller evaporator is 38°F and the water entering the evaporator is 55°F. As the warmer water passes over the tubes of the evaporator, the heat from the water is transferred from the water to the liquid refrigerant. As the refrigerant is heated, it immediately begins to boil because it is already at its boiling point of 38° (38° is the evaporator or suction temperature).

When the refrigerant enters the evaporator it is a liquid-vapor mixture. It is about 18% vapor and 82% liquid. The enthalpy, or heat content of this mixture is approximately 36 Btu per pound. As the refrigerant passes through the evaporator tube the boiling process continues. When all the refrigerant is vaporized it will have a heat content of 109 Btu per pound. Therefore, each pound of refrigerant passing through the evaporator absorbed 73 Btu from the water (109 − 36). This is called the "net refrigeration effect."

As long as the refrigerant is changing state from a liquid to a vapor, the temperature remains at 38°F. However, once all the liquid has been changed to a vapor (this occurs near the end of the evaporator), the vapor can now absorb additional heat. This heat will be *sensible heat*. The process is called "superheating the vapor." Only a vapor can be superheated. The refrigerant will enter the compressor at 50°F. The heat content of the vapor is 110 Btu per pound. This system has picked up about 12° of superheat (50 − 38) and 1 Btu per pound (110 − 109). Proper superheating helps to ensure that only a vapor will enter the compressor. The evaporation stage has removed heat from the water. In the process it has changed the low temperature (38°F), low pressure (80.81 psia, 66.11 psig) liquid-vapor mixture into a low temperature, low pressure vapor (38°F, 66.11 psig).

Stage 2: Compression

The refrigerant vapor leaves the evaporator and goes through the suction pipe. It then enters an electrically driven mechanical compressor. This is the compression stage. The compressor has two main functions in the refrigeration cycle. One function is to pump the refrigerant vapor from the evaporator so the desired temperature and pressure can be maintained in the evaporator. The second function is to increase the pressure of the refrigerant vapor through the process of compression and simultaneously increase the temperature of the vapor. Remember, increasing the pressure on a refrigerant vapor will increase its temperature. This change in pressures, or pressure rise, across the compressor causes the refrigerant to flow throughout the entire system.

Continuing with this example, the compressor increases the pressure of the vapor to about 212.6 psia. The corresponding temperature of the vapor is increased to about 100°F. This is the condensing temperature and is the temperature in the condenser. However, the actual temperature of the discharge vapor is about 140°F. The extra heat (40°F), is picked up in the compressor and is called the "heat of compression." This 40° is sensible heat. The heat content of the vapor at 140°F is 121 Btu per pound. In the heat of compression the vapor has picked up 11 Btu per pound (121 − 110). The refrigerant is now at a high temperature, high pressure as compared to the low temperature, low pressure condition found in the evaporator.

Stage 3: Condensation

The condensation stage is where the heat is rejected from the refrigeration system. The condensation stage has a heat exchanger called a condenser, a warm fluid (either air or water) passing through the condenser and a hot refrigerant (in a vapor-liquid mixture) flowing through the condenser. In order for the refrigerant to be able to pick up more heat from the evaporator, it must once again become a low temperature liquid. The removal of heat from the vapor in the condenser causes the refrigerant to change state from a vapor to a liquid. This process is called **condensation.**

In order for heat to flow there must be a difference in temperature. Heat always flows from a higher level or temperature to a lower level or temperature. The temperature of the water entering the condenser tubes is 85°F. The temperature of the vapor entering the condenser is 140°. Since the water through the condenser tubes is cooler than the refrigerant over the tubes, and heat travels from a higher temperature to a lower temperature, the heat will be picked up by the water. The refrigerant is cooled and the water is heated. The condenser is said to be discharging, or "rejecting" its heat into the water. The heat from the evaporator (73 Btu per pound) plus the heat from the compressor (11 Btu per pound) and any superheat (1 Btu per pound) is the total amount of heat (85 Btu per pound) that must be rejected.

In this example, the high pressure, high temperature vapor leaves the compressor at

140°F (121 Btu per pound) and goes through the discharge, or hot gas pipe to the condenser. In the discharge pipe and the beginning of the condenser the 140°F vapor will reject 40° of sensible heat (8 Btu per pound). In other words, the pressure of the vapor, 212.6 psia, will only support a temperature of 100°F. Therefore, 40° is given up. This is called *desuperheating*. Only a vapor can be desuperheated. The vapor in the condenser is now 100°F (212.6 psia). The heat content of the vapor at 100°F is 113 Btu per pound.

As the refrigerant vapor passes through the condenser the condensation process continues. As long as the refrigerant is changing state from a vapor to a liquid, the temperature remains at 100°. The heat rejected during the change of state is called the "latent heat of condensation." When the refrigerant is all liquid at 100°F the heat content is 41 Btu per pound. Each pound of refrigerant has given up 72 Btu of latent heat of condensation (113 − 41).

Once all the vapor has been changed to liquid, the liquid can reject additional heat. This will be *sensible heat* rejected in the condenser. This is called **subcooling.** Only a liquid can be subcooled.

As the refrigerant, which is now a high pressure, high temperature liquid (212.6 psia, 100°F), flows through the liquid line to the expansion stage it continues to give up heat. The liquid is subcooled to 86° and loses 14° of sensible heat (110 − 86). The liquid now has a heat content of 36 Btu per pound. Each pound of refrigerant passing through the condenser, discharge pipe and liquid line rejected approximately 8 Btu of sensible desuperheat (121 − 113), 72 Btu of latent heat of condensation (113 − 41) and 5 Btu of sensible subcooling (41 − 36) for a total heat of rejection of 85 Btu per pound.

Stage 4: Expansion

The expansion stage consists of a pressure reducing and metering device to reduce the refrigerant pressure and corresponding temperature. It also controls the flow of refrigerant into the evaporator coil.

In this example, the liquid refrigerant will enter the expansion stage's pressure reducing and metering device at approximately 86°F (still at approximately 212.6 psia). This reduction in pressure (from 212.6 psia to 80.81 psia) reduces the boiling point of the liquid refrigerant from 100°F to 38°F. However, the temperature of the liquid refrigerant is still above the new boiling point. It is approximately 86°F. Because the liquid refrigerant is hotter than its boiling point, a part of the liquid refrigerant begins to boil off. This boiling off of the liquid refrigerant is called *flashing*. The liquid refrigerant, which is boiled off or flashed, changes state to a vapor. This vapor is called "flash gas." When a part of the liquid refrigerant is flashed, it removes heat from the remaining refrigerant. Flashing continues until the remaining liquid refrigerant is cooled down to the boiling point which corresponds to the pressure on the liquid. In other words, the refrigerant in the example is now back down to 38°F at 80.81 psia. The refrigerant is once again a low temperature, low pressure liquid ready to pick up more heat in the evaporator and the cycle is repeated. Eighteen percent of the refrigerant has been flashed into vapor, leaving 82% of the total refrigerant available for heat removal.

Cooling Tower-Heat Rejection Stage

Water from the cooling tower (85°F) enters the condenser and picks up the heat from the refrigerant. The water leaves at 95°F. The water flow through the condenser and tower is

149.5 gpm. A variable speed fan in the tower draws air across the tower at a rate that will maintain the tower leaving water at 85°F. The air coming across the tower evaporates the water removing heat. The total heat rejected in the tower will be 747,500 Btuh.

$$THR \text{ (Btuh)} = \text{gpm} \times 500 \times \Delta t$$
$$THR \text{ (Btuh)} = 149.5 \times 500 \times 10$$
$$THR \text{ (Btuh)} = 747,500$$

$$\text{Tons of refrigeration} = \frac{\text{Btuh}}{12,000 \text{ Btuh per ton}}$$

$$\text{Tons of refrigeration} = \frac{747,500}{12,000 \text{ Btuh per ton}}$$

$$\text{Tons of refrigeration} = 62.3$$

REFRIGERATION COMPONENTS (FIG. 10.3)

Compressors

One function of the compressor is to pump refrigerant vapor from the evaporator to the condenser. The compressor must be capable of pumping the refrigerant vapor from the evaporator as fast as it vaporizes. If it doesn't, the accumulated refrigerant vapor will increase the pressure inside the evaporator. If this happens, the boiling point of the liquid refrigerant will be raised and the cooling process will stop. The second function of the compressor is to compress the refrigerant vapor from a lower pressure to a higher pressure. The process of compression adds heat to the vapor, changing it from a lower temperature vapor to a higher temperature vapor.

Three types of mechanical compressors are commonly used for HVAC refrigeration

FIGURE 10.3.   REFRIGERATION COMPONENTS

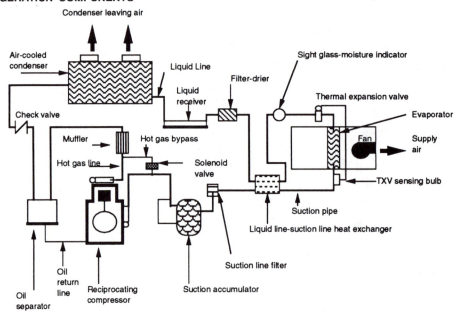

systems. They are: reciprocating, rotary and centrifugal. Positive displacement compressors (reciprocating and rotary) are more economical below 100 tons of refrigeration capacity. Centrifugal compressors are not economical for small capacity systems. They start at about 80 tons and go to several thousand tons. The larger the capacity, the more advantageous the centrifugal compressor becomes. Each of the compressor types have certain advantages and disadvantages. The type of compressor used for a certain application depends on the size and requirements of the application (Table 10.2). The operating pressures and performance of a compressor are determined by many factors, such as the temperature of the refrigerant in the evaporator and the condenser.

Reciprocating Compressors

Cylinder Arrangement

Reciprocating compressors come in a variety of sizes and designs. They may have one or more cylinders arranged in a "V" or "W" pattern, or the cylinders may be in-line or radial. Two and three cylinder compressors are generally arranged in-line. Compressors with four or more cylinders use the radial or the V or W patterns.

Pistons

The reciprocating compressor has a group of pistons operated by a rotating crankshaft. Most HVAC reciprocals use single-acting automotive type pistons which compress the vapor on the upstroke of the piston only once during each revolution. The four stages of the piston cycle are:

- The refrigerant leaves the evaporator and travels through the suction pipe. When the piston moves down (intake stage), refrigerant vapor is drawn into the cylinder through the intake or suction valve.

- As the piston moves up (compression stage), it begins to compress the vapor. The volume of the refrigerant vapor is reduced and the pressure of the vapor is increased.

- At the top of the piston stroke (discharge stage), the vapor exits through the discharge valve and enters the discharge pipe to the condenser.

- The piston starts its downward stroke (reexpansion stage) and the cycle is repeated.

Cylinder Valves

Cylinder suction valves and discharge valves operate on the difference in the pressure between the inside of the cylinder and the pressure in the suction and discharge pipes. The suction valve opens on the downstroke of the piston when the pressure inside the cylinder becomes less than the pressure in the suction line. On the upstroke of the piston, the pressure inside the cylinder is increased. This closes the suction valve. As the piston continues up-

TABLE 10.2.

Compressor Type	Positive Displacement	Compression Mechanism
Reciprocating	Yes	Reciprocating piston
Rotary	Yes	Rotating piston, vane, screw
Centrifugal	No	Impeller

ward, the cylinder pressure continues to increase. When the pressure inside the cylinder is greater than the pressure in the discharge pipe, the discharge valve opens and the compressed vapor flows into the discharge pipe.

Lubrication

Compressors require lubrication for the bearings, cylinders, pistons, gears and other moving parts. The lubricating oil is mixed in and circulates with the refrigerant. The oil also acts as a sealant in the space between the piston and the cylinder walls to force the refrigerant vapor out of the cylinder into the discharge pipe. Without the oil to seal this space the capacity of the system is reduced.

Splash Feed Lubrication System

Small, open compressors generally use a "splash feed" lubrication system. The crankcase is an oil sump and is filled with oil to a level about even with the bottom of the main bearings. As the crankshaft rotates, it dips into the oil in the crankcase. The oil is splashed around within the crankcase, lubricating the bearings, cylinder walls and other rubbing surfaces as the crankshaft rotates.

Forced Feed Lubrication System

Larger compressors use a forced feed system. The crankshaft drives a positive displacement oil pump which supplies oil to the crankshaft, bearings, cylinder walls and the other moving parts. An oil pressure regulator is used to prevent excessive oil pressure.

Capacity Controls

Many large HVAC reciprocating compressors are equipped with controls to provide variable capacity. The type and staging of capacity controls depends on the size of the compressor and the application. The types of capacity controls are:

Hot Gas Bypass. This type of capacity control uses a solenoid valve in the bypass pipe. On a call for capacity reduction the solenoid is energized and the valve opens. A portion of the hot gas discharge is bypassed back to the compressor suction pipe. The rest of the compressed discharge vapor goes to the condenser. This reduces the effective capacity of the compressor by the amount of gas bypassed. With this type of capacity control the reduction in horsepower and energy usage is minimal.

Cylinder Bypass. This type of capacity control also uses a solenoid valve in the bypass pipe. The solenoid is controlled from either temperature or pressure. On a call for capacity reduction the solenoid valve opens. The discharge vapor from one or more cylinders is bypassed back to the compressor suction pipe, where it mixes with the incoming suction vapor. As long as the suction pressure remains below the cut-in setting of the control, the discharge from the isolated cylinders continues to bypass to the suction pipe. When the suction pressure rises to the cut-out setting of the control, the solenoid is deenergized, the valve is closed and the compressor goes to full capacity. A check valve in the discharge pipe does not allow high pressure vapor into the isolated cylinders.

Cylinder Unloaders. Cylinder unloaders may be electrically, mechanically or hydraulically operated. For example, a hydraulic cylinder unloader is operated by a capacity control actuator which is mounted inside the crankcase. The actuator operates on a difference in pressure between the refrigerant suction pressure and atmospheric pressure. If the demand for refrigerant in the evaporator decreases, the suction pressure in the evaporator drops. The capacity control actuator senses the drop in pressure and reduces the oil pressure to a hydraulic mechanism. The mechanism opens the cylinder suction valve. The suction valve remains open until the oil pressure in the mechanism is increased. The piston continues through its four stages. However, with the suction valve being held open, there is no compression of the refrigerant vapor on the upstroke of the piston because the refrigerant vapor is forced back out the suction opening. Since there's no increase in pressure to overcome the pressure in the discharge line, the discharge valve stays closed. If the suction pressure continues to decrease the capacity control actuator, valving mechanism and unloader assembly all function to unload another cylinder. This process continues until the operating cylinders are matched with the cooling load from the conditioned space. Therefore, at partial loads (which is generally the case) not all the cylinders are pumping refrigerant—they are merely idling, needing only enough power to overcome friction. This means that the horsepower requirement is reduced. When the conditioned space becomes too warm, the space thermostat will send out a call for cooling. More refrigerant will be allowed into the evaporator, increasing the suction pressure. This increased pressure is sensed by the capacity control actuators. The actuators increase the hydraulic oil pressure to the cylinder unloaders. The unloaders are filled with oil and close the suction valve for normal operation. As the piston goes down the suction valve is opened, allowing in refrigerant vapor until the suction pressure and the pressure in the cylinder equalize. When this happens, the suction valve closes and the piston starts upward, compressing the refrigerant. There is significant horsepower and energy reduction when using unloaders for capacity control since no pumping work is done on the unloaded cylinders.

Speed Control. The fourth method of capacity control is to vary the speed of the compressor by varying the speed of the compressor driver. Two speeds are normally used and the compressor operates at either full capacity or 50% capacity. When more than two speeds are needed, a motor with two separate windings is used to have a four speed drive. Horsepower and energy is approximately reduced by the cube of the change in capacity.

Rotary Compressors

Rotary compressors operate smoother than reciprocating compressors because of the rotary motion of their compressing mechanism. A screw compressor uses two helically grooved rotors which intermesh to progressively reduce the space inside the compression cylinder. The cylinder has an inlet and outlet port. The refrigerant vapor enters the space between the rotors through the inlet port at the suction end of the screw cylinder. As the rotors turn, they close off the suction port. The screw action forces the refrigerant vapor to the discharge end of the cylinder, where it is compressed against a discharge plate. As the rotors continue to turn, they open a port in the discharge plate. The compressed vapor is forced out into the discharge pipe and the cycle is repeated. Screw compressors are generally used on systems of 50 tons of refrigeration or larger.

The rolling piston compressor uses a steel piston to progressively reduce the space inside the compression cylinder. The cylinder has a suction and discharge port. The piston rolls around the cylinder wall. Refrigerant vapor enters through the suction port and goes into the space between the roller and the cylinder wall. As the piston rolls around the cylinder, the space ahead of the piston gets smaller. The vapor trapped ahead of the roller is compressed and forced out the discharge pipe. The roller continues on and a new space is made for the next cycle.

The rotating vane operates in a similar manner to the rolling piston. Vanes are attached to the rotor and move back and forth in slots in the rotor as the rotor turns inside the cylinder wall. This maintains a positive seal against the walls. Refrigerant vapor enters through the suction port and goes into the space between the roller and the cylinder wall. As the rotor turns, the space in the cylinder is reduced and the vapor is compressed. It is then forced out the discharge port into the discharge pipe. The roller continues on, and a new space is made for the next cycle.

Capacity Control

Capacity control for rotary compressors is most often done with some type of bypass application. For example, capacity control on screw compressors is done with a slide valve in the housing wall underneath the rotors. The slide valve is hydraulically operated. When the system calls for a reduction process the valve is opened, allowing some vapor to recirculate in the cylinder without being compressed.

Centrifugal Compressors

Centrifugal compressors are high capacity machines moving large volumes of vapor. The pumping force is based on impeller size and rotating speed. Centrifugal compressors may be either open or hermetic.

Lubrication

Centrifugal compressors have no cylinders, valves or pistons. Therefore, there are fewer parts needing lubrication. For example, in the hermetic compressor, the only points needing lubrication are the main bearings supporting the drive shaft and the motor bearings. In open compressors, lubrication is required for the shaft seal as well.

Forced Feed Lubrication System

Centrifugals use a pressurized, forced feed lubrication system. The pressurized oil is fed to the compressor bearings by an oil pump submerged in an oil sump. The oil pump is driven by the compressor shaft or by a separate motor. The pump brings the oil pressure up to requirements before the compressor can be started. If the oil pressure is below the minimum limits, an oil pressure control senses the differential pressure across the bearings and prevents the compressor from starting. The oil pressure is regulated by a valve in the pump discharge.

The temperature of the oil in the sump is maintained at about 130°F through the use of an electrical heater. The oil is heated to prevent refrigerant from condensing in the sump and diluting the oil. When the oil leaves the sump, it is cooled down to the required bearing lubrication temperature (approximately 100°), by an oil cooler.

Motor Cooling

Depending on the size and manufacturer of the chiller, the compressor motor may be cooled by the liquid refrigerant or by chilled water. A jacket surrounds the motor windings. On some systems, liquid refrigerant flows by gravity from the economizer into the jacket. The heat from the motor causes the refrigerant to boil. This cools the motor. The refrigerant vapor is returned to the economizer where it is drawn off into the compressor. On other systems, chilled water is circulated through the jacket.

Capacity Control

Capacity control can be done by varying the speed of the compressor with an electronic variable frequency drive. Increasing the speed of the compressor increases capacity (and vice versa).

Varying the system capacity can also be accomplished by opening or closing guide vanes directly ahead of the impeller inlet. The guide vanes, which are also called "prerotation vanes," vary the amount and direction or rotation of flow of the refrigerant vapor immediately before it enters the impeller.

EVAPORATORS

HVAC evaporators are either a fin tube coil (such as the cooling coil in an air handling unit) or a bare tube in a shell and tube heat exchanger (such as in a chiller evaporator). Evaporators may be either dry expansion (cooling coil or chiller evaporator) or flooded (chiller evaporator).

The dry expansion evaporator, also called a direct expansion, or "DX" evaporator, has the refrigerant in the tubes. In a dry expansion evaporator there is no separation line between the liquid and vapor anywhere in the evaporator. There is also no recirculation of liquid or vapor refrigerant. By comparison, a flooded evaporator has the refrigerant surrounding the tubes. There is a separation line between the liquid and the vapor, and there is recirculation of liquid refrigerant. By controlling the liquid level and recirculating un-evaporated liquid, the flooded evaporator assures that virtually all the surface area of the coil is in contact with liquid refrigerant under any load condition. Generally, smaller chillers are DX evaporators because they are easier to control and they require less refrigerant. Larger chillers are flooded evaporators.

HVAC evaporator temperatures are usually between 34 and 45°F. Operating at less than 34°F increases the likelihood of frosting up a dry expansion refrigerant coil or freezing the water in the water chiller. Operating at the highest evaporator temperature for application also reduces the horsepower per ton ratio of the compressor. This means less energy will be used to operate the refrigeration system.

Cooling can also be done with an evaporative cooler, which is sometimes called a "swamp cooler." The evaporative cooler is essentially a box with water pads fitted over louvers in the box. A pump circulates or drips water over the pads. A fan in the box draws outside air into the box across the wetted pads. As the warmer outside air comes across the cooler pads, heat is rejected from the air into the water. The water evaporates. The air has been sensibly cooled to as much as 20°F. However, the air has picked up moisture. This raises the relative humidity of the air going into the conditioned space. Therefore, direct evaporative coolers are used mostly in climates where the outside air has low relative humidity (low wet bulb temperatures) and high dry bulb temperatures (such as the desert area of the Southwest). If evaporative coolers are used in areas with high relative humidities, adding more moisture to the air would make the conditioned space uncomfortably humid— even though the dry bulb temperature of the air might be reduced. The other reason for using evaporative coolers in relatively dry climates is that the capacity of the evaporative cooler depends on the heat content (wet bulb temperature) of the entering air.

CONDENSERS

HVAC condensers are generally either fin tube coil or shell and tube. HVAC condensers may be either air-cooled, water-cooled, or evaporative. The compressor's discharge pressure depends on how rapidly the condenser cooling medium—the air or the water—will carry away the heat of the refrigerant vapor. This heat transfer rate depends on both the temperature of the condenser cooling medium and the volume of flow of the medium across or around the heat transfer surfaces of the condenser.

Lower condenser temperatures are important. The lower the refrigerant temperature that can be maintained in the condenser, the lower the condenser pressure will be. The compressor horsepower per ton ratio will also be smaller. However, the rate of flow and the velocity of the condensing medium through the condenser must be such that it will produce turbulent flow and good heat transfer. It should not, however, be excessive, creating a large pressure drop and increasing fan or pump horsepower requirements.

The air-cooled condenser (fin tube coil) uses air as the condensing medium. The refrigerant vapor is in the tubes. Heat is rejected from the refrigerant into the air. Most often, air is induced or forced by the action of a fan across the condenser tubes. The fan and condenser coil may be mounted on a frame with the compressor (chassis mounted) or located separate of the compressor (remote). However, in all cases, the condenser should be located so there is an adequate supply of air across the coils.

The water-cooled condenser (shell and tube) uses water as the condensing medium. Water flows through the condenser tubes. The refrigerant vapor flows over the tube and heat is rejected from the refrigerant into the water. The shell and tube condenser consists of a cylindrical shell with a number of straight tubes in parallel. The condensing water flows through the tubes which are attached at each end to tube sheets. The refrigerant surrounds the tubes. The shell and tube condenser is available in capacity sizes ranging from several tons to hundreds of tons.

Water-cooled condensers may use either waste water or recirculated water as the condensing medium. In waste water systems, the water supply is usually taken from the city water main and dumped into the sewer after passing through the condenser. In recirculated systems, the water leaving the condenser is piped to a water cooling tower and back to the condenser.

The evaporative condenser (fin tube coil) uses both air and water as the condensing medium. The refrigerant vapor is in the tubes. Water is pumped from a pan in the base of the unit to sprayers over the coil. The water is sprayed on the condenser tubes while air is forced (or induced) across the coil. Heat is rejected from the hotter refrigerant into the cooler water. This evaporates the water. Eliminators are provided to remove any entrained water from the air. The water lost through evaporation or bleed off is made up through a water supply line to the pan in the unit. A float valve regulates the level of water. The capacity of the evaporative condenser depends on the heat content (wet bulb temperature) of the entering air. The higher the wet bulb temperature of the entering air, the lower the capacity of the unit.

RECEIVERS

A receiver is a temporary storage tank for liquid refrigerant. Refrigeration systems that have large load variations are not always called upon to remove heat at a constant rate. Such variations in load conditions may cause the refrigeration to accumulate in the condenser. The receiver stores the refrigerant not required in the system.

METERING DEVICES

A metering device controls the flow of refrigerant to the evaporator. The general types of metering devices are capillary tube, automatic expansion valve, thermal expansion valve, low side float, high side float, orifice or other devices.

COOLING TOWERS

Types

Cooling towers are classified according to the method of air circulation. A natural draft tower has air circulated across the tower by natural convection. A fan is used to circulate air across mechanical draft towers. Depending on the location of the fan, mechanical draft towers are classified as either induced draft or forced draft. If the air flows across the water, the tower is designated as a crossflow type. A counterflow tower has the water falling as the air is circulated up past the water.

Operation

Warm water from the condenser is pumped over the top of the tower by the condenser water pump (Fig. 10.1). The water is sprayed or falls down along the fill to a basin in the bottom of the tower. As the water falls on the fill it is broken down into small droplets. As the air moves across the tower some of the water is evaporated. This cools the water. The cooled water is then pumped from the basin to the condenser where it picks up more heat from the refrigerant vapor.

Although there may be some sensible cooling when the water temperature is warmer than the air temperature, most of the cooling of the water comes from the evaporation process. The evaporation process creates a warm water vapor. The air circulating through the tower carries the water vapor away. The sensible (or dry bulb temperature) and the moisture content (or wet bulb temperature) of the circulated air is increased. Since most of the cooling is accomplished by evaporation, the effectiveness of the tower depends to a large degree upon the wet bulb temperature of the entering air.

In theory, the lowest temperature to which the water can be cooled is the wet bulb temperature of the entering air. However, in practice, the temperature of the water leaving the tower will normally be 7 to 10° above the wet bulb temperature of the entering air. This difference in temperature between the leaving water temperature and the entering air wet bulb temperature is called the "tower approach." The decrease in temperature as the water passes through the tower is called the "range" of the tower. The load on the tower can be approximated by measuring the flow rate over the tower and the entering and leaving water temperatures.

Equation 10.3: Tower approach

$$A = LWT - EWB$$

A = tower approach (°F)
LWT = temperature (°F) of the water leaving the tower
EWB = wet bulb temperature (°F) of the air entering the tower

Equation 10.4: Tower range

$$R = EWT - LWT$$

R = tower range
EWT = entering water temperature (°F)
LWT = leaving water temperature (°F)

Equation 10.5: Tower load

$$L = \text{gpm} \times 500 \times TD$$

L = tower load (Btu per hour)
gpm = water flow rate through the tower in gallons per minute
500 = constant for water (8.33 pounds per gallon $\times$ 60 minutes per hour $\times$ 1 Btu/lb)
TD = temperature difference between the entering water and the leaving water

Example 10: Determine the tower range, approach and load of a cooling tower with a leaving water temperature of 84°F. The entering water temperature is 93°F and an entering air wet bulb temperature of 73°F. The water flow over the tower is 200 gpm.

$$R = EWT - LWT$$
$$R = 93 - 84$$
$$R = 9°F$$

The range is 9°.

$$A = LWT - EWB$$
$$A = 84 - 73$$
$$A = 11°F$$

The approach is 11°F.

$$L = \text{gpm} \times 500 \times TD$$
$$L = 200 \times 500 \times (93 - 84)$$
$$L = 200 \times 500 \times 9$$
$$L = 900,000 \text{ Btuh}$$

The load is 900,000 Btuh.

Drift and Bleed Off

Water that is entrained in and then carried away by the air circulating through the tower is called "drift." The amount of drift is a function of tower design and installation and environmental conditions. For example: tower height, type of fill, fan speed, tower location, wind velocity, etc.

Tower water contains a certain concentration of dissolved solids and other impurities which leave scale deposits as the water is evaporated. "Bleed off" is the continuous or intermittent draining of a percentage of the tower water to avoid the buildup of scale in the tower and the condenser. The amount of bleed off required to maintain the concentration of dissolved solids at low levels is a function of the tower design, the conditions of the water (as determined by the facility or design engineer) and the water chemical treatment specialist.

Make-Up Water

Water lost by evaporation, drift and bleed off is replaced by new water piped to the lower tower basin. A float valve system in the basin adjusts to maintain a constant water level in the basin.

CHAPTER 11 Water Chillers

This chapter will describe the two types of water chillers commonly used in HVAC refrigeration systems. The most often used chiller (a mechanical chiller) uses either a reciprocating, screw or centrifugal compressor. The other major components in the mechanical water chiller package are: condenser, evaporator, accessories, intercooler, purge system and controls.

The other type of chiller is the absorption chiller. The absorption chiller does not have a mechanical compressor, but instead uses a generator, an absorber, a condenser, an evaporator, a purge system, a heat exchanger, accessories and various controls.

MECHANICAL CHILLERS

Centrifugal Chiller

A centrifugal chiller (Figs. 11.1 and 11.2) system will be used to explain the fundamentals of mechanical chiller operation.

Compressor

To maintain the centrifugal force needed to compress the vapor, the compressor impeller rotates at very high speeds. In fact, speeds to 25,000 rpm are common. However, centrifugal compressors don't build up as much pressure as do positive displacement reciprocal and rotary compressors. Therefore, several impellers are put in series to increase the pressure of the vapor. Commonly, centrifugal compressors will have two, three or four impellers. Each impeller is a stage of compression. After the vapor leaves an impeller, it's directed into another impeller or into the discharge line. The system application will dictate the capacity of the compressor, number of stages and compressor speed.

Low pressure, low temperature, low velocity refrigerant vapor is drawn from the evaporator through inlet vanes and into the eye of the impeller. The impeller is in a casing near the center of the compressor. The vapor enters the inlet of the impeller and goes through several stages of compression. As the impeller spins, the vapor is discharged at

FIGURE 11.1.   CENTRIFUGAL WATER CHILLER

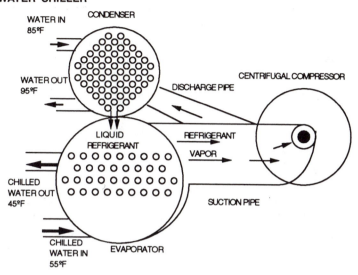

FIGURE 11.2.   CENTRIFUGAL WATER CHILLER

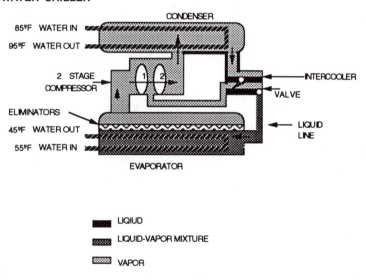

high velocities to the outside of the casing. The vapor is at a relatively high temperature and pressure.

Condenser

The hot vapor enters the condenser. A condenser pump circulates water from the cooling tower through the condenser (Chap. 10, Fig. 10.1). The temperature of the water going into the condenser is about 85°F. As the water passes through tubes in the condenser it picks up heat from the refrigerant vapor. The water, which is now about 95°F, goes back to the cooling tower to release into the air the heat that it has picked up in the condenser. The hot liquid refrigerant drains from the condenser into a pressure chamber, which is called an "economizer" or "intercooler."

Intercooler

The liquid refrigerant goes into the high pressure chamber of the intercooler. It then passes through a high side float valve into the intermediate chamber of the intercooler. As the refrigerant goes through the valve the pressure on the refrigerant is reduced. This reduction in pressure reduces the boiling point of the refrigerant. However, the temperature of the liquid refrigerant is still above the new boiling point. Because the liquid refrigerant is hotter than its boiling point, a part of the liquid refrigerant begins to boil off. This boiling off of the liquid refrigerant is called "flashing." One of the purposes of the intercooler is to pre-flash liquid refrigerant in its intermediate pressure chamber to reduce the temperature of the liquid refrigerant to the lower temperature corresponding to the pressure in the intermediate chamber.

Another purpose of the intercooler is to take the preflashed vapor from the intermediate chamber and send it to the suction side of the second stage compressor to be compressed. The pressure of this vapor is above the evaporator pressure. Therefore, the power required to compress it to the condensing pressure is less. Also, the temperature of this vapor is cooler than the temperature of discharged vapor from the first stage compressor. When the two vapors are mixed, the temperature of the vapor going into the second stage compressor is reduced, thereby increasing the capacity and efficiency of the system.

From the intermediate chamber, the lower temperature liquid refrigerant passes through a float valve into the evaporator. The refrigerant is now at an intermediate pressure. Its temperature is somewhere between the higher pressure and temperature of the compressor and the lower pressure and temperature of the evaporator. As the liquid refrigerant passes through the intermediate float, the pressure is reduced to the evaporator pressure. Some of the liquid flashes, cooling the remainder of the liquid to the evaporator temperature. The liquid-vapor refrigerant goes to the evaporator through the liquid line.

The intercooler reduced the total volume of flash gas required to cool the refrigerant to its evaporator temperature. This reduction in the volume of flash gas in the intercooler means that more of the liquid refrigerant is available for use in the evaporator. This makes the chiller system more efficient by increasing net refrigerant effect. It also means that there is less load on the first stage of the compressor, thereby reducing the horsepower requirements.

Evaporator

The refrigerant liquid-vapor mixture goes through the liquid line into a flooded evaporator. The temperature of the refrigerant mixture is about 40°F. The temperature of the water entering the evaporator is about 55°F. As the water flows through the evaporator tubes it is cooled down 10° (to approximately 45°F).

Refrigerant

Depending on the type of refrigerant used, the system may operate at pressures that are less than atmospheric. For instance, a system using refrigerant R-11 will have an evaporator pressure of about 15 inches of mercury. This is called "vacuum pressure," and would be equal to about 7 psi below standard atmospheric pressure of 14.7 psi.

Purge System

The function of the purge system (Fig. 11.3) is to remove noncondensable and condensable gases from the condenser. For example, if the system is below atmospheric pres-

FIGURE 11.3.   CENTRIFUGAL CHILLER PURGE UNIT

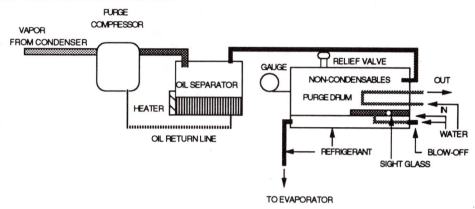

sure it is possible that air (a noncondensable gas) and moisture (water vapor is a condensable gas) will leak into the system and become trapped in the condenser. If allowed to accumulate, these gases will increase the condenser pressure and lower the efficiency of the system.

The purge system draws the condensable and noncondensable gases, including refrigerant vapor, from the condenser into a purge compressor. The gases become entrained with the compressor's lubricating oil. The oil and gas mixture is then discharged into an oil separator tank. The tank has a heater which heats the gases and separates them from the oil. The oil collects at the bottom of the tank and is returned to the compressor crankcase through an oil return line. The heated gases are released from the oil and travel to the purge drum.

Inside the purge drum is a chilled water coil. The heated water vapor and refrigerant vapor condense on the coil. The condensed refrigerant liquid collects on the bottom of the purge drum. When a set amount of refrigerant has collected, a float valve will open and allow the refrigerant to reenter the system through the side of the evaporator. Any accumulated water floats on top of the refrigerant. A sight glass shows the amount of water collected and a manual blow off valve can be opened to remove the accumulated water. The noncondensable gases rise to the top of the drum. When the pressure inside the drum exceeds a setting on the pressure relief valve, the valve opens, releasing the noncondensable gases into the atmosphere.

Chiller Controls

A centrifugal chiller control system will consist of various components including a transmitter/sensor in the return water pipe, a controller, a compressor switch and a load limiting relay. In a typical sequence of operation, a rising return chilled water temperature indicates that the conditioned space is becoming warmer and the system load is increasing. This condition would cause a direct-acting transmitter in the return water pipe to send an increasing signal to the controller. The controller receives the signal from the transmitter and sends out an increasing signal to a compressor start/stop switch.

When the signal from the controller exceeds the cut-in set point on the compressor switch, the switch closes and the compressor motor is started. The output signal from the controller also goes to the operator of the compressor inlet vanes. The vanes are normally closed. As the signal increases, the vanes start to open, allowing more refrigerant vapor into the compressor.

As long as the compressor motor current is less than a set percentage of maximum load, a load limiting relay (*LLR*) passes the signal on to the inlet vane operator. If the motor current exceeds the set point, the LLR interrupts the signal to the vane operator, closing the inlet vanes. Closing the vane reduces the flow of refrigerant to the compressor and reduces the load on the motor. The load limiting relay remains in control until the load on the compressor is reduced below the setting on the relay. At this point, the LLR once again allows the control signal to pass on to the vane operator, and control of the chiller is returned to the chilled water temperature control system.

A manual or automatic demand limiter sets the limits of chiller operation. Typical load limit settings are 40, 60, 80 or 100% of full load.

ABSORPTION CHILLERS

Absorption chillers are similar to mechanical chillers in that:

- They use heat to vaporize a volatile refrigerant at low pressures in the evaporator.
- They take a low pressure refrigerant vapor from the evaporator and deliver a high pressure refrigerant vapor to the condenser.
- They condense the refrigerant vapor in the condenser.
- They recycle the refrigerant.

The main differences are:

- The absorption chiller does not use a mechanical compressor. Instead, heat from various sources is used to change the concentration of the refrigerant-absorbent solution. As the concentrations change, the pressures in the various components will change. This pressure difference moves the refrigerant around the system.
- A generator and absorber replaces the mechanical compressor.
- An absorbent is used. Usually water or lithium bromide.
- The refrigerant is usually either ammonia or water.
- The energy input comes from heat (steam or hot water) supplied directly to the generator. The heat may come from boilers or furnaces. In some cases, heat from other processes may be used. These include low pressure steam or hot water from industrial plant work, waste heat recovered from exhaust gases of gas engines or turbines, or low pressure steam from steam turbine exhaust.

The drawing of the absorption chiller in Fig. 11.4 will be used to explain the fundamentals of HVAC absorption chiller operation. Generally, commercial HVAC absorption systems will have capacities from 25 to over 1,000 tons. The chiller consists of the following components.

- Generator
- Condenser

FIGURE 11.4. ABSORPTION CHILLER

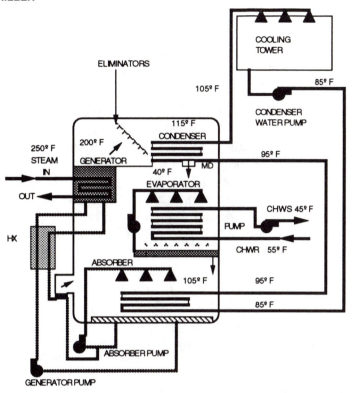

- Evaporator
- Absorber
- Fluid pumps
- Purge unit
- Controls

The evaporator and the absorber are on the low pressure side of the system in a vacuum of about 0.12 psia (0.248 inches of mercury). The generator and the condenser are also in a vacuum on the high pressure side of the system. The pressure on the high side is 1.5 psia (3.06 inches of mercury), or approximately one-tenth of atmospheric pressure (14.7 psia). There are two working fluids in an absorption system—a refrigerant and an absorbent. The refrigerant in this chiller is water and the absorbent is lithium bromide. The refrigerant flow is:

From	To	State
Condenser	Evaporator	Liquid
Evaporator	Absorber	Vapor
Absorber	Generator	Liquid
Generator	Condenser	Vapor

The absorbent flow is:

From	To	Solution
Absorber	Generator	Diluted to concentrated
Generator	Absorber	Concentrated to diluted

Absorbent

The operation of the absorption chiller depends on an absorbent that has a great attraction for the refrigerant. For use in large capacity absorption chillers, lithium bromide, a salt, is such an absorbent. Lithium bromide, when dry, is in a crystal form. The amount of lithium bromide dissolved in the water to form the lithium bromide solution is measured by weight and not by volume. The concentration of the lithium bromide solution is stated in terms of the amount of lithium bromide in the solution. For example, a 100 pound solution of lithium bromide and water may have 65 pounds of lithium bromide and 35 pounds of water. This would be called a 65% concentration because 65% of the total weight is lithium bromide.

Refrigerant

The refrigerant is water. The evaporator is maintained in a vacuum of about 0.248 inches of mercury (0.12 psia). The temperature of the refrigerant corresponding to this pressure is approximately 40°F.

Evaporator

High pressure liquid refrigerant from the condenser passes through a metering device into the evaporator. The pressure on the refrigerant is reduced, thus lowering the temperature of the refrigerant to approximately 40°F. In addition to the refrigerant from the condenser, a pump recirculates liquid refrigerant collected in the bottom of the evaporator. This refrigerant goes through a spray header over the evaporator tubes. The sprayers help keep the tubes wet at all times. They also break the liquid refrigerant into small droplets, allowing it to vaporize more easily. This improves heat transfer and makes maximum use of the refrigerant. As the refrigerant comes in contact with the warmer evaporator tubes, part of the refrigerant is vaporized and absorbs latent heat from the material being cooled. In this example, the material being cooled is return water from the air handling units. The water enters the evaporator tubes at 55°F. The refrigerant cools the water down to a leaving temperature of 45°F. Any liquid refrigerant that is not vaporized collects under the tubes and is recirculated. The low pressure refrigerant vapor from the evaporator passes through eliminators and removes any liquid refrigerant.

The refrigerant vapor flows from the evaporator to the absorber. This occurs because the vapor pressure of the lithium bromide solution in the absorber is lower than the vapor pressure of the refrigerant vapor in the evaporator. The higher the concentration and the lower the temperature of the lithium bromide, the lower the vapor pressure of the solution. The refrigerant vapor from the evaporator is drawn into the low pressure area created by controlling the temperature and concentration of the refrigerant-absorbent solution.

Absorber

An absorber pump circulates a concentrated solution of lithium bromide and water through a spray header over the absorber tubes. The lithium bromide solution from the sprayers mixes with and absorbs the refrigeration vapor from the evaporator. Heat is generated during the mixing process. To maintain the temperature in the absorber, water from a cooling tower flows through the absorber tubes. The temperature of the water entering the absorber tubes is approximately 85°F. The water in the tubes picks up the heat in the absorber and leaves at 95°F. If the heat is not removed, the temperature and pressure in the absorber will rise and the vapor flow from the evaporator will stop.

When the mixing takes place, the refrigerant vapor condenses into a liquid to create a diluted solution of refrigerant and lithium bromide. This is called the *dilute solution*. This dilute solution then drops into the bottom of the absorber shell, from where it is pumped through a heat exchanger to the generator. The solution must be reconcentrated in order for the refrigeration process to continue.

Generator

The dilute solution flows through the heat exchanger and picks up heat from the concentrated lithium bromide solution returning to the absorber. After passing through the heat exchanger, the dilute solution flows over the generator tubes. Low pressure steam flows through the tubes. The heat from the steam vaporizes a portion of the refrigerant (which reduces the water content of the lithium bromide solution), making it a concentrated solution. The vaporization of the refrigerant is possible because it has a lower boiling point than the absorbent and the generator is never at a temperature high enough for the absorbent to boil.

The refrigerant vapor in the generator is about 200°F. It passes through eliminators (which remove any entrained lithium bromide) and flows into the condenser. Approximately 3.06 inches of mercury (1.5 psia) is maintained in the high side, which means that the refrigerant vapor will condense on the condenser tubes at approximately 115°F.

Condenser

Water from the absorber is pumped through the condenser tubes and then into the cooling tower. The water enters the condenser at 95°F and picks up the latent heat of condensation. The water leaves the condenser at about 105°F. The water goes to the cooling tower, where it is cooled down to 85°F and is recirculated back to the absorber.

The condensed refrigerant liquid flows by pressure differential into the evaporator through the metering device, and the operating cycle is repeated.

Heat Exchanger

The concentrated solution is pumped from the generator through the heat exchanger to the absorber. The heat exchanger transfers heat from the warmer concentrated lithium bromide solution, leaving the generator, to the cooler dilute lithium bromide solution coming from the absorber. The efficiency of the absorption system is improved by using the heat exchanger. The temperature of the dilute solution is increased, thus reducing the amount of heat needed in the generator. The temperature of the concentrated solution is reduced, resulting in less heat having to be removed in the absorber by the water from the cooling tower.

The concentrated solution leaves the heat exchanger and enters a metering device which lowers the pressure of the refrigerant and a small part of the remaining refrigerant liquid flashes. Flashing cools the rest of the refrigerant. The refrigerant vapor enters the absorber. Dilute solution pumped by the absorber pump from the bottom of the absorber mixes with the cooler concentrated solution from the generator and goes through the sprayers. This solution is now ready to absorb refrigerant vapor from the evaporator.

Purge system

Noncondensable gases will tend to collect on the surface of the dilute lithium bromide solution in the bottom of the absorber. If the noncondensable gases are not removed by purging, they will increase the pressure in the absorber to a point where the flow of refrigerant vapor from the evaporator will stop.

Chiller Controls

Control of the operating cycle is accomplished with an automatic control valve on the steam line. This control valve modulates the flow of steam to the generator tubes. The control valve operates from a temperature sensor in the chilled water line leaving the evaporator. For example, if the chilled water temperature was too cold, the control valve would close, reducing the quantity of steam supplied to the generator. The reduced amount of steam would boil a smaller quantity of refrigerant out of the generator and reduce the concentration of the lithium bromide solution being sprayed in the absorber. The solution would not be able to absorb as much refrigerant vapor and the cooling ability of the evaporator would be reduced.

Water flow rate through the condenser will be controlled with a bypass loop in the condenser piping.

CHAPTER 12 Verifying the Performance of HVAC Control Systems

In this chapter you will learn how to verify the performance of HVAC control systems and components including sensors, controllers, relays, switches, actuators and controlled devices.

Not all the documentation listed below will be available for every system, but the more information you get, the better your understanding of the system will be. With greater "system understanding" you will be able to make better recommendations for system optimization. Therefore, from the control contractor get:

Engineering drawings

Shop drawings

"As-built" drawings

Schematics

Control diagrams

In some cases the control drawings will have to be made in the field as part of the verification process.

From the control contractor or the component manufacturer get:

Equipment catalogs

Recommendations for testing controls

Operation and maintenance instructions

Study the control drawings, specifications and catalogs to become familiar with the control system and its design intent. For clarity, be sure that all controls are designated correctly on the drawings and the reports. While reviewing the documents, make note of any control system component or other condition that should be specifically investigated during the field verification of performance.

VERIFYING THE PERFORMANCE OF THE CONTROL SYSTEM

Put the control system in normal operation. Inspect the control system and verify its operation. Record the findings for information on the general operating condition of the system and for possible optimization opportunities. While the system is being tested you may find that some changes will need to be made during the verification process, while other changes will need to be considered after all information on the system is obtained.

General Procedure

Sequence of Operation

- Review the drawings for the sequences of operation for all control modes.
- Observe the sequence of operation for all control modes and verify that they are in accordance with the control drawings.

Sensors

- Check the location and installation of all sensors to determine if they will operate properly and sense only the intended temperatures, humidities or pressures.

Controllers

- Operate each controller in the system being tested.
- Review service records to verify calibration dates. If the control system is not maintained on a service contract, check calibration of individual controllers using manufacturer's recommended procedures.
- Using the information from the contract drawings, verify that each controller's set points meet design intent, including limiting safety controllers such as firestats or freezestats.

Relays and Switches

- Verify the settings, operation and adjustment of all relays, gradual switches, minimum position switches, end switches, mercury switches, solenoid valves, contactors, etc.

Actuators

- Check the operation of all actuators.
- Verify that all actuators are properly connected to the intended controlled device.
- Verify the operation of pilot positioners.

Controlled Devices

- Operate controllers and verify that all controlled devices are operated by the intended controller.
- Confirm the proper operation of all controlled devices as applicable to normally open or normally closed.
- Examine the span of control from a normally open position to a normally closed position, observing any deadbands, excessive pressures, leading or lagging of simultaneously or

sequentially controlled devices. Record any overlapping of operation of controlled devices.

- Test the fail safe modes of all controlled devices.
- Verify that all controlled devices are in the position indicated by the controller—open, closed or modulating.
- Observe that all controlled devices are located correctly and properly installed in the HVAC distribution systems. Check for valves installed backwards and dampers linked incorrectly.
- Verify that all controlled devices have free travel.
- Check controlled devices for tight shut-off and full open position.

Safeties

- Verify the proper operation of all safeties.
- Verify the proper operation of all lockout or interlock controls.

VERIFYING THE PERFORMANCE OF THE PNEUMATIC CONTROL SYSTEM

In addition to the items listed for the general procedure:

- Check the compressed main air supply for proper pressure(s) and observe the general operation of the compressor, dryer, pressure gauges, and cut-in and cut-out settings.
- Note the on- and off-times of the compressor.
- Inspect filters, drains and safeties. Note the general condition of the system.
- Inspect the system for air leaks.

Main Air

- Verify the operation of the main air supply system including the air compressor, air dryer and pressure controls.
- Inspect the system and confirm that the system is complete and that all components are installed correctly. If the system is not operating, confirm that the compressor is operational and that the air pressure control cut-in and cut-out points are correct.
- Observe several operating cycles of the compressor and record on-time and off-times.
- Verify that the air dryer is operating. Open the manual drain valve and observe output.

Controllers and Sensors

- Record the location and verify the operation of all controllers and sensors including thermostats, humidistats, pressurestats, firestats, freezestats, receiver-controllers, remote bulb sensors and transmitters.
- Calculate sensor and controller output pressures.
- Confirm that the controllers are set according to the control drawings.

- Check controller operation. This is done using the instruments and tools from a pneumatic calibration kit.

  —Connect the pneumatic tubing and test gauge to the calibration test point.

  —Observe output pressure.

  —If the controller is out of calibration, recalibrate according to manufacturer's instructions or contact a control contractor.

- Check throttling range, authority, and set points according to manufacturer's instructions or contact a control contractor.

- Apply normal control pressure input and observe controller pressure output. Confirm that the output pressure is correct over the full throttling range.

- Verify transmitter calibration.

  —Connect the pneumatic tubing and test gauge from the test kit to the calibration test point.

  —Read the actual value at the transmitter and compare actual output pressure with predicted output pressure.

Relays and Switches

- Record the location and verify the operation of all relays and switches.
- Connect the pneumatic tubing and test gauge from the test kit to the relay or switch according to manufacturer's instructions.

Amplifying or Retarding Relays

- Apply input pressure and verify that the output pressure remains at zero until the input pressure increases to the bias pressure setpoint. At this point, the output pressure should increase by the correct ratio (1:1, 1:1.5, 1:2, etc).
- Adjust or replace if needed.

Averaging Relays

- Use a manual transmitter and gauge with normal control pressure and apply a low input pressure and one high input pressure to the relay.
- Compare the output pressure with the average of the input pressures.
- The output pressure should be +/−0.5 psi of the arithmetic average of the input pressures.
- Replace if needed.

Pneumatic to Electric (P-E) Switches

- Verify the setting of P-E switches by applying an increasing input pressure with a gradual transmitter and precision gauge.
- Observe the actual cut-in and cut-out points and compare them to the points specified.
- Adjust if needed.

Diverting Relays

- Apply various input pressures to verify the operation of diverting relays.
- Increase the input pressure and note the pressure when the common port is switched from the normally open port to the normally closed port.
- Decrease the input pressure and note when the common port is switched back to normally open port.
- Compare the switching point pressures and the differential pressure with the control drawings.
- Adjust if needed.

Reversing Relays

- Verify proper operation by applying various input pressures with a gradual transmitter and precision gauge.
- Observe output pressures and compare them to the output pressures specified.
- Adjust the relay if needed to provide the required output pressure for a selected input pressure.

Selector Relays

- Use a manual transmitter and gauge with normal control pressure and apply a low input pressure and one high input pressure to the relay.
- Using the manufacturer's recommendations, cap off or jumper unused ports.
- Make one test for each input.
- Verify that the relay is selecting the higher or lower input pressure and switching to the output pressure as required by the control drawings.
- Replace if needed.

Limiting Relay

- Apply normal control pressure and observe the position of the actuator when the input pressure is below the minimum setting. Do it again when the input pressure is above the maximum setting.
- Adjust the relay if needed and mark.

Positive Positioning Relays

- Apply input pressure and verify that output pressure increases when the input pressure reaches setpoint.
- Adjust or replace if needed.

Gradual Switch

- Apply normal control pressure and observe that the output pressure is increased or decreased with adjustment of the switch.

- Adjust the switch to position the required setting and mark the switch.

Minimum Position Switch

- Apply normal control pressure and observe the position of the actuator when the input pressure is below the set point.
- Adjust the switch to position the actuator at the required minimum setting and mark the switch.

Pneumatic-Electric Transducers

- Apply input pressure and record electrical output in ohms, dc volts or milliamperes, as appropriate.
- Predict output values for various input pressures and compare.
- Replace if needed.

Controlled Devices and Actuators

- Record the location and verify the operation of all controlled devices and actuators.
- Observe the input pressure to each actuator.
- Verify that with zero input pressure to the actuator the controlled device is in the "normal position" according to the controlled specifications.
- Using the tubing and test gauge from the pneumatic test kit, apply input pressure and verify the spring range of the actuator.
- Observe the movement of the damper or valve linkage through the full range of the actuator.
- If the linkage adjustment is incorrect, make corrections so that the linkage provides the full range of motion between full open and full closed.

VERIFYING THE PERFORMANCE OF THE ELECTRIC, ELECTRONIC AND DIRECT DIGITAL CONTROL SYSTEM

In addition to the items listed for the general procedure, as applicable:

- Verify that the control voltage is correct.
- Confirm that control terminations are correct.
- Put the system in normal operation and test each control loop at both ends of its control range to verify that all control loops and their individual field points are responding correctly.
- Verify the calibration and response time of all transducers.
- Note if the system has lightning protection and battery backup.
- Confirm the application and accuracy of the software algorithms for each control loop.
- Verify the operation of the phone modem.

Main Power Source

- Review the control drawings and specifications.
- Verify the requirements for power supply to the control system.
- Check that 120 volt power circuits are properly protected and labeled for 24 hour operation.
- Inspect the wiring connections at sensors, controllers and controlled devices.

Controllers and Sensors

- Record the location and verify the operation of all controllers and transmitters.
- Predict sensor and controller output signals (explained later in this chapter).
- Confirm that the controllers are set according to the control drawings.
- Check controller operation. This is done using appropriate testing instruments such as a potentiometer, a selectable resistance device (decade box), a volt-ohm-milliamp meter (*VOM meter*), or a digital multi-meter (*DM meter*).
  —Using the information from the manufacturer, connect the instrument to the controller.
  —Observe output voltage at the calibration setpoint and over the full throttling range.
  —Calibrate controller by comparing expected controller output voltage with actual value. If the output voltage is not as predicted, recalibrate the controller as instructed by the manufacturer or contact a control contractor.
- Check throttling range, authority, and set points according to manufacturer's instructions or contact a control contractor.
- Apply input signal and observe controller pressure output. Confirm that the output pressure is correct over the full throttling range.
- Verify transmitter calibration.
  —Measure the actual condition of the controlled variable (temperature, pressure or humidity).
  —Connect the testing instruments and observe output signal.
  —Compare the expected transmitter output with the actual value. If the output signal is not as predicted, replace the transmitter or contact a control contractor.

Relays and Switches

- Record the location and verify the operation of all relays and switches.
- Connect the test instrument to the relay or switch according to manufacturer's instructions.

Amplifying or Retarding Relays

- Apply input signal and verify that the output signal remains at zero until the input signal increases to the bias pressure setpoint. At this point, the output signal should increase by the correct ratio.
- Adjust or replace if needed.

Averaging Relays

- Apply one low input voltage and one high input voltage to the relay.
- Compare the output voltage with the arithmetic average of the input voltages.
- Replace if needed.

Diverting Relays

- Apply various input signals to verify the operation of diverting relays.
- Increase the input signal and note the signal when the common port is switched from the normally open port to the normally closed port.
- Decrease the input signal and note when the common port is switched back to the normally open port.
- Compare the switching point signal and the differential pressure with the control drawings.
- Adjust if needed.

Reversing Relays

- Apply various input voltage signals and observe output signal.
- Adjust the relay as needed to provide the required output voltages for the various input signals.

Selector Relays

- Apply a low input voltage and one high input voltage to the relay.
- Using the manufacturer's recommendations, cap off or jumper unused ports.
- Make one test for each input.
- Verify that the relay is selecting the higher or lower input voltage and switching to the output voltage as required by the control specifications.
- Replace if needed.

Limiting Relay

- Apply control signal and observe the position of the actuator when the input pressure is below the minimum setting and again when the input pressure is above the maximum setting.
- Adjust the relay if needed and mark.

Gradual Switch

- Apply input signal and observe that the output signal is increased or decreased with adjustment of the switch.
- Adjust the switch to position the required setting and mark the switch.

Minimum Position Switch

- Apply input signal and observe the position of the actuator when the input signal is below the set point.
- Adjust the switch to position the actuator at the required minimum setting and mark the switch.

Electric-Pneumatic Transducers

- Apply input voltage and record output pressure.
- Predict output values for various input voltages and compare.
- Replace if needed.

Controlled Devices and Actuators

- Verify the input voltage signal to the controlled device.
- Verify the "normal" position, range of actuators and proper movement of all controlled devices.
- Record the position of the controlled device with a zero input signal.
- Vary the input signal with a manual transmitter and observe actuator motion from starting point to stopping point.
- Observe the movement of the controlled device as the input signal is varied.
- Observe the movement of the damper or valve linkage through the full range of the actuator.
- If the linkage adjustment is incorrect, make corrections so that the linkage provides the full range of motion between full open and full closed.

CONTROL OPTIMIZATION

After the control system verification of performance has been completed, a list of optimization opportunities may be made. Generally the list will include some of the following:

Calibrate

Clean

Repair

Replace

Redesign

PNEUMATIC SYSTEMS

General

- Adjust the pneumatic pressure switch on the air compressor to provide desired pressure range and operating time.

- Relocate sensors if the possibility of erroneous or erratic operation is indicated due to outside influences such as sunlight, drafts, sensors located on outside walls, etc.
- Repair air leaks.
- Tighten electrical connections.

Calibrating Pneumatic Thermostats and Humidistats

Pneumatic controls should be calibrated at least annually—more often if they are located in a dirty environment. There are just six basic steps to calibration of pneumatic room thermostats or humidistats. It is advisable to always consult manufacturer's instructions before doing any actual calibration of control components. The basic steps are:

1. Determine the actual value, temperature or humidity at the controller's sensing element. Use a thermometer for thermostats or a hygrometer for humidistats.
2. Turn the controller's set point adjustment to the actual value.
3. Install a pneumatic output pressure test gauge in the controller's test port.
4. Turn the calibration adjusting screw until the pressure read on the test gauge is midway between the spring range of the controlled device.
5. Remove the test gauge.
6. Turn the controller's set point dial to the desired set point.

Example 12.1: The desired set point on a pneumatic room thermostat is 74°F. The thermostat operates a hot water valve which has a spring range of 4 to 8 pounds. To check controller calibration, remove the thermostat cover and place a recently calibrated test thermometer on top of the thermostat. In this example the thermometer indicates 76°F. Next, set the thermostat to 76°F. Insert a pneumatic test gauge into the test port of the thermostat and turn the thermostat's adjustment screw until the test gauge indicates 6 pounds (halfway between 4 and 8 pounds). Finally, replace the cover and set the thermosatat to 74°F. The thermostat is set to 76°F. Insert the test gauge into the test port and turn the adjustment screw until the test gauge reads 6 pounds (halfway between 4 and 8 psi). The set point on the thermostat is turned to 74°F.

CHAPTER 13 Control Components

Automatic control systems (*ACS*) are installed in residential, commercial and industrial buildings. They start, stop or regulate the flow of air, water or steam and maintain the desired building environmental conditions (temperature, humidity and pressure) for comfort, process function, economy and safety. Understanding the control system is the key to understanding HVAC systems. This chapter explains the various types of control components. Chapter 15 will show typical control applications and ways to retrofit the control system.

An automatic control system is a group of components designed to interact with the other components to make the system self-regulating. Each component has a definite function. HVAC automatic control systems are comprised of pneumatic, electric and electronic-direct digital controls and are classified according to the source of power used to operate the various components.

System	Source of Power
Pneumatic	Compressed air
Electric	Low voltage electricity (normally 24 volts ac) or line voltage electricity (normally 110 to 220 volts ac);
Electronic	Low amperage or low voltage electricity (normally 4 to 20 milliamps dc or 0 to 15 volts dc).

PNEUMATIC CONTROL SYSTEMS

The power source for a pneumatic control system is compressed air. The typical components of a single or dual pressure pneumatic system are (Figs. 13.1, 13.2 and 13.3):

Air compressor	Control piping or tubing
Receiver tank	Sensors
Refrigerated air dryer	Controllers
Filters	Relays and switches
Pressure reducing valve	Actuators
Pressure relief valve	Controlled devices

FIGURE 13.1.   SINGLE PRESSURE PNEUMATIC AIR SYSTEM

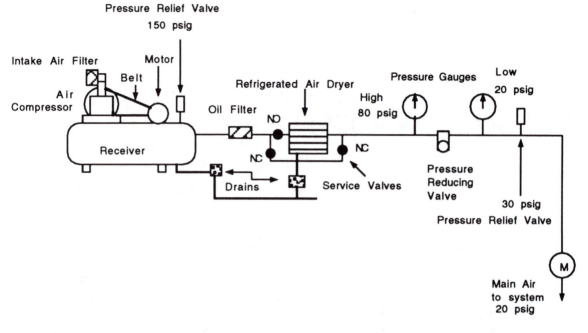

FIGURE 13.2.   DUAL PRESSURE PNEUMATIC AIR SYSTEM

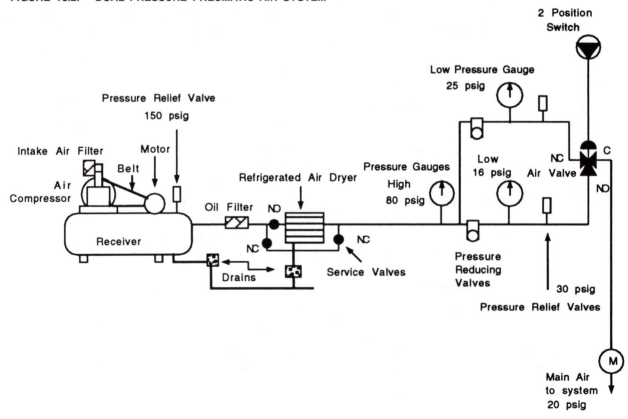

FIGURE 13.3.   TYPICAL SIMPLIFIED PNEUMATIC SYSTEM

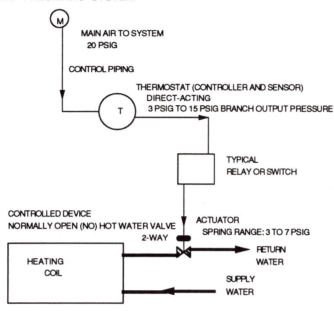

Each of these components is described in the following sections.

Air Compressor, Receiver Tank and Drain

The air compressor is an electrically driven, reciprocating machine generally sized about 25 horsepower or less. The compressor selected will normally have an operating schedule of about 35% on-time and 65% off-time. The operating time of the compressor is limited to extend compressor life. Since compression adds heat to the air, limiting operating time allows sufficient time to cool the air in the compressor's receiver tank. An automatic or manual drain is installed in the receiver tank to remove any accumulated water, oil, dirt or scale.

Filters and Air Dryers

The compressed air used to operate the system must be kept clean, dry and oil-free in order for the pneumatic components to function correctly. A number of devices are installed in the system to dry the air and remove oil, vapors, dirt and other contaminants. For instance, an air intake filter is installed on the compressor to keep dirt and oil vapors from entering and being passed through the compressor and then condensing into droplets in the air lines.

To ensure that air lines are oil and dirt free, a refrigerated air dryer equipped with an automatic drain is placed downstream of the receiver tank to remove any moisture carryover. An oil filter is installed upstream of the air dryer to collect any oil vapor or particles of dirt. A manual bypass valve is installed to service the refrigerated air dryer and downstream filter without interrupting the system operation.

Pressure Reducing Valve and Pressure Relief Valve

A single pressure pneumatic system has a pressure reducing valve (*PRV*) which maintains the system pressure between 18 and 20 psig. On either side of the PRV is a pressure gauge which indicates the tank pressure upstream and the line pressure downstream of the PRV. To protect the receiver tank from excessive pressures, a pressure relief valve is installed. Another pressure relief valve is installed downstream of the PRV to protect the system if the PRV fails. The maximum safe operating pressure for most pneumatic devices is 30 psig.

Dual Pressure Systems

A dual pressure pneumatic system has two different applications, either summer/winter or day/night. Each application requires a different main air pressure. The configuration of a dual pressure system is the same as the single pressure system up to the PRV. Since two different pressures are required, two pressure reducing valves are needed. Depending on the make of the controller, one PRV reduces pressures to 13 to 16 psig and the other PRV reduces pressures to 18 to 25 psig. The higher pressure PRV supplies air to the controller when the device is on the winter or night setting. The lower pressure is supplied to the controller for summer or day operation. Details on dual pressure thermostats are provided later in this chapter. Downstream from the pressure reducing valves is a three-way air valve and a manual (or automatic), two-position switch. The function of the switch is to change the ports on the three-way air valve.

Example 13.1: The two position switch is set for summer operation. On the three-way air valve the normally open (*NO*) port is open and the normally closed (*NC*) port is closed. Air from the lower pressure PRV is allowed to flow through the NO port into the common (*C*) port and out to the system. Air from the higher pressure PRV is blocked. When the switch is set for winter operation, the NO port on the three-way air valve closes and the NC port opens. This allows air from the higher pressure PRV to flow through the NC port into the common (*C*) port and out to the system. Air from the lower pressure PRV is blocked.

Example 13.2: The two-position switch is set for day operation. On the three-way air valve, the normally open (*NO*) port is open and the normally closed (*NC*) port is closed. Air from the lower pressure PRV is allowed to flow through the NO port into the common (*C*) port and out to the system. Air from the higher pressure PRV is blocked. When the switch is set for night operation, the NO port on the three-way air valve closes and the NC port opens. This allows air from the higher pressure PRV to flow through the NC port into the common (*C*) port and out to the system. Air from the lower pressure PRV is blocked.

Piping

The control piping or tubing conveys the compressed air (normally 3–25 psig) to various controllers such as thermostats, humidistats, pressurestats and receiver-controllers and then to the controlled devices.

Sensors (Pneumatic and Electric/Electronic Sensing Elements)

A sensing element, either internal or remote of the controller, measures the controlled variable (temperature, humidity or pressure) and sends a signal back to the controller.

Temperature Sensors. Temperature sensors sense a change in temperature. There are two general types of pneumatic temperature sensing elements.

- Thermal expansion elements such as bimetal or metal rod and tube elements, and vapor or liquid-filled elements such as sealed bellows, remote bulb or capillary fast response or averaging elements.

- Bimetal elements are commonly used in room thermostats.

- Rod and tube elements are generally used in insertion and immersion temperature controllers, such as those located in boilers or storage tanks.

- Sealed bellows elements are commonly used in room thermostats.

- Remote bulb elements are used where the temperature measuring point is a distance from the controller location, such as in a duct or pipe.

- Fast response and averaging capillary elements are used instead of the bulb in a remote bulb element. The fast response element is a tightly coiled capillary with a response time many times faster than the standard remote bulb. The averaging element is a capillary evenly strung across a duct to obtain the average temperature in the duct.

Humidity Sensors. Humidity sensors are used to measure relative humidity or dew point of the air. Mechanical-pneumatic humidistats (hygrometers) operate on the principle that hygroscopic materials such as nylon, silk, wood, leather and human hair will expand when exposed to moisture. The change in the material is detected mechanically and converted to a pneumatic signal.

Pressure Sensors. Mechanical-pneumatic pressure sensors convert changes in absolute, gauge or differential pressures using bellows, diaphragms or Bourdon tubes. The change is detected mechanically and converted to a pneumatic signal.

Controllers

The pneumatic sensor sends out a signal which is proportional to the valve of the variable being measured. The pneumatic controller compares the signal from the sensor with the desired pressure and sends out a control signal based on this comparison. A pneumatic controller continuously receives and acts on input data.

Controllers may be either direct- or reverse-acting. *Direct-acting* means that a rise in the controlled variable (such as space temperature) causes a rise in the branch pressure output of the controller. *Reverse-acting* means that a rise in the controlled variable (such as space temperature) causes a decrease in the branch pressure output of the controller.

Controllers can also be classified as either one-pipe, bleed-type controller, or two-pipe, relay-type controller. The bleed-type controller has only a one pipe connection. Main air is introduced through a restrictor into the branch line between the controller and the controlled device. The branch control air output is a function of the amount of air flowing into the controller in response to a change in the controlled device. For example, on a direct-acting thermostat, the flow of air into the thermostat is restricted on a rise in space temperature, which increases the output pressure to the controlled device.

The two-pipe controller has two connections (branch and main) and receives main air directly. This type of controller will feed air to the branch and/or exhaust it through a leak port in response to the controlled variable. The two-pipe controller provides a greater volume of air to the controlled device which produces a faster response to a change in the controlled variable.

Thermostat (Temperature Controller)

The standard thermostat has a temperature range of 55 to 85°F and a 3 to 15 psi output range. The throttling range may be adjusted between 2 and 12°F.

Example 13.3: A direct-acting thermostat set for 73°F and a 6°F throttling range (70°F to 76°F) will have a branch output of 3 psig when the room temperature is 70°. When the room is 73°F the output will be 9 psig. At 76°F or above the branch output will be 15 psig.

*Space Temperature (°F)	Direct-Acting Output (psig)
70	3
73	9
76	15

Example 13.4: A room has a reverse-acting thermostat with a 6°F throttling range and a set point of 72°F.

Space Temperature (°F)	Reverse-Acting Output (psig)
69	15
72	9
75	3

High Limit (HL) and Low Limit (LL) Thermostats. High and low limit thermostats monitor the condition of the controlled medium (air, water, or steam) and interrupt system operation if the monitored condition becomes excessive (high limit) or drops below the desired minimum value (low limit). For example, on a conventional air economizer, the mixed air plenum controller is a low limit thermostat. The outside controller is a high limit thermostat. Another application for a low limit thermostat is for protection of the coils against freezing. The sensing element of this low limit thermostat, generally called a "freezestat," is strung across the discharge side of the heating coil. When at least one foot of the element senses set point temperature, the system goes to its fail safe condition, which is typically: fan off, outside air dampers closed, return air dampers open and heating valve open.

Deadband Thermostats. A deadband thermostat is used for energy conservation to eliminate simultaneous heating and cooling. It is a two-pipe controller that operates similar to a single pressure, single temperature thermostat. The exception is a temperature span or "deadband" between the heating and cooling set points when no heating or cooling occurs. The deadband thermostat uses two bimetal strips—one for heating and one for cooling. The heating bimetal modulates the output pressure between 3 psig and the deadband pressure. The deadband pressure is the output pressure at which no heating or cooling occurs. The cooling bimetal modulates the output pressure between the deadband pressure and 15 psig.

Example 13.5: A deadband thermostat allows heating below 68°F and cooling above 74°F. The deadband pressure is 8°F and the temperature span is 6°F (74–68). The deadband is adjustable within limits depending on the heating and cooling set points selected.

Space Temperature (°F)	Direct-Acting Output (psig)	
68 or below	3 to 7	heating occurs
between 68 and 74	8	no heating or cooling
74 or above	9 to 15	cooling occurs

Dual Pressure Summer/Winter Thermostat. The summer/winter system provides for the seasonal requirements of either cooling or heating. Depending on the season, either chilled water or hot water is supplied to the coil. Since the valve controlling the flow of hot or chilled water remains the same (either normally open or normally closed, but not both), the system must have a thermostat which can be both direct-acting and reverse-acting. The bimetallic strip in the thermostat is changed from direct-acting to reverse-acting by a change in the main air pressure.

Example 13.6: A dual pressure summer/winter thermostat controls a normally open, two-way valve. The system is in the winter condition and the higher main air pressure is sent to the thermostat. This makes the thermostat direct-acting. As the space temperature rises, an increased branch pressure is sent to the valve causing it to close. This allows less hot water into the coil, and the space begins to cool.

Example 13.7: The system is switched to summer conditions and the lower main air pressure is sent to the thermostat, which changes it over to reverse-acting. On a decrease in space temperature, an increasing pressure is sent to the valve causing it to close, reducing the chilled water to the coil.

Dual Pressure Day/Night Thermostat. The day/night dual pressure thermostat allows for setting and controlling space temperature at different points for the day and night or varying loads. A day/night thermostat is essentially the same as a summer/winter thermostat except that the day/night has two bimetal strips and both are either direct-acting or reverse-acting. The two bimetal strips have separate set points. When the higher main air pressure is sent to the thermostat, the night bimetal strip is in control. When the lower pressure is sent, the day bimetal is in control.

Example 13.8: A reverse-acting day/night thermostat controls a two-way, normally open cooling valve. During the day the lower main air pressure is sent to the thermostat. The thermostat's day set point is 72°F. Any temperature above 72°F will send a decreasing branch pressure to the actuator causing the valve to open. At 72°F or below the thermostat will send an increasing branch pressure to the actuator causing the valve to close. This allows less chilled water into the coil.

Example 13.9: A direct-acting day/night thermostat controls a two-way, normally open heating valve. The main pressure is switched at night, sending the higher main air pressure to the thermostat. The thermostat will now modulate the branch pressure to the valve actuator based on the night bimetal strip which is set for 60°F. At 60°F or below the thermostat will send a decreasing branch pressure to the valve causing it to open. This allows more hot water into the coil. Any temperature above 60°F will send an increasing branch pressure to the valve causing it to close.

Humidistat (Humidity Controller)

A humidistat uses a hygroscopic material (such as nylon) to sense moisture and control the relative humidity in a space.

Example 13.10: A reverse-acting humidistat controls a normally closed, two-way steam valve. As the room's relative humidity drops, the pressure to the steam valve actuator is increased. This opens the valve and allows steam to enter the humidifier.

Space Humidity (%)	Reverse-Acting Output (psig)
35	15
45	9
55	3

Master/Submaster Controller

The master controller sends its branch output pressure to the reset port on a second controller—the submaster controller—instead of a controlled device. Both master and submaster are piped with main air. The submaster's set point changes as the signal pressure from the master controller changes. There are two types of reset—direct and reverse. When the change is direct-acting it is called a "direct reset" type of control. When the change is reverse-acting it is called a "reverse reset" type of control.

Direct Reset

Example 13.11: The master controller in this example is a space thermostat. The submaster controller is a thermostat with a remote sensing element located in the discharge air duct. The branch output from the submaster is piped to a two-way, normally open, hot water valve. As the space thermostat (master controller) senses an increase in room temperature, an increased pressure is sent to the submaster controller to reset its set point lower. The submaster then senses discharge air temperature. If the discharge air temperature is above the submaster set point, the submaster controller sends a signal to the hot water valve to close.

The direct reset schedule is:

Space Temp (°F)	Master Controller Output (psig)	Submaster Set Point (°F)	Discharge Air Temp (°F)	Hot Water Valve Position
70	3	100	100	Open
73	9	80	80	Modulating
76	15	60	60	Closed

Reverse Reset. Another application of a master/submaster controller is to reset hot water (submaster controller) from outside air temperature (master controller). This is a reverse reset.

Example 13.12: The sensor for the outside air master controller is set for 70°. The submaster controller operates the boiler hot water valve. The set point on the submaster is 80°. As the outside air temperature falls, the submaster set point is reset upwards and the hot water temperature is increased. The relationship between the master and submaster in this example is 1:1.5 (for every degree the outside air drops the water temperature is reset upwards by 1.5°).

The reverse reset schedule is:

Outside Air Temp (°F)	Master Controller Output	Heating Water Temp (°F)	Hot Water Valve Position
70	3	80	Closed
35	9	132.5	Modulating
0	15	185	Open

Receiver-Controller and Transmitter

A receiver-control consists of two main parts—the controller and the transmitter.

Controller. A receiver-controller, like the other controllers, receives a signal from a sensing element and then varies it's branch output pressure to a controlled device or another receiver-controller. Single input receiver-controllers function the same as the controllers described before. Dual input pressure receiver-controllers are used in master/submaster reset applications.

Transmitter. The sensing element in a receiver-controller is called a transmitter. Transmitters are one-pipe, direct-acting, bleed-type devices which use a restrictor in the supply line to help maintain the proper volume of compressed air between the transmitter and the receiver-controller. Transmitters are used to sense temperature, humidity or pressure and send a varying pneumatic signal back to the receiver-controller. All transmitters have a pressure output span of 12 psi, but, they come in a variety of transmitter spans (Table 13.1). The sensitivity of a transmitter is its output span divided by its transmitter span. For example, a transmitter with a 100° span has a sensitivity of 0.12 psi per degree (12 psi divided by 100°F).

Relays and Switches

The number of applications for pneumatic relays and switches is virtually unlimited. The following is a list and basic applications of ten of the more common relays and switches.

Air Motion Relay. An air motion relay senses air movement across a fan or coil to verify air flow. It is used as a safety device.

Amplify or Retard Relay. An amplifying or retarding relay is used to change the output start point. It is also called a *bias start relay* or *ratio relay*. A typical application is when a retarding relay is installed to eliminate the simultaneous heating and cooling that occurs when a heating valve and cooling valve are operating from the same controller and have an overlapping spring range.

TABLE 13.1.   TYPICAL TRANSMITTER SPANS AND PRESSURE OUTPUT SPANS

Transmitter Type	Transmitter Range	Transmitter Span	Pressure Output Span (psi)
Temperature	0 to 100°	100°	12
Temperature	-25 to 125°	150°	12
Humidity	30% to 80% RH	50% relative humidity	12
Pressure	0 to 7 inches water gauge	7 inches water gauge	12

Example 13.13: A normally open heating valve has a spring range of 3 to 7 psi. The normally closed cooling valve's spring range is 7 to 11 psi. A retarding relay is installed between the valves. The sequence is for the output pressure from the thermostat to go to the heating valve actuator and then to the relay. The relay then sends a signal to the cooling valve actuator. If the thermostat output pressure is 7 psi, the heating valve actuator sees 7 psi, as does the input to the relay. If this relay is set for a 2 psi retard bias, the input to the relay is 7 psi, but the output of the relay to the cooling valve actuator is 5 psi. Therefore, the cooling valve would not start to open until the output from the thermostat was 9 psi (the cooling valve sees 7 psi).

Averaging Relay. An averaging relay is used to reset a controller or operate a controlled device from the average signal of two or more controllers. For example, two direct-acting thermostats each send a signal to an averaging relay. One signal is 4 psi and the other is 8 psi. The output signal from the relay to the controlled device is 6 psi (4 + 8 divided by 2).

Electric-Pneumatic (E-P) Relay or Switch. An electric-pneumatic (E-P) relay or switch is an electrically operated solenoid three-way air diverting valve. It is used to control a pneumatically operated device from an electric circuit.

Example 13.14: A typical application is when the outside air dampers are interlocked with the operation of the fan. The sequence is that when the fan is turned on the E-P relay, which is wired to the fan, is energized. This allows control air piped at the normally closed (NC) port to connect to the common (C) port and go on to the damper actuator, opening the dampers. When the fan is turned off, an internal plunger blocks the NC port and connects the C port to the normally open (NO) port. This allows air in the actuator to bleed off through the NO port, closing the dampers.

Diverting Relay. A diverting relay is used to switch air signals. A diverting relay is a three-way air valve used primarily to convert a signal, at a predetermined set point, into a signal for a controlled device. A typical application for a diverting relay is to use it as either a high limit or low limit control in an economizer application.

Example 13.15: A high limit diverting relay in the outside air is set at 72°. This allows the mixed air controller to control the outside air (0A) and return air (RA) dampers up to 72°F. The output from the mixed air controller is piped into the normally open (NO) port of the diverting relay. As long as the outside air temperature is below 72°F this signal is passed along to the dampers through the common (C) port to the damper actuators. The outside dampers are open and the return air dampers are closed. When the OA temperature reaches 72°F, the diverting relay switches and blocks the NO port and connects the C port to the normally closed (NC) exhaust port to allow the air pressure to be exhausted from the damper actuators. This closes the outside air dampers and opens the return air dampers.

Pneumatic-Electric (P-E) Relay or Switch. A pneumatic-electric (P-E) relay or switch is an air actuated device used to make or break electrical contacts. Pneumatic-electric relays are used to start or stop fans, pumps or other electrically driven equipment and can be wired either normally open or normally closed. Note—when using electrical terms, normally open means the circuit is deenergized. Normally closed means that it is energized.

Reversing Relays. Reversing relays are used to reverse a signal from a controller.

Example 13.16: A direct-acting space thermostat is controlling a normally open heating valve and a normally open cooling valve. A reversing relay is installed between the heating valve and the cooling valve. The branch pressure from the controller is piped into the heating valve and then to the reversing relay. The signal from the reversing relay is piped to the cooling valve. On an increase in pressure from the thermostat, the heating valve sees this increase while the cooling valve sees a decreasing pressure. This sequence closes the heating valve and opens the cooling valve.

Input Pressure To the Relay (psig)	Output Pressure To the Cooling Valve (psig)
3	15
4	14
5	13
6	12
7	11
8	10
9	9
10	8
11	7
12	6
13	5
14	4
15	3

Selector Relays. A selector relay is used to compare, select and transmit pneumatic signals. The relay may be either a low select, high select or a high-low select. The relay receives two or more signals, compares them and selects and transmits either the lowest signal, the highest signal or both the lowest and the highest signals. It is also called a *discriminating relay*.

Example 13.17: A high-low select on a multizone air handling unit receives the input from 7 direct-acting zone thermostats. The highest pressure (12 psig) is sent to the cooling valve to allow only enough chilled water into the coil to cool the zone (zone 6) with the greatest requirement. The lowest pressure (4 psig) is sent to the heating valve to allow only enough hot water into the coil to heat the zone (zone 4) with the greatest requirement.

Zone	Pressures (psig)
1	5
2	7
3	9
4	4
5	11
6	12
7	8

Gradual Switches. A gradual switch is a manually operated device. It is used to select a branch air pressure (0 to 20 psig) to be delivered to a controller or controlled device. A typical application is for the gradual switch to receive a pneumatic signal from a transmitter.

The output pressure from the gradual switch can then be manually increased or decreased to raise or lower the set point on a receiver-controller.

Minimum Position Switches. A minimum position switch (*MPS*) is a gradual switch with a built-in high pressure selector relay. A typical application is to use a minimum position switch to position the outside air dampers on an economizer system. The minimum position switch is piped with main air. When the supply fan is energized, an electric signal is sent to the E-P relay. The E-P relay is energized and allows control air to pass to the mixed air controller. The mixed air controller sends a branch signal to the minimum position switch, which sends the signal on to the outside air dampers. The MPS pressure has been manually set (in this example, set point is 5 psig) to maintain the outside air dampers open to the required minimum position whenever the fan is operating. When the output pressure from the mixed air controller is less than the MPS set point, the switch will provide minimum pressure (5 psig) to the outside air damper actuator. This allows for the minimum volume of outside air required by local ventilation codes. When the fan is stopped, control air to the mixed air controller is blocked by the E-P and branch air is exhausted from the controller, MPS and outside air damper actuator.

Actuators

Control air pressure from a controller is used to position actuators. The actuator, sometimes called an *operator* or *motor,* consists of a cylindrical housing, air connection port, rubber diaphragm, piston, spring and a connector rod to the controlled device. The operation of the actuator is such that as the control air pressure is introduced into the actuator, the diaphragm begins to expand. As the diaphragm expands it forces the piston outward against the spring, driving the connector rod out. As air pressure is increased, the connector rod is forced to the maximum of the spring range. When air is removed from the actuator, the spring's tension returns the piston to its normal position.

Example 13.18: An actuator has a spring range of 3 to 7 psig. The actuator connector rod is in its normal position when the air pressure is 3 psig or less. Between 3 and 7 psig, the stroke of the connector rod is proportional to the air pressure (5 psig would mean that the rod is halfway extended). Above 7 psig the maximum stroke is achieved.

Controlled Devices

The automatic controlled devices used in HVAC systems are dampers (or air valves) for air flow and temperature control and water (or steam) valves for temperature and water or steam flow control.

Dampers and Air Valves

Dampers or air valves are either normally open or normally closed. The damper position (air valve) is determined by the way the damper (air valve) is connected to the actuator. If the damper (air valve) opens when the actuator is at minimum stroke, the damper (air valve) is normally open. If, on the other hand, the damper (air valve) closes when the actuator is at minimum stroke, the damper (air valve) will be normally closed. Damper (air valve) actuators may be directly or remotely connected to the damper (air valve).

Multiblade Temperature Control Dampers. Some systems will use multibladed automatic face and bypass dampers for temperature control of coils. Most large air conditioning systems have multibladed automatic temperature control dampers to regulate the mixture of outside air and return air. The operation of these dampers is controlled by temperature requirements of the system, and not by airflow requirements. These multibladed dampers are either opposed blade or parallel blade dampers (see Chap. 4, Air Distribution Components). The terms "opposed" and "parallel" refer to the movement of the adjacent blades. The opposed blade damper has a linkage which causes the adjacent blades to rotate in opposite directions, resulting in a series of slots that become increasingly narrow as the damper closes. This type of blade action results in a straight, relatively uniform airflow pattern sometimes called "nondiverting."

The parallel blade damper blades all rotate parallel to each other. A parallel blade damper has a "diverting" pattern, because when closing, the damper blades have a tendency to divert the air sideways. This type of pattern is beneficial when properly used to mix incoming outside air and return air. However, a diverted flow pattern may adversely affect coil or fan performance if the damper is located too closely upstream. All temperature control dampers should be tight shut-off.

Water Valves

Automatic temperature control water valves (see Chap. 7, Water Distribution Components) are used to control flow rate or to mix or divert water streams. Valves are classified according to body design, control action and flow characteristics.

Body Design. Valves are constructed for either two-way or three-way operation. Two-way valves may be either single-seated or double-seated. The single-seated valve is the most common. The valve must be installed with the direction of flow opposing the closing action of the valve so the flow and pressure tend to hold the valve open. If the valve is installed the opposite way it will cause chatter, because as the valve modulates towards the closed position, the velocity of the water around the plug becomes very high. At some point near closing, the velocity pressure overcomes the spring resistance and forces the valve closed. Then, when flow is stopped, the velocity pressure goes to zero and the spring takes over and opens the valve. The cycle is repeated and chattering is the result.

The double-seated (or balanced valve) is used when high differential pressures are encountered and tight shut-off is not required. The flow direction through this valve tends to close one port while opening the other port. This design creates a balanced thrust condition which enables the valve to close off smoothly without water hammer, regardless of the differential pressure.

Three-way valves are classified according to their internal construction and not by their application. The internal difference is necessary so the valve will seat against flow. The classifications are mixing and diverting. A mixing valve has two inlets and one outlet. The diverting valve has one inlet and two outlets. Either valve may perform flow control in a bypass application or temperature control in a mixing application, depending on where the valve is installed in the system. These valves combine the two water streams. Diverting valves should not be substituted for mixing valves and vice versa. Using either design in the wrong application would tend to cause chatter.

Another type of three-way modulating valve also has two inlets and one outlet but does not mix or divert the water streams. This type of three-way valve is used in several applications. One application is in the supply line to coils in a three-pipe system. One inlet is supplied with heating water and the other is supplied with chilled water. The valve varies the amount of heating and chilled water. Depending on the temperature in the occupied space, the valve opens to allow either heating water only or chilled water only into the coil. This nonmixing three-way valve is also used in the supply line of four-pipe, one coil systems. The return line also has a three-way valve. This valve has one inlet and two outlets and is two position. The water leaving the coil enters the valve and is diverted to either the heating water return or the chilled water return, depending on the temperature of the water entering the coil. For example, if the supply valve is allowing chilled water into the coil, the return valve will divert this water to the chilled water return line.

Control Action. The control action (or valve position) is either normally open or normally closed and is determined by the way the valve is connected to the actuator. If the valve closes when the actuator is at minimum stroke, the valve is normally closed. If, on the other hand, the valve opens when the actuator is at minimum stroke, the valve will be normally open. Generally, normally open valves with low operating ranges are used in heating applications. Normally closed valves with higher operating ranges are used in cooling applications. This will allow for sequencing valves without simultaneous heating and cooling and for the system to fail safe to heating.

Flow Characteristics. A valve's flow characteristic refers to the relationship between the percent of plug lift to the percent of flow. Control valves have three basic types of seat plug configurations: quick opening, linear and equal percentage (Fig. 13.4).

- *Quick opening valve:* The two-position quick opening valve has a flat plug which delivers nearly maximum flow at about 20% lift. A typical application for a quick opening valve might be on a stream or water preheat coil where it is important to have maximum fluid flow as quick as possible.

- *Linear valve:* In a linear valve, the percent of plug lift and percent of flow are proportional. For instance, if the lift is 30%, the flow is approximately 30%. This type of valve is used in chilled water systems.

- *Equal percentage:* For heating or hot water systems where a small amount of flow results in a large heating capacity, the equal percentage valve should be used. With this type of valve each equal increment of plug lift increases the flow by an equal percentage and will provide a better relationship between lift and capacity.

FIGURE 13.4.   WATER VALVE FLOW CHARACTERISTICS

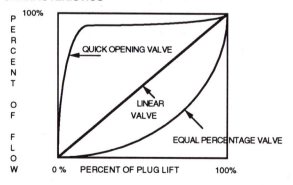

ELECTRIC/ELECTRONIC CONTROL SYSTEMS

The power source for an electric control system is low voltage electricity (normally 24 volts ac) or line voltage electricity (normally 110 to 220 volts ac). The power source for an electronic control system is low amperage or low voltage electricity (normally 4 to 20 milliamps dc or 0 to 15 volts dc). The typical components of an electric or electronic system are:

Sensors
Control wiring
Controllers
Relays and switches
Actuators
Controlled devices

Sensors

A sensing element, either internal or remote of the controller, measures the controlled variable (temperature, humidity and pressure) and sends a signal back to the controller.

Temperature Sensors. Temperature sensors sense a change in temperature. There are two general types of electronic temperature sensing elements.

- Thermistors or resistance temperature detectors (*RTD*) which sense a change in temperature with a change in electrical resistance.
- Thermocouples, which sense a change in temperature with a change in voltage.

Humidity Sensors. Humidity sensors (or hygrometers) are used to measure relative humidity or dew point. Electronic hygrometers sense changes in humidity from either changes in capacitance or resistance in the electronic circuitry.

Pressure Sensors. Electronic pressure sensors sense changes in pressures using mechanical devices and then convert this signal to produce current or voltage.

Wiring

The control wiring conveys the electricity to the various controllers. For electrical control systems the voltage is 24 volts ac (low voltage), or 110/220 volts ac (line voltage). Electronic control systems use a voltage of 0 to 15 volts dc, or a current of 4 to 20 milliamps dc.

Controllers

Types of Control. Electric and electronic controllers can control flow using any of the following methods: proportional (modulating), two-position, two-position timed, floating control. They may be single-pole, double-throw (SPDT), or single-pole, single-throw (SPST).

- Proportional control uses a reversible motor with a feedback potentiometer.

- Two-position control is used simply to start or stop a device or control a spring-return motor.

- Two-position timed control uses SPDT circuits to actuate unidirectional motors.

- Floating control uses a SPDT circuit with a reversible motor.

Electronic Direct Digital Controllers. Electronic-direct digital controllers use digital computers to receive electronic signals from sensors and converts the signals to numbers. The digital computer (microprocessor or microcomputer) compares the numbers to design conditions. Based on this comparison, the controller then sends out an electronic or pneumatic signal to the actuator.

Direct digital control (*DDC*) differs from pneumatic or electric/electronic control in that the controller's algorithm (sequence of operation) is stored as a set of program instructions in a software memory bank. The controller itself calculates the proper control signal digitally, rather than using an analog circuit or mechanical change. Interface hardware allows the digital computer to receive input signals from sensors. The computer then takes the input data, and in conjunction with the stored algorithms, calculates the changes required. It then sends output signals to relays or actuators to position the controlled devices.

Direct digital controllers are classified as preprogrammed or operator-programmable control. Preprogrammed control restricts the number of parameters, set points and limits that can be changed by the operator. Operator-programmable control allows the algorithms to be changed by the operator. Either hand-held or console type terminals allow the operator to communicate with and, where applicable, change the controller's programming.

Actuators

There are two general categories of electric/electronic actuators. One is the solenoid type of actuator. It consists of a magnetic coil operating a moveable plunger. It is limited to the operation of smaller controlled devices. Most solenoid actuators are two-position.

The other category of actuators is the electric/electronic motor. These motors are further classified as unidirectional, spring-return or reversible.

Unidirectional Motors. The unidirectional (or one direction) motor drives the controlled device open or closed during one-half revolution of the shaft. It is a two-position control. Once started, the motor will continue opening or closing the controlled device until it goes through its half cycle. This is regardless of any secondary action by the controller. A limit switch is installed to stop the motor at the end of each stroke. If the controller is not satisfied, the motor will hold the controlled device at this position until the controller is satisfied. At this time, the motor will drive the controlled device to the opposite position.

Spring-return Motor. The spring-return motor is also two-position. The motor drives the controlled device to one position and holds it there. When the circuit is broken the spring returns the controlled device to its normal position.

Reversible Motor. The reversible motor is used for proportional or floating operation. It can run in either direction and can stop in any position. It is sometimes equipped with a spring return so that if control energy (electricity) is removed, the spring will drive the controlled device to its normal condition.

In proportional control applications the motor assembly also contains a balancing relay (in some cases the balancing relay is external) and a feedback potentiometer. The potentiometer's wiper is mechanically connected to the motor shaft and moves with it. On a change in the controlled variable, the wiper on a potentiometer in the controller moves and changes the resistance between the wiper and either end of the potentiometer. This causes the balancing relay to pivot to one side, completing a circuit which drives the motor shaft. As the shaft moves, so does the wiper on the feedback potentiometer. This changes the resistance in the feedback potentiometer in the opposite direction of the controller potentiometer. When the resistances balance out, the motor stops.

CHAPTER 14 Establishing Goals for Optimizing and Retrofitting the HVAC System

The goal of an HVAC system is to provide a high degree of occupancy comfort and safety, and to maintain conditions for any process function while holding operating costs to a minimum.

HVAC systems need optimizing because they become obsolete. Systems or components become obsolete for three basic reasons:

- They simply wear out. This is commonplace but unnecessary. If a machine is well maintained, in theory, it would never wear out. Predictive and preventative maintenance (PPM) means scheduling maintenance and replacing parts in a timely manner so the system will remain operational.

- Newer technology creates better ways to achieve results.

- The system does not perform well, either energy-wise or comfort-wise. This may be because of deficiencies in the initial design or installation, or inadequate or improper maintenance.

EVALUATING THE HVAC SYSTEM TO DETERMINE WHETHER IT NEEDS OPTIMIZING

First, let's address whether the HVAC system needs optimization. There are many ways to evaluate whether or not the system is in need of optimization. For instance, general appearance, equipment down-time, maintenance records, maintenance costs and occupant complaints are items that need to be taken into consideration. However, one of the best ways to evaluate how the system is performing, if energy usage costs are a major concern, is by comparing energy used over several years. To make this comparison you will need to develop a Building Energy Usage Number (BEUN).

To develop the BEUN (Fig. 14.1), gather the utility records for the past 24 months. If you are not familiar with the rates charged by your energy supplier, you will need to learn about commodity rates, demand rates, discounts, taxes, on- and off-peak rates, power factor rates, ratchet charges, etc. Determine the total energy used by the building's HVAC systems for one year. This will include electricity, natural gas and any other commodity (such as liquid petroleum gas or oil) used to operate the HVAC system. Convert the energy used to Btu per year. Divide the Btu per year by the square feet of the building's conditioned space only. This is your BEUN for the base year in Btu per year per square foot. You can now compare your BEUN with other similarly constructed and used buildings in the area. You

FIGURE 14.1.   BUILDING ENERGY USAGE NUMBER

BUILDING				YEAR		
Energy	**Energy Units**	**Energy**	**Conversion**	**Btu**	**Space Area**	**Btu per S. F.**
Source	**Used Per Year**	**Units**	**Factor**	**Per Year**	**Square Feet**	**Per Year**
Electricity		KWH	3,413			
Natural Gas		THERMS	100,000			
Natural Gas		MCF	*			
Fuel Oil		GAL	*			
LP Gas		GAL	*			
Coal		TN	*			
Steam		MLB				
Other						
Total		******	******			

Definitions
kWh = kilowatt hours
mcf =1000 cubic feet
gal = gallons
tn = tons
mlb = 1000 pounds
BTU/yr = energy units used per year × conversion factor
BTU/sf/yr = BTU/yr divided by sf

*Consult supplier for conversion factor.

may obtain BEUN information for other buildings from federal, state or local government agencies or the utility company. To determine if a retrofit of the system is advisable, you may also need other information including occupancy data, such as:

- the hours the building is occupied
- the type of work performed
- the number of people per shift, etc.

You will also need weather data for the base year and present year, data on how the building is constructed, and operation and maintenance logs and manuals. Next, identify the type of systems and their interaction with each other. Then determine system performance. Evaluate system maintenance. After gathering all the information needed to develop the BEUN and evaluate the HVAC systems, you can set goals for optimization. *Note:* Only collect as many facts and opinions as needed to solve the problem/opportunity. Don't use data collection as an excuse for procrastinating. Be as objective as possible.

ESTABLISHING OPTIMIZATION AND RETROFIT GOALS

First, set up the survey team, consisting of staff, maintenance personnel, consultants, contractors and the utility company. Then, survey the facility to determine system performance and the retrofit opportunities. Define the optimization/retrofit opportunity or problem. This may sound simple, but it is not necessarily easy to do. Most of us gloss over the real problem/opportunity and deal with the symptoms. We never really solve the problem or take advantage of the opportunity. What should you do? Ask yourself:

- What is the real problem/opportunity?
- Am I stating the problem/opportunity or just a symptom of the problem/opportunity?
- Am I stating the problem/opportunity correctly?
- Am I stating the problem/opportunity in specific terms?
- Is there more than one problem/opportunity? Can I separate them?

After answering these questions, you can then set challenging but realistic optimization/retrofit goals. Next you should divide large goals into smaller (and more manageable) specific goals. For example: The goal is to reduce the energy consumed by the air handling motors by 10% in the next twelve months. This will be accomplished by cleaning filters and coils, repairing leaks in the duct and air balancing the systems.

Determine what resources you will need to achieve the goals, in terms of staffing and money. Then, assign responsibilities for achieving the goals. Finally, set a time schedule to reach these goals.

The next step is to act upon your goals. Plan your goal and then take action. Plan your work and work your plan. Action cures fear. Finally, monitor the results. Review and revise your objectives as needed.

How to Prioritize Optimization and Retrofit Opportunities

The six ways to prioritize optimization and retrofit goals are: financially, energy usage, safety conditions, comfort conditions, scheduling and system interaction.

Financially

Management can evaluate the retrofit proposal as they would any financial investment based on tangible benefits—benefits on which dollar values can be placed.

- Increase revenue.
- Reduce or eliminate an existing expense (cost displacement).
- Avoid future expenses (cost avoidance).

Energy Usage

Evaluate the energy used by each of the three systems that make up the facility.

- Energized systems (the HVAC systems).
- Nonenergized systems (the building envelope).
- Occupancy (the people).

Safety Conditions

Safety should always be a main concern.

Comfort Conditions

Evaluate the comfort level in the facility and/or the number and nature of the complaints. Comfort may be classified as an intangible benefit—a benefit not qualified in financial terms.

Scheduling

Some retrofits may need to be accomplished before others are started. If work is done out of sequence it may cause a delay in the completion of the retrofit, a reduction in energy savings and/or use of additional energy. Schedule for time of year and/or typical weather concerns.

System Interaction

Some retrofit measures may be cost effective when viewed by themselves, but become counterproductive when analyzed in relation to other systems. Always evaluate the system or component to be retrofitted in relation to the operation of other components or systems.

Choose the Best Solution for the Optimization/Retrofit Opportunity

Evaluate the information gathered.

Brainstorm.

Combine ideas.

List the advantages and disadvantages of each solution presented.

Select the solution that will produce the expected result.

Never accept anything to be true if you are not certain of it.

Divide each objective into as many parts as necessary for the solution.

Start with the simplest solution and proceed to the more complex.

Be as thorough as possible.

Review objectives and solutions until you know that nothing has been omitted.

Incorporate Your Goals into a Long-Range Plan for Optimizing or Retrofitting the HVAC System

The written plan is updated periodically and should state the following:

A statement of goals and objectives which include:

Intitial cost of the retrofit

Yearly costs to maintain the system

Payback period

Yearly savings

Return On Investment (ROI)

Increased occupant comfort and safety

Increased equipment reliability

Increased equipment life

Increased efficiency

Start time for the project

The Building Energy Usage Number (BEUN).

Reports on the HVAC system performance and condition.

A procedure for monitoring the retrofit.

A program for training operation and maintenance personnel and building occupants as applicable.

Detailed description of each planned retrofit including:

Priority

Cost

Projected results

Current status

Schedule

Detailed description of each implemented retrofit including:

Priority

Cost

Current results

Current status

Schedule

Detailed description of each completed retrofit including:

Cost

Results

Status

Figure 14.2 provides a sample format for an energy management plan.

FIGURE 14.2.   **SAMPLE FORMAT: ENERGY MANAGEMENT PLAN**

I.   GOALS
Statement of energy management goals.

II.   ENERGY USAGE
The Building Energy Usage Number (BEUN) documenting energy consumption for the previous 12 to 24 months.
Summary of the building's energy history and trends.
Forecast of future energy usage and energy cost.

III.   ENERGY SURVEY
A list of the HVAC systems and hours of usage.
System performance and equipment condition.

IV.   FACILITY DESCRIPTION
General information
Construction
Size
Specific information
Geographic location
Building orientation
Occupancy

V.   RETROFIT MEASURES
PROPOSED RETROFIT MEASURES
Retrofit Description
Priority
Objective
Cost
Status
Expected Completion Date
CURRENT RETROFIT MEASURES
Retrofit Description
Priority
Objective
Cost
Current results
Current status
Schedule

VI.   COMPLETED RETROFIT MEASURES
Retrofit Description
Objective
Total Cost
Results—Total Savings—Payback
Current status

VII.   PROCEDURE FOR MONITORING AND TRACKING RETROFITS

VIII.   TRAINING PROGRAM
An outline of the energy management and maintenance training program for the operations and maintenance personnel.
An outline of the energy management training program for the building occupants.

CHAPTER 15 HVAC Unit Operation, Maintenance, Optimization and Retrofit

This chapter will provide information on the operation and maintenance of some of the components in an HVAC system and ways to increase component and system efficiency. When working on your systems, not only for your safety, the safety of others and the safety of the equipment, always refer to the manufacturers published information for instructions on proper maintenance and operation.

CHECKLIST FOR AIR SIDE OPERATION, MAINTENANCE AND OPTIMIZATION

- Check that the fan is rotating in the correct direction.
- Clean the fan blades.
- Clean the filters and coils.
- Repair leaks in the duct system.
- Avoid installing restrictive ductwork on the inlets and discharges of fans.
- Install balancing dampers.
- Proportionally balance the system.
- When reducing airflow, change fan speed instead of closing main dampers.
- Design or retrofit the system for lower air flows. Horsepower is proportional to the cube of the air flow. Change fan speed.
- Design for the lowest pressure needed.
- Install variable volume systems where applicable.
- Reduce equipment on-time.
- Use air side economizers where applicable.
- Insulate duct.
- Reduce resistance in system.
- Maintain systems including controls.

Maintaining and Optimizing Filters

How often you'll need to change filters will depend on the type of HVAC used and the cleanliness of the environment in which the system is located. In order for filters to function properly they must be installed correctly. To help you with this, most filters will have an arrow showing direction of airflow. When changing filters, always turn the fan off. Not only does this make the job easier, but more importantly, the coil is still protected from dirt and debris which might be drawn in. To help you to determine when the filters need changing, a Magnehelic® gauge can be installed with measuring points on either side of the filters. This will give a pressure drop across the filters. When the pressure drop reaches a certain point, you'll know that the filters need to be changed. Also inspect the filter frame. It should be tight against all sides of the filter housing so that no air bypasses the filters.

Maintaining Coils

Coils need to be kept clean. A coil is a heat exchanger. If the coil is dirty or has debris impinged upon it, the coil loses some of its heat transferring surface and will become ineffective. This means not only a loss in efficiency, but also an increase in energy as the coil tries to maintain the normal operating conditions. Also check the fins on the coil. If the fins are mashed together, the surface area of the coil is reduced, which reduces the effectiveness of the coil. If the fins have been pushed in, they need to be straightened.

Maintaining Fans

Inspecting for System Effect

System effect is the term used to describe any condition in the fan plenum or distribution system that adversely affects the aerodynamic characteristics of the fan and reduces fan performance. Inspect the fan and the ductwork attached to (or around) the fan for any of the following outlet conditions which can cause system effect:

- *Straight Duct:* The length of straight duct on the fan outlet should be at least one duct diameter for each 1,000 feet per minute of outlet velocity, with a minimum length of two and a half duct diameters. For example, a fan with an outlet velocity of 1,500 fpm would need a straight duct of 2.5 duct diameters, while a fan with an outlet velocity of 3,000 fpm would need a straight length of duct of 3 duct diameters.

- *Elbows:* If an elbow must be installed closer than one duct diameter for each 1,000 fpm of outlet velocity, the ratio of the elbow radius to duct diameter should be at least 1.5 to 1.

- *Duct size:* The size of the outlet duct should be within plus or minus 10% of the fan outlet area.

- *Duct slope:* The slope of an outlet transition should not be greater than 15% for converging transitions or greater than 7% for diverging transitions.

- *Volume dampers:* Parallel-bladed (Chap. 4) are not recommended as volume dampers. When parallel-bladed dampers are installed at or near the fan outlet they may, when partially closed, divert the air to one side of the duct, resulting in a nonuniform velocity profile beyond the damper. This can create airflow problems for takeoffs close to the damper. Opposed-bladed dampers (Chap. 4) are recommended for volume control at the fan outlet.

- *Cutoff:* Check the fan cutoff plate for integrity and proper positioning.

Then, inspect the fan and the ductwork attached to or around the fan for any of the following inlet conditions which can cause system effect:

- *Duct size:* Inlet duct should be within plus or minus 10% of the fan inlet area.
- *Duct slope:* The slope of the inlet transition should not be greater than 15% for converging transitions or greater than 7% for diverging transitions.
- *Inlet duct:* The inlet duct or fan inlet should have a smooth, rounded entry to reduce inlet losses. A converging taper entry or a flat flange on the end of the duct will also reduce inlet losses. Inlet conditions can be improved by installing splitters or airflow straighteners in the duct.
- *Elbows:* An unvaned elbow at the fan inlet will create uneven airflow distribution into the fan. Losses can be reduced if a straight length of duct and turning vanes are added to the elbow.
- *Inlet cone:* Inspect the inlet cone for integrity and proper positioning.

Fan performance can also be reduced if the space between the fan inlet and the fan plenum wall is too close. There should be at least one half the fan wheel diameter between the plenum walls and the fan inlet. There should be at least one wheel diameter between inlets of fans in parallel.

Check the installation and operation of parallel fans. If one of the fans has a restricted inlet it may handle less air than the nonrestricted fan. This can result in fan pulsations which may reduce fan performance. The fan pulsations can also cause noise and vibration problems. The vibrations, if excessive, can damage the fan or the ductwork.

Lubricating Fan Bearings

Fan bearings must be lubricated to the manufacturer's specifications. Do not lubricate to excess.

Maintaining the Fan Wheel

The fan wheel should be kept clean. An accumulation of dirt on the wheel can reduce the performance characteristics of the fan. It can also put the fan wheel in an unbalanced condition, causing noise or vibration problems.

Checking Fan Speed

Check fan speed. The fan speed must be correct for the safety of the equipment. For example, if the fan is turning too slow, the evaporator (refrigerant) coil in a fan-coil unit may ice over. Icing of the coil can occur when the moisture in the air condenses on the coil and there is not enough warm air moving across the coil. The coil temperature drops to freezing and ice begins to form on the coil. Icing can also occur if the fan belt breaks, is thrown off or removed.

If the fan is turning too fast there could be damage to the fan wheel or the shaft and bearings. Calculate fan blade tip speed (Chap. 2) and compare with the manufacturer's maximum tip speed chart if there is a problem.

If fans are not operating at the proper speed, the conditioned space will not be able to maintain the correct temperatures and air changes per hour.

Changing Fan Speed. If you determine that a fan speed change is needed, use the following drive equations to determine the size of the sheaves needed to get the correct fan speed and airflow. You'll notice that the pitch diameter (Chap. 1) is used in the calculations. For field calculations, use the outside diameter of a fixed sheave for pitch diameter. For adjustable sheaves, when the belt is riding down in the groove, an approximation of the pitch diameter will be used for calculation purposes. Also, notice that increasing the size of a fixed pitch motor sheave, or adjusting the belts to ride higher in an adjustable motor sheave, will mean an increase in fan speed. Decreasing the size of a fixed pitch motor sheave, or adjusting the belts to ride lower in an adjustable motor sheave, will result in a decrease in fan speed. The opposite is true if you're changing the fan sheave instead of the motor sheave. In other words, increasing the pitch diameter of the fan sheave decreases the fan speed, while decreasing the pitch diameter of the fan sheave increases the fan speed.

The drive equations are:

$$rpm_m = rpm_f \times \frac{D_f}{D_m}$$

$$Dm = rpm_f \times \frac{D_F}{rpm_m}$$

$$rpm_f = rpm_m \times \frac{D_m}{D_f}$$

$$Df = rpm_m \times \frac{Dm}{rpm_f}$$

rpm_m = speed of the motor shaft
Dm = pitch diameter of the motor sheave
rpm_f = speed of the fan shaft
Df = pitch diameter of the fan sheave

After the size of the sheave is calculated, calculate the belt length needed for the drives to determine if the belts will also need changing. If it's necessary to change any of the belts on a multiple groove sheave, a matched set should be used. Because belt lengths and tension strengths vary, some belts could end up being too tight and others too loose, resulting in excessive wear.

The equation for finding belt length is:

$$L = 2C + 1.57(D + d) + \frac{(D - d)^2}{4C}$$

L = belt pitch length
C = center to center distance of the shafts
D = pitch diameter of the large sheave
d = pitch diameter of the small sheave
1.57 = constant (Pi/2)

Inspecting the Drives

Inspect the drives (sheaves and belts) for integrity, alignment and proper size. Verify the installed drive if a change is needed. Replace worn or broken drive components with a new component of the same size. If a different size component is needed, calculate the size of the new drive component and install.

Ordering Sheaves. You will need the following information to order the proper sheave:

- Motor and fan shaft diameter. To help you in measuring, remember that motor shaft diameters are in increments of 1/8 inch and fan shaft diameters are in increments of 1/16 inch.
- Brushing sizes: Sheaves may have a fixed bore which will fit the exact size of the shaft, or they may have a larger bore to accept bushings of various bore diameters that will fit different shaft sizes.
- Number of belt grooves.
- If the motor is mounted on an adjustable base, measure the amount of motor movement on the motor slide rail or adjustment rod to allow for adjustment of belt tension.

Installing Sheaves. To change the sheaves, first loosen and slide the motor forward for easier removal of the belts. Never force the belts over the sheaves. For proper removal or mounting of sheaves or adjustment of adjustable sheaves, consult the manufacturer's published data. Caution: Before trying to remove or adjust the pitch diameter of an adjustable sheave, be sure to loosen all locking screws. Be sure to lock all locking screws after adjustments are finished.

Aligning Sheaves. To prevent unnecessary belt wear or the possibility of a belt jumping off the sheaves, the motor and fan shafts should be parallel to each other and the motor and fan sheaves in alignment. To align the motor and fan sheaves use an electronic alignment instrument or:

- Place a straightedge from the fan sheave to the motor sheave. The straightedge is on the outside flange of the sheaves.
- Move the motor or the sheaves to an equal distance from the straightedge and the center of both fan and motor sheaves.

Installing Belts. After the sheaves are in place, install the proper size belts:

- Loosen and slide the motor forward.
- Put the belts on the sheaves and move the motor back to adjust the belts for proper tension.
- Secure the motor.
- After the belts have been installed, recheck the sheave alignment. After the first day's operation and again a few days later, recheck the sheaves, drive alignment and belt tension. Belts shouldn't be too tight or too loose. Slack belts will squeal on start-up. They wear out quicker and deliver less power. Belts with excessive tension will also wear

faster, cause excessive wear on shaft bearings and possibly overload the motor and drive. The correct operating tension is defined as the lowest tension at which the belts will perform without slipping under peak load conditions. A belt tension checker is available from some belt manufacturers.

Maintaining Belts. Monthly to quarterly, inspect the drive belts. Check for wear and proper tension. The tension can be checked by either using a tension gauge or using light hand pressure (depressing each belt at its middle). The proper tension, as determined by hand pressure, is when the belt will depress approximately one half to three quarters of an inch. An indication that a belt is too loose is if it squeals when the motor is started. Another indication that a belt is too loose is when belt dust is found on the inside of the belt guard. If the belt is too loose, not only will this mean premature failure of the belt, but the fan or other component being driven will not operate at the proper speed. If the belt is too tight, it could cause damage to the motor, fan shaft or bearings (in addition to wearing out the belts quickly).

CHECKLIST FOR WATER SIDE OPERATION, MAINTENANCE AND OPTIMIZATION

- Check the pump. Make sure it is rotating in the correct direction.
- Clean the strainer and coils.
- Avoid installing restrictive piping on the inlets and discharges of pumps.
- Install flow meters and provide balancing valves.
- Proportionally balance the system.
- When reducing water flow, change pump speed instead of closing main valves.
- Design or retrofit the system for lower water flows. Horsepower is proportional to the cube of the water flow. Trim impeller.
- Design for the lowest pressure needed.
- Reduce time equipment is on.
- Use water side economizers where applicable.
- Keep water coils, condensers, evaporators and cooling towers clean.
- Keep debris and other obstructions away from coils and towers.
- Use two speed or variable speed fans on cooling towers.
- Raise ΔT on heat exchangers.
- Use primary-secondary circuits and variable flow systems where applicable.
- Repair leaks in the pipe system.
- Repair and replace leaking valves.
- Repipe crossed-over piping.
- Insulate pipe.
- Reduce resistance in system. Clean strainers and restricted valves.
- Maintain systems (including controls).

Checking Pump Rotation

Start and stop the motor to "bump" the pump just enough to determine the direction of rotation. If the rotation is incorrect, reverse the rotation on a three-phase motor by changing any two of the three power leads at the motor control center or disconnect. On single-phase motors change rotation, as applicable, by switching the internal motor leads within the terminal box.

Inspecting for Pump Cavitation

Cavitation can occur in a water pump when the pressure at the inlet of the impeller falls below the vapor pressure of the water, causing the water to vaporize and form bubbles. The bubbles are entrained in the water and are carried through the pump impeller inlet to a zone of higher pressure where they implode with terrific force. Snapping and crackling noises at the pump inlet, severe vibration, a drop in pressure, a drop in brake horsepower, a reduction in water flow or no water flow are all symptoms of a cavitating pump. Cavitation usually results in pitting and erosion of the impeller vane tips or inlet.

The problem of cavitation can be eliminated by maintaining a minimum suction pressure at the pump inlet to overcome the pump's internal losses. This minimum pressure is called net positive suction head (Chap. 6). Generally, if pumping is limited to air-conditioning chilled water closed circuit systems, there will be enough pressure for good pump operation. Cavitation is not ordinarily a factor in open systems or hot water systems unless there's considerable friction loss in the pipe, or the water source is well below the pump and the suction lift is excessive. However, if the friction losses in the system are too great (meaning a lower than designed pressure at the pump suction), or the water temperature is high, the pressure at some point inside the pump may fall below the operating vapor pressure of the water, allowing the water to vaporize.

If the pump is cavitating, or the measured NPSHA is insufficient, check the suction line for undersized pipe, too many fittings, throttled valves or clogged strainers.

Using Water Side Economizers

As with air side economizers, the purpose of water side economizers is, when practical, to reduce or eliminate the need for mechanical refrigeration. "Free cooling" is available with open cooling towers when the outside ambient air temperature is low enough that a water side economizer system can be used to directly or indirectly cool the space load. There are two basic types of open cooling tower water side economizers, described in the following paragraphs.

Open Cooling Tower with Strainer

In warm weather, the strainer cycle system is a conventional vapor-compression refrigeration water-to-water (Chap. 10, Fig. 10.1) or water- (cooling tower/water-cooled condenser) to-air (evaporator coil in fan unit) system. In cooler weather, when the ambient outside air temperature is low enough, the cooling tower water is diverted from the condenser and circulated directly through the chilled water circuit (Fig. 15.1). Special strainers and water treatments are incorporated in the piping to minimize corrosion and fouling in the

FIGURE 15.1.

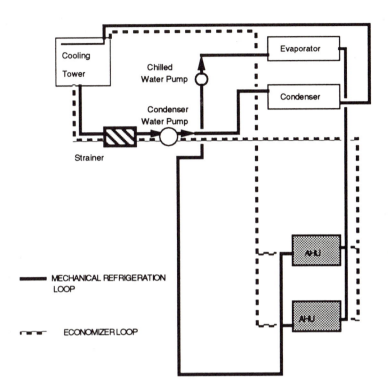

chilled water circuit. This type of water side economizer system has the greatest potential for energy savings.

Open Cooling Tower with Heat Exchanger

This water side economizer system is also a conventional vapor-compression refrigeration system during warm weather conditions. In cooler weather, when the ambient outside air temperature is low enough, the cooling tower water is diverted from the condenser through a plate-type heat exchanger (Fig. 15.2). The closed loop chilled water circuit is also diverted through the heat exchanger. There is an exchange of heat from the "warm" water in the chilled water loop to the "cooler" water in the condenser/cooling tower loop. The heat exchanger keeps the condenser water loop separate from the chilled water loop. Therefore, there is less potential for fouling problems than with the strainer cycle system, but there is also less potential for energy savings.

Using Variable Flow Systems

Variable flow systems are used to achieve either full- or part-load heating or cooling conditions while reducing the energy consumed by the pump. Variable flow systems may use a constant speed pump with two-way automatic control valves or a variable speed pump with either two- or three-way automatic control valves. Both types of systems reduce flow to maintain a constant water temperature difference across the terminal. The equation is $Q = $ gpm $\times 500 \times \Delta T$. As the heat load ($Q$) in the conditioned space changes, the flow (gpm) changes to maintain constant temperature (ΔT) across the terminal.

Variable flow systems that use a constant speed pump and two-way valves do not

FIGURE 15.2.

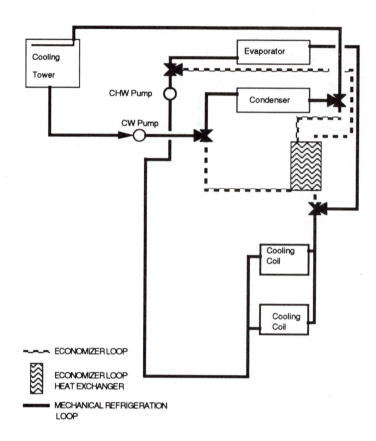

reduce the energy consumed by the pump nearly as much as variable speed systems. In a typical constant speed system, a temperature control device in the conditioned space sends a signal to a terminal's two-way valve to reduce or increase flow. If the valve closes to reduce flow there is an increase in resistance in the system. This increased resistance "backs the pump up on its curve", reducing water flow through the pump and decreasing the horsepower and the energy required by the pump. For example, a pump that uses 33.2 brake horsepower than circulating 1,250 gallons per minute of water at 82 feet of head will require only 29.9 bhp when pumping 1,000 gpm. The flow and the horsepower are reduced because the two-way valves in the system partially close to increase the resistance to 100 feet of head.

Systems that use variable speed pumping to control the water flow rate use an electronic device called a "variable frequency drive" to vary the speed of the pump motor and the pump. This type of system will reduce pumping horsepower approximately by the cube of the pump speed change. In the example above, if the system had variable speed pumping, the horsepower would be reduced to approximately 17 bhp with a reduction in flow from 1,250 to 1,000 gpm.

A differential pressure (*DP*) device is installed in the piping to control the speed of the pump. In a typical commercial system, as the temperature controls are satisfied, the automatic control valves for the terminals begin to close. The differential pressure devices sense the increase in system pressure and the pump speed is reduced to meet actual system demand.

To set the differential pressure on the DP device, the pump and the system are first set for full flow. The DP is then set to maintain the required differential pressure at that point in the system. For instance, if the device is placed in the ends of the supply and return mains, it is set for the required pressure drop across the last terminal in the system. This pressure drop includes the terminal's associated valves and piping. When the differential pressure device is located at the end of the system, the pump will operate at maximum energy savings.

If the DP device is placed closer to the pump, some energy savings will be lost because the pump will have to operate at a higher speed to overcome the pressure losses in the piping and terminals that are past the device. In some systems the DP may have to be installed closer to the pump to maintain control of the system. The DP is installed in the correct location when it will control the variable frequency drive so that the pump operates at the lowest speed to maintain required demand to all terminals.

Figure 15.3 shows a variable flow system. The system cooling coils are variable flow/constant temperature. The system maximum load is 2,000 gpm. The system pumps are sized for maximum load. These pumps are variable speed. As the load changes in the system, the differential pressure (DP) sensor senses the change and sends a signal to the pump motors' variable frequency drive (VFD). The chiller pumps are constant speed/constant volume and are sized for the chiller load. The temperature leaving the chiller is 45°F. The temperature entering the chiller is 55°F.

FIGURE 15.3.   VARIABLE FLOW WATER SYSTEM

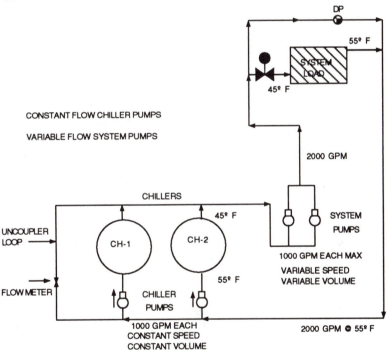

Example 15.1: Figure 15.4. The system load drops. The two-way (ATC) valves on the coils begin to close to reduce the water flow through the coils to maintain a constant Δ T (10°F). The DP sensor senses the increase in pressure in the system and sends a signal to the VFD to reduce the speed of the motors and respective pumps. To match the load, the system pumps reduce flow to 1,500 gpm. The water is entering the system coils at 45°F and leaving at 55°F. The water coming into the chiller loop is 1,500 gpm at 55°F. The pump for chiller two (CH-2) is the first to receive the water. This pump is constant speed and constant volume. It has been set up for 1,000 gpm. It must pump 1,000 gpm. Of the 1,500 gpm at

FIGURE 15.4.   VARIABLE FLOW WATER SYSTEM

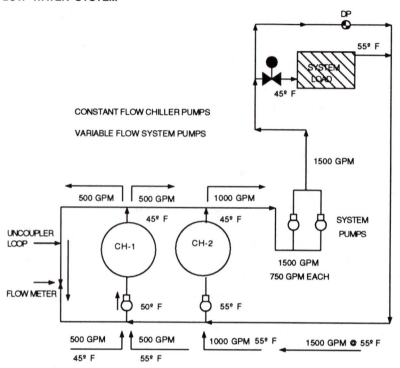

55°F coming into the chiller loop, 1,000 gpm at 55°F goes to CH-2. The other 500 gpm at 55°F goes to chiller one (CH-1).

The pump for chiller one is also constant speed and constant volume. It has been set up for 1,000 gpm. It must pump 1,000 gpm. This pump receives 500 gpm at 55°F from the system. It also receives 500 gpm at 45°F from the uncoupler loop. The result of this mixture is that the water entering CH-1 is 50°F. The chiller will cut back because the load on it has been reduced to a 5° Δ (50-45). The water flows through the uncoupler loop because that is the path of least resistance.

Example 15.2: Figure 15.5. The system load continues to drop. The two-way (*ATC*) valves on the coils begin to close more to reduce the water flow through the coils to maintain a constant Δ T (10°F). The DP sensor senses the increase in pressure in the system and sends a signal to the VFD to reduce the speed of the motors and respective pumps. To match the load, the system pumps reduce flow to 1,200 gpm. The water is entering the system coils at 45°F and leaving at 55°F. The water coming into the chiller loop is 1,200 gpm at 55°F.

Of the 1,200 gpm at 55°F coming into the chiller loop. 1,000 gpm at 55°F goes to CH-2. The other 200 gpm at 55°F goes to CH-1. CH-1 receives 800 gpm at 45°F from the uncoupler loop. The result of this mixture is that the water entering CH-1 is 47°F. The chiller will cut back because the load on it has been reduced to a 2° Δ. At this point (or some other set point), CH-1 sends a signal to stop its pump because the water temperature coming into the chiller is too low.

The chiller closest to the uncoupler loop shuts down first. If there were four chillers in this system, numbered from left to right, the sequence of shut down would be: CH-1, CH-2, CH-3 and CH-4. When the chillers are restarted the sequence is 4, 3, 2, 1.

The system pumps (Fig. 15.6) are moving 1,200 gpm. 1,000 gpm at 45°F from CH-2

FIGURE 15.5.   VARIABLE FLOW WATER SYSTEM

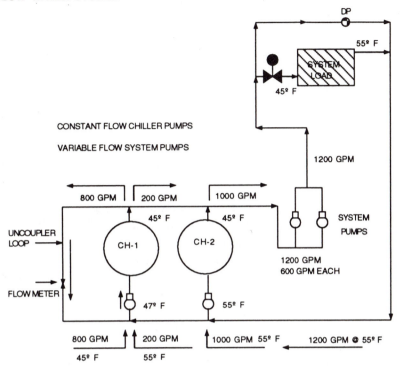

FIGURE 15.6.   VARIABLE FLOW WATER SYSTEM

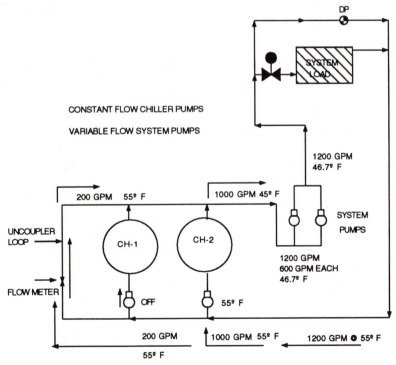

and 200 gpm at 55°F from the uncoupler loop. The flow through the uncoupler loop has been reversed (easiest path). The temperature of the water going out into the system is approximately 46.7°F. When the load on the system increases, the two-way valves open and there is a reduction in system pressure. The DP acts to maintain its set point and sends a signal to the VFD to increase water flow. The flow meter in the uncoupler senses an increase in flow above its set point (in this example, 200 gpm). It then sends a signal to start the pump for CH-1. When the chiller controls verify that the pump is operating at 1,000 gpm, the chiller will come on (Chap. 10).

MOTORS

The motors used on HVAC equipment will, in most cases, be either single-phase or three-phase alternating current induction motors. The number of motor phases will generally be determined by size. Smaller motors, ½ horsepower and below, will normally be single-phase, while larger motors will use three-phase current. The advantages of three-phase motors over single-phase motors are: (1) Three-phase motors have a capacity approximately 150% of that of single-phase motors of the same physical size, and (2) they are lower in initial cost and maintenance and they generally perform better than single-phase motors.

On most motors you will find an attached nameplate. The exceptions to this will be some very small motors (such as found on toilet exhaust fans. Of course, some nameplates will be missing, inaccessible, disfigured or painted over. The motor nameplate information that will be most important to you is: manufacturer, horsepower, phase, voltage, amperage, rpm and service factor.

Motor Terms. Let's take time to explain some of these motor terms. *Horsepower* (hp), is a unit of power. One horsepower equals 746 watts. *Brake horsepower* (bhp), is the total horsepower applied to the drive shaft of any piece of rotating equipment. *Phase* (PH) is the number of separate voltages supplying the motor. The *nameplate voltage* (v or volts) and *nameplate amperage* (*current*), or *full load amperage,* (*a, amps* or *fla*), are the rated operating voltage and full load operating current at rated horsepower. Many of the motors on HVAC units will be dual voltage motors. This means that depending on how the motor is wired, it is capable of operating from either of the voltages listed on the motor nameplate. For example, single-phase motors may be listed at 110/220 volts or 115/230 volts. A three-phase motor may list dual voltages of 220/440 volts, 230/460 volts or 240/480 volts. A dual voltage motor will also list dual amperage on the nameplate. For example, a three-phase, 60 horsepower, dual voltage motor has the following ratings: 230/460 volts, 140/70 amps. This means that if the motor is wired for 230 volts, the full load amps will be 140. Notice that volts and amps are inversely proportional. In other words, when the voltage doubles the amperage reduces by one half.

The rated rpm on the nameplate is the speed that the motor will turn when it's operating at nameplate horsepower. The rpm may vary, but only a small amount when more or less than rated voltage is applied. Some motors will have two to four different speeds available. The motor speed can be changed by switching the winding connections. Wiring diagrams are usually provided on the motor.

Motors basically consist of two windings. One is a stationary outside ring made of laminated steel and is called the **stator.** The other is the rotor which is a rotating cylindrical core inside, but separated from, the stator by a uniform air gap. Another term that you'll see on compressor motors more often than on fan motors is *locked rotor amperage* (or amps), abbreviated (LRA). Locked rotor amps is the starting current drawn from the line with the rotor locked and with rated voltage supplied to the motor. This amperage will generally be 5 to 6 times the full load amps. For this reason it is a good idea, when taking current readings with an analog ammeter that has a number of scales, to start with the highest scale and work down until the amperage reading is in the top half of the scale. The same is true for voltage readings with an analog voltmeter.

The *service factor* (SF), is the number by which the horsepower or amperage rating is multiplied to determine the maximum safe load that a motor may be expected to carry continuously at its rated voltage and frequency. For example, a 50 horsepower motor might have a service factor of 1.10. The service factor would allow this motor to operate safely at 55 horsepower (50 × 1.10). Typical service factors are: 1.0, 1.10, 1.15 and 1.25.

You may have heard the term *single phasing*. This is the condition which results when one phase of a three-phase motor circuit is broken or opened. Motors will not start under this condition, but if already running when it goes into the single phase condition, the motor will continue to run with a lower power output and possible overheating.

Motor Enclosure. The type of motor enclosure is another important factor to understand about HVAC motors. The "open protected" (or "open") motor is open for air passage through the windings for cooling purposes. The "totally enclosed" motor has a special housing design to prevent water damage to a motor installed in a wet atmosphere (such as a cooling tower). An "explosion proof" motor is designed to prevent a spark from the motor entering an area where it could cause an explosion.

Fully loaded, open motors are generally designed for a maximum temperature rise of 104°F above ambient air temperature. Totally enclosed and explosion proof motors are generally designed to operate at a maximum temperature rise of approximately 131°F. There's also an internal maximum allowable temperature rise that is indicated on the motor nameplate. Maximum internal temperatures can be over 200°F. However, even though the motor is not overheated and the surface temperature is well below 200°F, the motor could still be too hot to touch. Therefore, not being able to hold one's hand on the motor does not necessarily mean that the motor is overheated.

Maintaining Motors

Maintenance on motors is generally limited to proper lubrication of bearings and protection from dirt, debris and moisture.

Taking Power Readings

It's important to know how to take power readings on the motor for general information keeping and determining if the motor is in an overloaded condition. Hand-held portable multimeters capable of measuring current or amperage, voltage and resistance or ohms are used for electrical measurements in the field. There are several types of electrical multimeters that combine the functions of an ammeter, voltmeter and ohmmeter. Although all the various multimeters can measure voltage, current and resistance, the clamp-on volt-ohm-ammeters are generally used to measure voltage and current. The volt-ohmmeter is used to measure voltage and ohms. In addition to motor voltage and current, you may also need to take power factor readings. Power factor measurements are taken with a hand-held clamp-on power factor meter.

Measuring Voltage

You'll take voltage measurements using a voltmeter, which measures the difference in potential (another term for voltage) between phases for three-phase circuits. For a single-phase circuit it measures the potential between phase and neutral. A good technique is to read the phase-to-phase voltage from left to right (Fig. 15.7). For example, L(line)1 to L2, L1 to L3, and L2 to L3. The most accurate voltage will be read at the motor terminal box. However, it's usually much safer to take readings at the motor control center or at the disconnect box. The measured voltage should be plus or minus 10% of the motor nameplate voltage. For example, a motor is rated at 230 volts. The readings are: 228, 235 and 225 volts. This motor is acceptable because it is within the limits of plus or minus 10% of 230 (207 to 253 volts). If the motor voltage is not within the acceptable range, notify the authority having jurisdiction.

FIGURE 15.7.

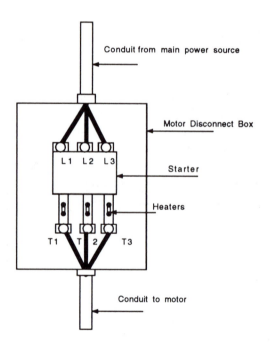

If the voltage isn't the same from phase-to-phase on a three-phase circuit—this is generally the case—there's a voltage unbalance. When there's a voltage unbalance, there's also a current unbalance, which can be as much as 10 times the percent of voltage unbalance. This means that the motor runs hotter than design. If the unbalance is large enough, it can reduce the life of the motor. Therefore the maximum allowable phase voltage unbalance for a three-phase motor is 2%. For more information, see Chap. 8.

Measuring Motor Current

Motor current is usually measured using a clamp-on ammeter. When reading current, as with voltage, all phases should be measured. Once again, when reading a three-phase motor, a good technique is to read from left to right (Fig. 15.7). For example, T(terminal)1 to T2, T1 to T3, and T2 to T3. Only one reading is required when measuring a single-phase circuit. That reading can be on either the hot wire or the neutral wire. The most accurate amperage will be read at the motor terminal box, but it's usually much safer to take readings at the motor control center or at the disconnect box.

The amperage measured on any phase shouldn't exceed the motor nameplate amperage. However, if the operating amperage reading is over the nameplate amperage, take one of the following steps to correct the problem.

Fan Motor

• If the reading is over the nameplate amperage but within the service factor and voltage limits, reduce the speed of the fan or close the main discharge damper until the amperage reading is down to nameplate or below.

• If the reading is over the nameplate amperage and outside the service factor limit, immediately turn the fan off and inform the person responsible for the fan. An exception is if the fan has been running and it's serving a critical area (such as a clean room or surgical room). If this is the case, leave the fan operating and immediately notify the person responsible for that fan's operation.

Pump Motor

• If the reading is over the nameplate amperage but within the service factor and voltage limits, close the main discharge valve until the amperage reading is down to nameplate or below.

• If the reading is over the nameplate amperage and outside the service factor limit, immediately turn the pump off and inform the person responsible for the pump. An exception is if the pump has been running and is a critical service. Immediately notify the person responsible for that pump's operation.

Power Factor

Power factor is generally read at the motor control center or at the disconnect box using a portable clamp-on digital or analog power factor meter. Power factor is measured on only one phase of a single-phase system and on each phase of a three-phase induction system. When reading a three-phase motor, a good technique is to read power factor from left to right.

Motor Control

The controllers for starting and stopping HVAC motors can be grouped into three categories: manual starters, contactors and magnetic starters.

• Manual starters are basically motor-rated switches that have provisions for overload protection. They are generally limited to motors of 10 horsepower or less. Manual starters are normally located close to the motor they control.

- Contactors are electro-mechanical devices that "open" or "close" contacts to control motors. Unlike manual starters, they can be remotely or automatically operated.

- Magnetic starters (Fig. 15.7) are contactors with an overload protection relay and are sometimes referred to as "mags" or "mag starters".

Protecting Motors From Overloading

If a motor becomes overloaded or one phase of a three-phase circuit fails (single phasing) there will be an increase in current through the motor. If this increased current draw through the motor lasts for any appreciable time and it is greatly in excess of the full load current rating, the windings will overheat. Damage may occur to the insulation, resulting in a burned out motor. To prevent motors from overheating, thermal overload devices must be installed. Single-phase motors often have internal thermal overload protection. This device senses the increased heat load and breaks the circuit, stopping the motor. After the thermal overload relays have cooled down, a manual or automatic reset is provided to restart the motor. Other single-phase and three-phase motors require external overload protection.

Installed external thermal overload devices, sometimes called "heaters" or "thermals" (Fig. 15.7) should be checked as part of the initial start-up of the motor and cataloging of equipment. Installed thermals will have a letter and/or number on them which is used for sizing. A chart identifying the thermals and their amperage ratings is usually found on the inside of the disconnect cover (Chap. 8).

CHAPTER 16

Retrofitting Single Zone, Terminal Reheat, Multizone and Dual Duct Systems

This chapter describes central fan systems, air side economizer systems, terminal reheat systems, multizone systems and dual duct systems including design intent. It provides details of how the systems operate, including airflow and control, water flow and control and refrigeration control. Finally, it outlines optimization and retrofit opportunities for increasing system performance for occupancy comfort and energy usage reduction. Items included will be reducing or eliminating: simultaneous heating and cooling, duct leakage, overpowered systems and thermal losses. Retrofitting to VAV system is also covered. Since each building and its HVAC system is unique, there may be other possible retrofit measures. Look for other opportunities by following the general guidelines in this chapter.

GENERAL GUIDELINES FOR ENERGY RETROFIT

- Understand the system.
- Survey system for ways to do the five "T's":
    Turn off, Turn down, Tune up, Turn around or Tear out
  For example:
    Turn off—Install a time clock.
    Turn down—Decrease fan speed.
    Tune up—Add capacitors.
    Turn around—Install a heart reclaim system.
    Tear out—Install a new system or component.

  Some examples of using the five "T's" are:

- Install a time clock or energy management system to turn off the fans when the space is vacant or the load can be shed or cycled.
- Adjust all dampers for tight shut-off. Install gaskets, end stops and side stops as necessary.
- Replace damper operators if they are not strong enough to fully drive the dampers closed.
- Adjust the minimum position switch to supply only the outside air that is needed. This switch can often be bypassed because the outside air damper leakage will satisfy the

minimum needs for ventilation. Another option is to modify the outside air duct to provide the minimum outside air quantity required. Install separate outside air dampers.

- Recalculate the actual required airflow.
- Reduce system effect by modifying the fan inlet and outlet connections to more nearly duplicate test rating conditions.
- Clean coils and comb the fins.
- Lower pressure drop across the filters. Clean filters. Install lower pressure drop filters.
- If outside and return air streams stratify, modify the dampers or install baffles to force mixing of the air streams.
- Calibrate and repair all automatic control components.
- Replace space thermostats with deadband thermostats to control the lowest heating and the warmest cooling temperatures to maintain comfort.
- Confirm that spring ranges on the hot water and chilled water valve operators will prevent simultaneous heating and cooling.
- Replace valve operators or install controls if the spring ranges overlap.

OPTIMIZING VENTILATION CONTROL

Occupied commercial and industrial buildings require a specified quantity of outside air for ventilation. Depending on the usage of the building, the outside air quantity will be approximately 15 to 40 cubic feet per minute (cfm) per person. In some buildings such as hospitals or chemical labs, where there exists the possibility of hazardous airborne materials, the HVAC system is supplied with 100% outside air.

Although the ventilation system may be ducted directly into the conditioned space, most systems are designed to combine outside ventilation air with the return air (Fig. 16.1). This conserves the energy needed to condition the air entering the heating and cooling coils.

FIGURE 16.1.

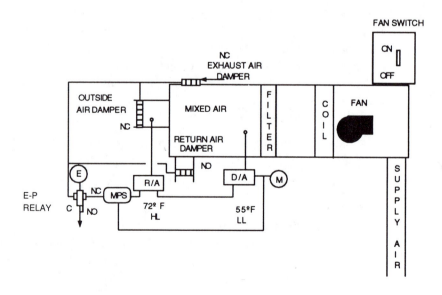

The combination of the return and outside air is called mixed air (*MA*). Any return air not used in the mixed air is exhausted to the outside and is called exhaust air (*EA*). The control of the return air, outside air, exhaust air and mixed air is called the "mixed air control" or "the economizer control." A mixed air, low limit thermostat (typically set for 55° F) modulates the outside air, return air and exhaust air dampers to maintain the mixed air temperature. Other controls such as a minimum position switch, outside air high limit and morning warm up low limit may be added to make the mixed air economizer system function more economically with better temperature control.

When properly controlled, the outside ventilation air can aid the heating, cooling and humidifying of the building spaces. It can also provide a positive static pressure in the conditioned spaces. This positive pressure reduces the amount of air infiltration. Commercial buildings will generally be pressurized at about 0.03 to 0.05 inches of water column static pressure. Automatic dampers are used to control the amount of outside ventilation air entering the building. Generally, pneumatic actuators are used to position the dampers.

Control Sequence

When the fan on-off switch is turned on the electric-pneumatic (E-P) relay is energized. This allows main air to travel through the normally closed (NC) and common (C) ports to the outside air damper actuator, driving the dampers open. When the fan switch is turned off, the E-P relay is deenergized and the air on the damper actuator is then exhausted through the E-P normally open (NO) port, closing the outside air dampers. The exhaust air and return air dampers open.

If freezing of water coils is a possibility, a preheat coil is installed between the outside air dampers and the other water coils (Fig. 16.2).

Hot water flows through the preheat coil. A direct-acting outside air low limit (OA-LL) controller operates a two-position, two-way valve. The controller's sensor is located in the outside air and is set for 40° F. When the outside air temperature drops to 40° F, the OA-LL opens the hot water valve (V-1) to the preheat coil. A reverse-acting high limit controller (HL) on the discharge of the preheat coil will close the valve when the tempera-

FIGURE 16.2.

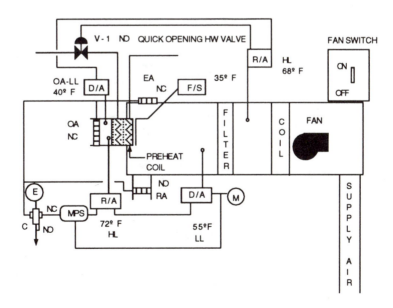

ture reaches 68° F so the plenum does not become overheated. A freezestat (FS) installed across the discharge of the preheat coil is set at 35° F. If the discharge air temperature drops to 35° F the fan is stopped, deenergizing the E-P relay and closing the outside air dampers.

Optimization Opportunities

- Install a 3-way valve (V-1).
- Install a recirculating pump in the pipe between the valve (V-1) and the preheat coil.

Figure 16.3 shows a preheating coil with a face and bypass damper. A low limit controller located in the air downstream of the coil varies the amount of air through the preheat coil. The set point for this controller is 60° F.

An outside air low limit (OA-LL) controls the valve (V-1) to the preheat coil. When the outside air temperature falls to 40° F, the valve opens. When the air temperature off the coil rises above the low limit set point of the downstream controller, the controller modulates the face dampers closed and the bypass dampers open to maintain set point temperature.

Optimization Opportunities

- Ensure tight sealing of dampers.
- Ensure that the dampers operate properly from full closed to full open.
- Ensure that stratification of the air downstream of the preheat coil does not occur. Install mixing baffles as required. Sufficient distance from the preheat coil to the water coils must also be available for adequate mixing.
- Consider retrofitting the system by removing the face and bypass dampers and adding a three-way mixing valve and recirculating hot water pump. The hot water is continuously

FIGURE 16.3.

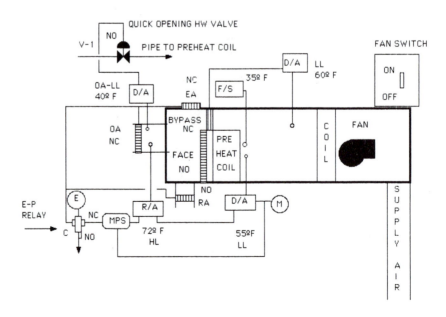

circulated through the coil to vary temperatures to meet requirements (see three-way valves in Chap. 7).

ENERGY RETROFIT FOR SINGLE ZONE SYSTEMS

The single zone system that will be described consists of a fan-coil central HVAC unit supplying air to a single conditioned space controlled by a thermostat. The basic intent of single zone systems is to heat and ventilate, or heat, cool and ventilate the space to design conditions. Humidity control may also be required.

Optimizing Heating Control

A basic heating application is shown in Fig. 16.4. The two-way heating water valve (*V-1*) is controlled from the space thermostat (*T-1*). The spring range of the actuator is 3 to 7 psig. The valve is normally open. The space thermostat is direct-acting and is set at 75° F.

Control Sequence

As the space temperature goes above thermostat set point, the valve closes, reducing water flow to the coil.

Optimization Opportunities.

- Lower thermostat set point temperature.

- Deactivate the thermostat and replace with a temperature controller in the return air duct as close to the space as possible (e.g., return air inlet).

- Make sure the spring range is correct.

FIGURE 16.4.

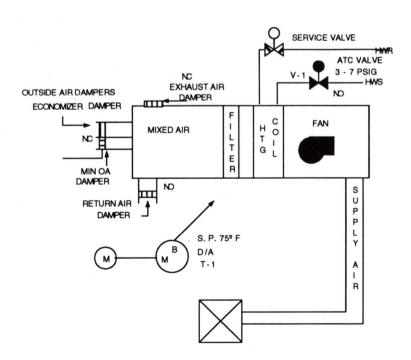

- Make sure the valve closes completely. Close the coil service valve and lower the thermostat set point below space temperature. This should close the valve. Check flow through the valve directly or by temperature.

- Install a high limit discharge thermostat to act as a feedback or anticipator control to decrease the operating differential between the duct temperature and the space temperature.

- Clean coil.

Figure 16.5 shows a heating system using face and bypass dampers controlled in sequence with the heating water valve by the space thermostat. The face and bypass dampers are controlled by an actuator with a spring range of 3 to 13 psig.

Control Sequence

If the air pressure from the thermostat is less than 3 pounds, the normally closed bypass dampers are closed. The normally open face dampers are open and the normally open water valve is open. As the space temperature rises above thermostat set point, the bypass dampers start to open and the face dampers start to close. The valve (8–13 psig) is full open. At 8 pounds, the face and bypass dampers are both at midpoint in their travel. The valve is still full open. As the temperature in the room continues to rise, the bypass dampers continue to open while the face dampers close. The valve is now starting to close. At 13 pounds, the bypass dampers are full open and the face dampers and the valve are full closed.

Optimization Opportunities

- Lower thermostat set point temperature.

FIGURE 16.5.

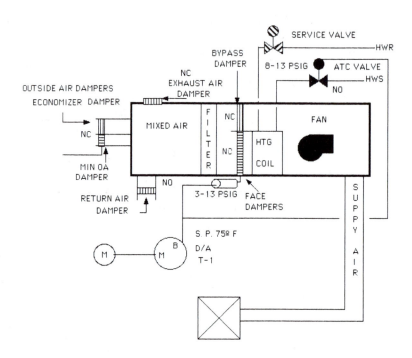

- Deactivate the thermostat and replace with a temperature controller in the return air duct as close to space as possible (e.g., return air inlet).

- Make sure the spring range is correct.

- Make sure the ATC valve closes completely. Close the coil service valve and lower the thermostat set point below space temperature. This should close the ATC valve. Check flow through the ATC valve directly or by temperature.

- Ensure that dampers open and close completely. This can be done visually or by measuring temperature or pressure differential.

- Install a high limit discharge thermostat to act as a feedback or anticipator control to decrease the operating differential between the duct temperature and the space temperature.

- Clean coil.

Optimizing Heating and Cooling Control

A basic heating and cooling application is shown in Fig. 16.6. The two-way heating water valve (V-1) to the heating coil (HC) is controlled from the space thermostat (T-1). The spring range of the heating valve actuator is 3 to 7 psig. The valve is normally open. The two-way cooling water valve (V-2) to cooling coil (CC) is also controlled from the space thermostat (T-1). The spring range of its actuator is 9 to 13 psig. The valve is normally closed. The space thermostat is direct-acting.

Control sequence: As the space temperature rises above the thermostat set point, the controller sends out an increasing signal to the valves. When the pressure reaches 7 psig, both the heating and cooling valves will be closed. At 9 psig the heating valve will remain

FIGURE 16.6.

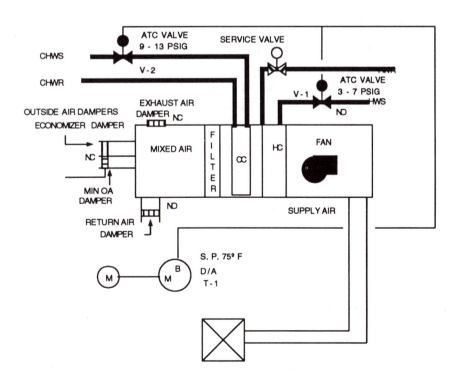

closed and the cooling valve will start to open. It will continue to open until fully open at 13 psig.

Between 7 and 9 psig, both valves will be closed and the system will be neither heating nor cooling. This is called a deadband and eliminates waste of energy from simultaneous heating and cooling.

Optimization Opportunities

- Deactivate the thermostat and replace with a temperature controller in the return air duct as close to the space as possible (e.g., return air inlet).

- Replace thermostat with a deadband thermostat set to control the coolest heating and warmest cooling temperature possible to maintain comfort level.

- Set the existing space thermostat as low as possible to maintain comfort. Install a direct-acting temperature controller in the return air to modulate the cooling water valve.

- Make sure the spring ranges are correct.

- Make sure the valves close completely. Close the heating coil service valve and turn the thermostat to a call for heating. This should close the cooling valve and open the heating valve. Check flow through the cooling valve directly or by temperature. If there is flow, or the temperature off the coil is cool, the cooling valve is leaking. Next, close the cooling coil service valve and turn the thermostat to a call for cooling. This should close the heating valve and open the cooling valve. Check flow through the heating valve directly or by temperature. Once again, if there is flow, or the temperature off the coil is warm, the heating valve is leaking.

- If the valves are overlapping (there is no deadband) install a retarding relay in the line to the cooling valve (see Chap. 13, Amplify or Retard Relay).

- Clean coils.

Optimizing Economizer Control

An air side economizer is used when it is advantageous to use outside air instead of mechanical refrigeration to cool the space. Figure 16.7 shows the basic dry bulb economizer cycle. The outside air (OA) normally closed, return air (RA) normally open, and exhaust air (EA) normally closed dampers are controlled by the mixed air low limit (MA-LL) controller. This controller is direct-acting and set for 55° F. The second control for this economizer is a reverse-acting outside air high limit (OA-HL) controller set for 72° F.

Control Sequence

When the air temperature in the mixed air plenum rises above the MA-LL set point (55° F, in this example), the controller will send an increasing branch signal to the OA, RA and EA dampers, moving them away from their normal positions. The OA and EA dampers will go open and the RA damper will go closed. This sequence will continue as long as the mixed air temperature is above set point and the outside air temperature is below set point (72° F). When the outside air reaches set point, the OA-HL will reverse the branch signal going to the dampers. The dampers will see a decreasing pressure which will cause them to return to their normal positions.

FIGURE 16.7.

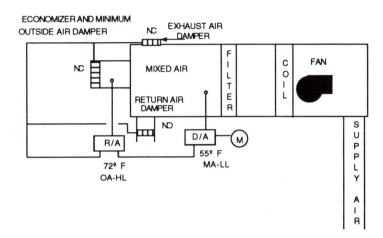

This economizer control scheme, sometimes called a "wild" or "partial" economizer can, depending on the circumstances, increase energy use and decrease comfort.

Example 16.1: It is a fall Monday morning. The outside air temperature is 50° F. The space temperature is 60° F. It is 6:00 A.M. and the time clock turns the fan on. The RA damper opens full and the OA damper opens to minimum. The air is circulated through the unit. The MA-LL senses a 60° F temperature. Since it is set for 55° F, it opens the OA and EA dampers and closes the RA dampers. Now, because the space temperature is below the thermostat set point, it sends a signal to the heating valve to open and we are heating up 50° F outside air.

A "partial" economizer will have some or all of the following controls:

- A direct-acting mixed air controller set at cooling temperature. This controller opens the outside dampers and the exhaust air dampers while closing the return air dampers when the mixed air temperature is above set point.

- A reverse-acting outside air lockout controller. This controller overrides the mixed air controller when the outside air is above its set point. The dampers return to their normal position.

- A minimum position switch. When the main air is on and the E-P is energized, the minimum outside air damper motor will receive enough air pressure so the outside air dampers will never close past the minimum required for ventilation and or space pressurization.

- An electric-pneumatic (E-P) three-way air valve which is energized when the fan is turned on. It allows main air to pass through the normally closed port into the common port and on to the dampers. When the fan is turned off, the normally closed port closes and the normally open port opens, which exhausts any air from the dampers and allows them to go to their normal positions.

To convert to "full" economizer cycle:

- Add a return air controller to modulate the outside, return and exhaust dampers to maintain space temperature. This direct-acting controller overrides both the mixed air and outside air controllers.

FIGURE 16.8.

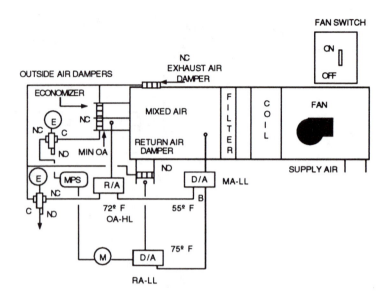

Optimization Opportunities

• Install a direct-acting return air low limit (*RA-LL*) morning warm-up control (Figure 16.8). Continuing with Ex. 16.1, the RA-LL would be set for 75° F. If the temperature in the return duct was below the set point, the RA-LL would bleed the air pressure from the damper actuators and the dampers would return to their normal positions—OA and EA closed, RA open. The dampers would remain at this condition until the air temperature in the return duct was above 75°. At this time, the mixed air controller and outside air controller would take over and modulate the dampers according to mixed air and outside air temperatures.

• Instead of operating one large damper for both minimum and maximum outside airflow install a separate minimum outside air damper.

• Install a minimum position switch (MPS)(see Chap. 13).

• Install an electric-pneumatic (E-P) switch to close the outside air dampers when the fan is off (see Chap. 13) .

• Make sure dampers have a tight seal. To get an idea of how much energy can be wasted by not optimizing dampers, let's look at the next example using outside air dampers.

Example 16.2: The minimum outside dampers are open to allow in 10% of the total supply flow. This means that return air is 90% of total supply air. If OA temperature is 60° F and RA temperature is 75° F, then the mixed air temperature is equal to 73.5° F. Mixed air temperatures:

$$MAT = (\%OA \times OAT) + (\%RA \times RAT)$$
$$MAT = (0.1 \times 60) + (0.9 \times 75)$$
$$MAT = 6 + 67.5$$
$$MAT = 73.5$$

When the OA temperature is 30° F and the RA temperature is 75° F, the mixed air temperature is 70.5° F.

$$MAT = (\%OA \times OAT) + (\%RA \times RAT)$$
$$MAT = (0.1 \times 30) + (0.9 \times 75)$$
$$MAT = 3 + 67.5$$
$$MAT = 70.5$$

This system has an airflow of 10,000 cfm. If the design discharge air temperature is 105°F, the heat transfer required will be 372,600 Btuh.

$$Btuh = cfm \times 1.08 \times TD$$
$$Btuh = 10,000 \times 1.08 \times 105 - 70.5$$
$$Btuh = 372,600$$

Next, the outside air dampers are open to allow in 40% rather than 10% outside air. The mixed air temperature is 57° F.

$$MAT = (\%OA \times OAT) + (\%RA \times RAT)$$
$$MAT = (0.4 \times 30) + (0.6 \times 75)$$
$$MAT = 12 + 45$$
$$MAT = 57$$

Now, the heat transfer required is 490,800 Btuh—an increase of 32%.

$$Btuh = cfm \times 1.08 \times TD$$
$$Btuh = 10,000 \times 1.08 \times 105 - 57$$
$$Btuh = 490,800$$

Let's see what happens on the cooling side. If the OA temperature is 90° F and the return air temperature is 75° F, then the mixed air temperature is 76.5° F.

$$MAT = (\%OA \times OAT) + (\%RA \times RAT)$$
$$MAT = (0.1 \times 90) + (0.9 \times 75)$$
$$MAT = 9 + 67.5$$
$$MAT = 76.5$$

Next, the outside air dampers are open to allow in 40% rather than 10% outside air. The mixed air temperature is 81° F.

$$MAT = (\%OA \times OAT) + (\%RA \times RAT)$$
$$MAT = (0.4 \times 90) + (0.6 \times 75)$$
$$MAT = 36 + 45$$
$$MAT = 81$$

If this is the case, then over four tons of extra refrigeration (just in sensible cooling) must be produced to maintain design conditions.

$$Btuh = cfm \times 1.08 \times TD$$
$$Btuh = 10,000 \times 1.08 \times 81 - 76.5$$
$$Btuh = 48,600$$

$$\text{Tons of refrigeration} = \frac{Btuh}{12,000 \text{ Btuh per ton}}$$

$$\text{Tons of refrigeration} = \frac{48,600}{12,000 \text{ Btuh per ton}}$$

$$\text{Tons of refrigeration} = 4.05$$

- Be sure the economizer dampers operate properly from full closed to full open. First, take a single point static pressure reading in the mixed air plenum. Then slowly operate all the economizer dampers. Observe the static pressure. It should remain relatively constant while one set of dampers is opening and the other is closing. If the static pressure varies greatly, one set of dampers is leading while the other is lagging. If the dampers are not tracking together it will result in wasted energy, improper airflow and reduction of system efficiency. See Chap. 19.

- Verify that the outside air dampers are set for minimum ventilation requirements. This can be done by measuring the outside air directly with a Pitot tube traverse of the OA duct, or with mixed air temperatures.

- Determine the outside air controller setpoint. One method to select a starting set point is to plot the average outdoor wet bulb and dry bulb temperatures for the economizer season on a psychrometric chart. Then plot the design room conditions (wet bulb and dry bulb) on the same chart. The starting set point for the OA controller is where the wet bulb line for the room condition intersects the outdoor average wet and dry bulb temperature line.

- In most cases a dry bulb controller will provide adequate energy savings—as long as the system is properly maintained. However, in areas of high latent load (wet bulb), you may want to consider converting to an enthalpy controlled economizer (Fig. 16.9). An enthalpy controller has a dry bulb and wet bulb/dew point sensor in the outside air and a similar sensor in the return air. If the enthalpy (total heat content) of the return air is greater than the enthalpy of the outside air, the enthalpy controller will pass the branch signal along to the mixed air controller. The economizer dampers will start to modulate the outside and exhaust air dampers open and the return damper closed. If the return enthalpy is lower than the enthalpy of the outside air, the economizer dampers will go to their normal positions. This type of control requires a considerable amount of maintenance. It is recommended that the sensors be monitored monthly. Extra sensors should be on hand for replacement while weathered sensors are returned to the manufacturer for reconditioning.

FIGURE 16.9.   ENTHALPY CONTROLLED ECONOMIZER

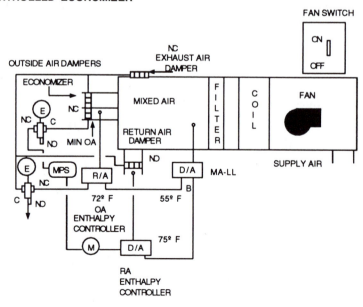

Controlling Humidity

There are a number of control schemes for decreasing or increasing the amount of moisture in the air. Decreasing the moisture in the air (dehumidifying) is accomplished when water is removed from the air as it passes over an evaporator or chilled water coil which is colder than the dew point of the air. The moisture that condenses out of the air is a byproduct of the cooling process and is a somewhat uncontrolled function. For controlled dehumidification, chilled water can be sprayed on the evaporator or cooling coil as the air passes over it, or by using an air washer which sprays chilled water directly into the air stream.

Humidification can be accomplished by using a steam humidifier, pan humidifier or an air washer. A steam humidifier ejects steam directly into the air stream. A pan humidifier evaporates heated water from a pan and an air washer sprays heated water directly into the air stream. Figure 16.10 shows a steam humidifier used to humidify an electronic assembly area where the control of static electricity is very important. The reverse-acting low limit room relative humidity controller (RH-LL) is set at 45% relative humidity. A reverse-acting high limit relative humidity controller (RH-LL) in the duct is set at 80% relative humidity. To prevent flooding in the duct if the control system fails, the steam valve is normally closed.

Control Sequence

As the relative humidity drops below the set point on the room RH-LL controller, an increasing signal is sent to the steam valve. When the relative humidity in the discharge duct reaches the set point of the high limit controller, this controller takes over control of the valve and maintains set point until the room controller is satisfied. The room controller then sends out a decreasing signal to close the steam valve.

FIGURE 16.10.   STEAM HUMIDIFIER

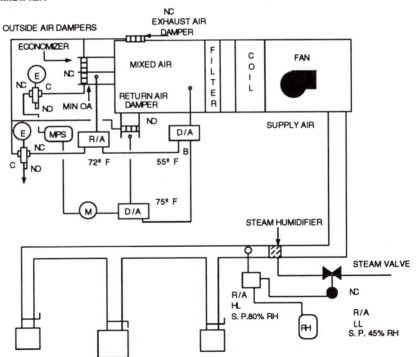

Optimization Opportunities

- Maintain humidifiers. Keep mineral deposits to a minimum. Use distilled or deionized water.

TERMINAL REHEAT SYSTEM

A terminal reheat system may be configured:

- *Air distribution*
  - —Single zone
  - —Multizone
- *Central unit*
  - —Cooling coil only—face and bypass damper
  - —Cooling and heating coils
- *Cooling coil*
  - —Chilled water coil
  - —Evaporator refrigerant coil
- *Heating coil*
  - —Steam
  - —Hot water
  - —Electric
- *Reheat coil*
  - —Steam
  - —Hot water
  - —Electric

Reheat systems are single path systems (the coils are in series) that provide a high degree of comfort control when operating correctly. However, even when operating properly, they are great energy wasters because the systems are heating and cooling simultaneously much of the time.

The terminal reheat system in Fig. 16.11 has a cooling coil in the central unit and hot water coils in each of its zones. The chilled water coil is controlled by a normally closed two-way valve with an actuator spring range of 8 to 13 psig and a direct-acting discharge air controller set for 58° F. The reheat coils are controlled by a direct-acting thermostat in each zone. The hot water valves are normally open with an actuator spring range of 3 to 7 psig.

Control Sequence

Chilled Water. When the discharge air temperature rises above set point, the controller sends an increasing signal to the chilled water valve. When the signal reaches 8 psig the valve starts to open. At 13 psig the valve is full open.

FIGURE 16.11.   TERMINAL REHEAT SYSTEM

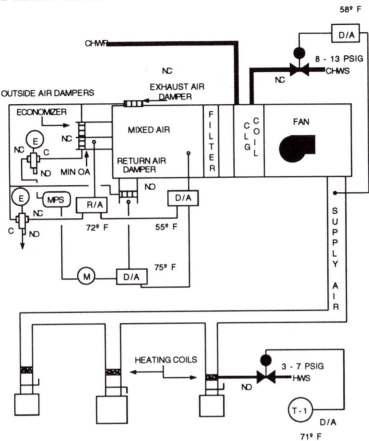

Hot Water Reheat. When the room temperature falls below set point, a decreasing signal is sent to the hot water steam valve. At 7 psig the valve starts to open. At 3 psig it is full open.

Electric Reheat. Electric strip heaters (Fig. 16.12) are usually controlled in steps of capacity required in response to the space thermostat. They may be controlled by either:

- Two-position control
- Timed two-position control
- Proportional control

Safety controls for electrical reheat include high limits (*HL*) and airflow switches (*FS*). Both a manual and an automatic reset may be required to reset the high limit control. The space thermostat (*T-1*) sends out a signal on a call for heat. If the minimum quantity of air is flowing across the heater, the flow switch (*FS*) allows the signal to pass to the contactor, and the heater is energized. If the heater overheats, a high limit (*HL*) will shut the heater off.

Optimization Opportunities

- Move the controller for the chilled water coil to the return air. This allows the controller to sense a temperature closer to the actual conditions in the space. Set the controller for 74° F.

FIGURE 16.12.   ELECTRIC STRIP HEATER

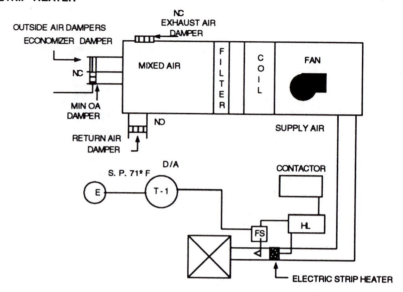

- Set the economizer morning warm up for 73° F.
- Repair any leaks in the distribution system.
  —Check joints.
  —Check vibration connection (flex connection, canvas connection).
- Clean the duct. Remove any restrictions (such as liner impinged against turning vanes).
- Reduce friction loss in the ductwork.
  —Straighten duct runs.
  —Keep flex duct to a minimum. Do not use flex duct runs longer than 7 feet.
  —Keep aspect ratio at 3:1 or less.
- Add turning vanes in supply and return elbows.
- Remove all unnecessary air diverting devices (such as pickups, extractors or splitters).
- Remove dampers near the face of grilles, registers and diffusers.
- Install single-bladed balancing dampers in all takeoffs.
- Clean all coils.
- Ensure that the flow switch on electric reheats will deactivate the heater with low airflow.
- Ensure that the high limit on electric reheats will deenergize the heater at excessive temperatures.
- Ensure that the reset(s) on electric reheats operate properly.
- Convert to variable air volume (VAV) system. Option 1:
  —Remove reheat coils and install VAV boxes. Some zones may need boxes with reheat coils.
  —Install variable frequency drive (VFD) on the fan motor.

—Install static pressure controls.

- Convert to variable air volume (VAV) system. Option 2:

    —Clean reheat coils.

    —Install automatic dampers after the reheat coils. All dampers will be normally closed.

    —Close the heating coil service valve to any zone which requires cooling only (generally interior zones).

    —Add a stop so that the dampers close to a minimum of 50% to any zone which requires heating and cooling (generally exterior zones).

    —Install variable frequency drive (VFD) on fan motor.

    —Install static pressure controls.

Control Sequence

- *Zones with reheat:* A direct-acting thermostat controls the heating valve and the VAV dampers.
- *Cooling only zones:* A direct-acting thermostat controls the VAV dampers.
- A static pressure sensor in the duct controls the VFD.

Example 16.3: The temperature in a cooling only zone falls below set point. The damper actuator is supplied with 13 psig. The thermostat sends out a decreasing signal. The damper starts to close. At 8 psig the damper is at minimum flow position (this may be as much as complete shut off—but it is not recommended). When the space temperature rises above set point, the thermostat sends out an increasing signal. The damper starts to open. At 13 psig the damper is at maximum flow position.

The temperature in a heating and cooling zone falls below set point. The damper actuator is supplied with 13 psig. The thermostat sends out a decreasing signal. The normally closed damper starts to close. At 8 psig the damper is at minimum flow position (50% open). If the space is still below set point, the branch pressure continues to fall. At 7 psig the heating valve starts to open. At 3 psig it is full open. When the space temperature rises above set point, the thermostat sends out an increasing signal. The valve starts to close. At 7 psig it is full closed. At 8 psig the damper starts to open from its 50% position. At 13 psig the damper is at maximum flow position and the space is supplied with 100%, 55° F temperature air.

As the dampers open and close, the static pressure controller sends a signal to the VFD which increases or decreases motor and fan speed to maintain system static pressure.

MULTIZONE SYSTEMS

A multizone system consists of a central fan unit with a heating coil and a cooling coil. The heating coil may be steam, hot water or electric. The cooling coil may be a chilled water coil or an evaporator coil. Some systems do not have a heating coil but will instead bypass mixed air around the cooling coil when the system is in a heating mode. A multizone system is a dual path system (the coils are in parallel) which maintains a constant airflow while supplying the space with the correct mixture of heated and cooled air to satisfy the space thermostat. Multizone systems provide a high degree of comfort control when operating properly. However, even when operating correctly, they are energy wasters because the systems are heating and cooling simultaneously much of the time.

The multizone system in Figs. 16.13A and 16.13B has a chilled water coil and a hot water coil. A two-way valve controls the water flow through the cooling coil. The heating coil has a secondary pump with a three-way mixing valve piped in a mixing application. The water flow through the cooling coil is controlled from the cold deck discharge air temperature set for 55° F. The controller is direct-acting. The chilled water valve has a spring range of 8 to 13 psig. The water flow through the heating coil is controlled from the hot deck air temperature and reset by the outside air temperature. The reset schedule is 1:1. For each degree that the outside air temperature drops, the hot deck temperature is reset upwards one degree.

RESET SCHEDULE

Outside Temperature °F	Hot Deck Temperature °F
70	70
50	90
30	110

At the central unit there are mixing dampers for each of the zone ducts. Generally, a multizone system will have 5 to 12 zones. This system has five zones. The duct where the heating coil is located is called the "hot deck." The "cold deck" is where the cooling coil is located. The mixing dampers are located in hot decks and cold decks after the coils. The hot and cold duct dampers are connected to each other. The hot deck dampers are normally open. The cold deck dampers are normally closed. There is one damper actuator with a spring range of 8 to 12 psig. A direct-acting thermostat is located in each zone.

Control sequence: As the temperature in the space (zone) rises above the set point of the zone thermostat, an increasing signal will be sent to the zone's mixing damper actuator. At 8 psig the hot deck dampers are full open and the cold deck dampers are full closed. At 10 psig the hot and cold deck dampers are at mid-range. At 12 psig the hot deck dampers are full closed and the cold deck dampers are full open.

FIGURE 16.13A.

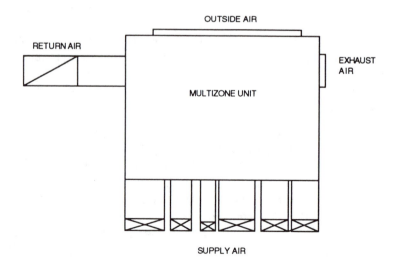

FIGURE 16.13B.

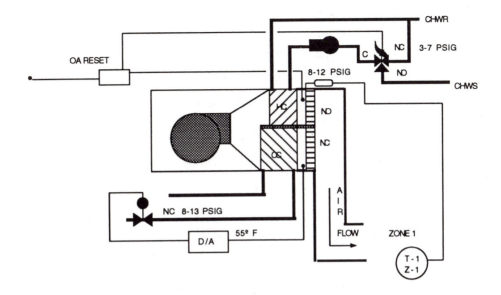

Optimization Opportunities

- Disconnect zone hot and cold deck dampers. Have separate actuators for both hot deck dampers and cold deck dampers.
- Install a discriminator or hi-lo select relay (Chap. 13) to control the hot and cold deck temperatures.
- Replace zone thermostats with deadband thermostats.
- Set thermometers to control the coolest heating and warmest cooling temperature possible to maintain comfort level.
- Blank off the hot deck side of any zone that requires cooling only.
- Lower zone thermostat set point temperature.
- Make sure spring ranges are correct.
- Make sure valves close completely.
- Clean coils.
- Ensure that dampers open and close completely.
- Be sure dampers are tightly sealed.
- Control the heating and cooling coils with a selector relay (Chap. 13).
- Repair any leaks in the distribution system.
    - Check transverse and longitudinal joints.
    - Check vibration connection (flex connection, canvas connection).
- Clean the duct. Remove any restrictions (such as liner impinged against turning vanes).
- Reduce friction loss in the ductwork.
    - Straighten duct runs.
    - Keep flex duct to a minimum. Do not use flex duct runs longer than 7 feet.

—Keep aspect ratio at 3:1 or less.

- Add turning vanes in supply and return elbows.

- Remove all unnecessary air diverting devices such as pick-ups, extractors, splitters.

- Remove dampers near the face of grilles, registers and diffusers.

- Install single-bladed balancing dampers in all takeoffs.

- Convert to VAV system.

DUAL DUCT SYSTEM

The dual duct system consists of a central unit with a heating coil and a cooling coil. As with the multizone system, the cooling coil may be either a chilled water coil or an evaporator coil. The heating coil may be steam, heated water or electric. The dual duct system maintains a constant airflow while supplying the space with a mixture of heated and cooled air to satisfy the space thermostat. This type of system wastes energy because it is heating and cooling simultaneously much of the time.

Figure 16.14 shows a dual duct system with a chilled water coil and a hot water coil. A three-way mixing valve controls the water flow through the cooling coil. This mixing valve is piped for a bypass application. The water flow through the heating coil is controlled by a two-way valve. The cooling coil valve is controlled from the cold deck discharge air temperature set for 55° F. The heating coil valve is controlled from the hot deck air temperature and reset by the outside air temperature.

The hot duct and the cold duct dampers in the mixing boxes are controlled by a thermostat in the space. The thermostat modulates the cold duct dampers from full open to full closed as the hot duct dampers are changing from full closed to full open. The box maintains

FIGURE 16.14.

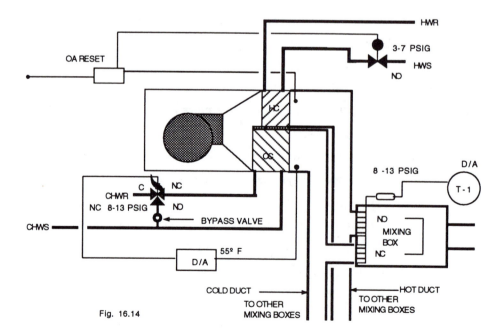

Fig. 16.14

a constant airflow while supplying the space with the correct mixture of heated and cooled air to satisfy the thermostat.

Optimization Opportunities

- Close the valve in the cooling coil bypass. This will make the cooling coil three-way valve operate as a two-way valve. A two-way valve is a variable flow system.
- Control cold deck discharge air temperature from return air temperature.
- Set thermometers to control the coolest heating and warmest cooling temperature possible to maintain comfort level.
- Blank off the hot deck side of any zone that requires cooling only.
- Install a deadband where the cold duct dampers in the mixing boxes will not open until the hot duct dampers are fully closed. These systems are a modified variable air volume system.
- Convert to a VAV system.
- Install a hot deck fan. The hot deck fan will only receive return air. The outside air will only be ducted to the cold deck fan. The cold deck fan will receive both return air and outside air.

CHAPTER 17 Testing the Air Side of Constant Volume Systems

This chapter covers field inspection and testing of the air side of constant volume systems. The field inspection consists of inspecting the building, air handling equipment and air distribution system. Testing the system includes recording relevant information on the system's fans, drives, motor, taking various system measurements and setting system controls.

Before beginning the field work, assemble and review various system documents. This work can be done in the office.

OFFICE PREPARATION FOR FIELD INSPECTION AND TESTING

First, *assemble all applicable contract documents, specifications, catalogs and previous reports,* including:

- Shop drawings
- "As-built" drawings
- Schematics
- Automatic temperature control drawings
- Manufacturers' catalogs and performance curves
  - —Fan description and capacity
  - —Fan performance curve
  - —Terminal box description and capacities
- Manufacturers' data and recommendations
  - —Testing of fans, boxes, inlets and outlets
- Equipment operation and maintenance instructions
- Air balance reports

Then *review the drawings, specifications, catalogs and reports.* Note any equipment, system component or condition that might change the system or component air volume, shut down the system, change the sequence of balancing the system or in any way affect the balance procedure.

Evaluate and make plans for the inspection or arrangement of any of the following conditions:

- Devices, controls or system features which might shut down the system or contribute to an unbalanced condition.
- General accessibility
  —Equipment that is difficult to get to
  —Limited access areas such as security areas, clean rooms, hotel
  —rooms, meeting rooms, etc.
- Time delays
  —High ceilings
  —Spline ceilings
- Sequence of balancing systems and means of generating an "out of season" heating or cooling load.
- Marking the final settings of balancing dampers.
- Arrangements for witnessing of the balancing.

Next, *check the instruments* to be used for balancing the system:

- Verify that the instruments selected for the balancing project meet the calibration criteria stated in the balancing specifications:
- Verify that the instruments have been calibrated within the last six months, or to manufacturer's recommendations.
- Include a list of the various instruments used in the test and balance report.
- Determine what instruments will be needed.
- Determine what measurements will be taken, and where.
- Determine if a correction for air density and air measurements will be needed because of altitude and/or temperatures.

Finally, *prepare test report sheets:*

- Gather the data and test sheets needed for each system being tested. The test and balance report may include any or all of the following.
  Air handling equipment data and test sheets (see Fig. 17.1)
  Drive data and test sheets (see Fig. 17.2)
  Motor data and test sheets (see Fig. 17.3)
  Duct traverse sheets (see Figs. 17.4 and 17.5)

FIGURE 17.1. AIR HANDLING EQUIPMENT DATA AND TEST SHEET

Project:_____   Engineer/Contact: _____

	Specified	Actual	Specified	Actual
Fan Designation				
Location				
Area Served				

	Specified	Actual	Specified	Actual
Fan Information				
Manufacturer				
Serial #				
Model #				
Fan Speed rpm				
Rotation				
Capacity cfm				
Efficiency				
Type				
Wheel Size				
Tip Speed				

Fan Pressure: Fan SP, Fan TP, Total SP or External SP

	Specified	Actual	Specified	Actual
Inlet TP				
Outlet SP				
Fan SP				
Inlet TP				
Outlet TP				
Fan TP				
Inlet SP				
Outlet SP				
Total SP				
External Inlet SP				
External Outlet SP				
External SP				

Pressure Differential: Filter, Heating Coil, Cooling Coil

	Specified	Actual	Specified	Actual
Filter SP Entering				
Filter SP Leaving				
Filter SP Diff.				
Htg. Coil SP Entering				
Htg. Coil SP Leaving				
Htg. Coil SP Diff.				
Clg. Coil Wet or Dry				
Clg. Coil SP Entering				
Clg. Coil SP Leaving				
Clg. Coil SP Diff.				

FIGURE 17.1. (*Continued*)

Project:_____   **Engineer/Contact:**_____

Fan Designation _____

Air Volume: Cubic Feet Per Minute _____

	Specified	*Actual*	*Specified*	*Actual*
Fan Total				
Outlet Total				
Outside Air Total				
Return Air Total				

Air Temperatures: Dry Bulb, Wet Bulb; Relative Humidity (%RH)

	Specified	*Actual*	*Specified*	*Actual*
Supply Air DB				
WB Supply Air				
SA Relative Humidity				
Return Air DB				
WB Return Air				
RA Relative Humidity				
Outside Air DB				
WB Outside Air				
OA Relative Humidity				
Mixed Air DB				
WB Mixed Air				
MA Relative Humidity				
Entering Clg. Coil DB				
Leaving Clg. Coil DB				
WB Entering Clg. Coil				
WB Leaving Clg. Coil				
Cond. Space % RH				

System Condition

Fan				
Duct				

Notes:

_____
_____
_____
_____
_____
_____
_____
_____
_____
_____
_____
_____
_____
_____

FIGURE 17.2.   DRIVE DATA SHEET

Project:_____   Engineer/Contact: _____

	Specified	Actual	Specified	Actual

Fan Designation

Drive Information

Fan Shaft Size				
Motor Shaft Size				
Shaft Center Distance				
Fan Sheave Size				
Manufacturer				
Fixed or Adjustable				
Motor Sheave Size				
Manufacturer				
Fixed or Adjustable				
Belt Manufacturer				
Number of Belts				
Belt Size				
Amt. of Motor Adjust.				
Alignment				
Tension				

Notes:

FIGURE 17.3.   MOTOR DATA AND TEST SHEET

Project:_____   Engineer/Contact: _____

	Specified	Actual	Specified	Actual

Fan Designation

Motor Information

Manufacturer				
Frame				
Horsepower				
Phase				
Hertz				
Motor Speed rpm				
Service Factor				
Voltage				
Amperage				
Power Factor				
Efficiency				
Brake Horsepower				
Starter Size				
Thermal OLP				

Notes:

FIGURE 17.4. RECTANGULAR DUCT TRAVERSE

Project: _____   **Engineer/Contact:** _____

System Data

System Designation _____
Traverse Designation _____

	Specified	Actual
Duct Size (inches)	_____	_____
Duct Area (square feet)	_____	_____
Volume (cubic feet per minute)	_____	_____
Average Velocity (feet per minute)	_____	_____
Center Line Static Pressure	_____	_____
Density (pounds per cubic foot)	_____	_____
Instrument Correction for Density	_____	_____

$$V = 4,005 \sqrt{VP}$$

No.	VP	Vel	No.	VP	Vel	No.	VP	Vel	No.	VP	Vel
1			18			35			52		
2			19			36			53		
3			20			37			54		
4			21			38			55		
5			22			39			56		
6			23			40			57		
7			24			41			58		
8			25			42			59		
9			26			43			60		
10			27			44			61		
11			28			45			62		
12			29			46			63		
13			30			47			64		
14			31			48					
15			32			49					
16			33			50					
17			34			51					
SUBTOTAL			SUBTOTAL			SUBTOTAL			SUBTOTAL		
									TOTAL		

Total velocity (fpm) ÷ No. of Readings = Average velocity (fpm)

Average fpm × Area = Total cfm (nearest 5 cfm)

Notes:

_____
_____
_____
_____
_____
_____
_____
_____
_____
_____
_____

FIGURE 17.5A.   ROUND DUCT TRAVERSE SHEET—12″ OR LARGER DIAMETER

Project: _____   **Engineer/Contact:** _____

System Data

System Designation _____
Traverse Designation _____

	Specified	*Actual*
Duct Size (inches)	_____	_____
Duct Area (square feet)	_____	_____
Volume (cubic feet per minute)	_____	_____
Average Velocity (feet per minute)	_____	_____
Center Line Static Pressure	_____	_____
Density (pounds per cubic foot)	_____	_____
Instrument Correction for Density	_____	_____

$$V = 4{,}005 \sqrt{VP}$$

No.	Factor	Inches	VP	Vel	No.	Factor	Inches	VP	Vel
1	0.052				11	0.052			
2	0.165				12	0.165			
3	0.293				13	0.293			
4	0.454				14	0.454			
5	0.684				15	0.684			
6	1.316				16	1.316			
7	1.547				17	1.547			
8	1.707				18	1.707			
9	1.835				19	1.835			
10	1.948				20	1.948			
			SUBTOTAL					SUBTOTAL	
								TOTAL	

Total velocity (fpm) ÷ 20 Readings = Average velocity (fpm)

Average fpm × Area = Total cfm

Notes:

_____
_____
_____
_____
_____
_____
_____
_____
_____
_____
_____
_____
_____
_____
_____
_____
_____
_____

FIGURE 17.5B.   ROUND DUCT TRAVERSE SHEET 10″ OR SMALLER DIAMETER

Project: _____   Engineer/Contact: _____

System Data

System Designation _____
Traverse Designation _____

	Specified	Actual
Duct Size (inches)	_____	_____
Duct Area (square feet)	_____	_____
Average Velocity (feet per minute)	_____	_____
Volume (cubic feet per minute)	_____	_____
Center Line Static Pressure	_____	_____
Density (pounds per cubic foot)	_____	_____
Instrument Correction for Density	_____	_____

No.	Factor	Inches	VP	Vel
1	0.088			
2	0.293			
3	0.592			
4	1.408			
5	1.707			
6	1.913			
7	0.088			
8	0.293			
9	0.592			
10	1.408			
11	1.707			
12	1.913			
		TOTAL		

Total velocity (fpm) ÷ 12 Readings = Average velocity (fpm)

Average fpm × Area = Total cfm

Notes:

_____
_____
_____
_____
_____
_____
_____
_____
_____
_____
_____
_____
_____
_____
_____
_____
_____

Traverse summary sheets (see Fig. 17.6)

Zone summary sheets (see Fig. 17.7)

Terminal box data and test sheets (see Fig. 17.8)

Air distribution sheets (see Fig. 17.9)

- Enter design information on the appropriate data and test sheet. Design data will include air quantities, fan information, motor information and air distribution information. The test and balance report is a complete record of design, preliminary, and final test data. It reflects the actual tested and observed conditions of all systems and components during the balance. The report lists any discrepancies between the design and the test data and possible reasons for the differences.

- Make a schematic drawing of each system. Show the central system with pressure and temperature drops (rises) across components such as filters, coils and fans. The schematic should show the general location of the air distribution devices (terminal boxes, outlets, volume dampers, etc.).

FIELD INSPECTION

Inspecting the Building

All walls, windows, doors and ceilings should be installed before beginning the balance. If the conditioned space isn't architecturally sealed, abnormal pressures and temperatures will adversely affect the system balance. Note any conditions that will affect the balance.

Inspecting the Air Handling Equipment

Check all equipment to make sure that:

- Motors, fans, drives, etc., are mechanically and electrically ready. Note any problems.
- Filters are clean and correctly installed
- Filter frame is properly installed and sealed
- Coils are clean, properly installed and sealed
- Drive components are installed
- Sheaves are properly aligned and tight on their shafts
- Belts are adjusted for the correct tension
- Belt guard is properly installed
- Fan vortex dampers are functional
- Fan housings are installed and properly sealed
- Fan wheel is aligned properly with adequate clearance in the housing
- Fan bearings are lubricated
- There is adequate distance from the fan housing to the fan inlet
- There is adequate distance between fans in parallel
- Flexible connections are installed properly

FIGURE 17.6.   TRAVERSE SUMMARY SHEET

Project:_____   **Engineer/Contact:**_____

System:_____

No.	Designation Duct Size	Sq. Ft.	SP	Cubic feet/minute			Notes
				Design	Actual	%D	

Notes:

_____

_____

_____

_____

Legend: SP = static pressure
    % = percent of design

FIGURE 17.7.   ZONE SUMMARY

Project:_____   Engineer/Contact:_____

System:_____

Zone	Duct Size	Sq. Ft.	cfm_d	cfm_m	%D	SP1	SP2	cfm_f

Notes:

_____

_____

_____

_____

Legend: cfm_d = design cubic feet per minute
cfm_m = measured cubic feet per minute
cfm_f = final cubic feet per minute
%D = percent of design (cfm)
SP_1 = initial static pressure
SP_2 = final static pressure

FIGURE 17.8.   TERMINAL BOX DATA AND TEST SHEET

Project: _____   **Engineer/Contact:** _____

Terminal Type: () **Constant** () **Variable** () **Single** () **Dual** () **Induction**

					Minimum		Maximum	
No.	Location	Mod. No.	Size	Min SP	cfm$_d$	cfm$_m$	cfm$_d$	cfm$_m$

Notes:

_____

_____

_____

_____

Legend: Moo. No. = model number
   Min SP = minimum static pressure required
   cfm$_d$ = design cubic feet per minute
   cfm$_m$ = measured cubic feet per minute

FIGURE 17.9.   AIR DISTRIBUTION TEST SHEET

Project: _____   **Engineer/Contact:** _____

Area Served	No.	Terminal Type	Size	Ak	Design cfm	Initial cfm	Initial %D	Proportioned cfm	%D	Final cfm	%D

Notes:
_____
_____
_____
_____
_____
_____
_____

Legend: A_k = air distribution correction factor

Inspecting the Air Distribution System

Make sure that:

- Ducts are clean, intact, sealed and leak tested according to specifications
- Access doors are installed and properly secured
- Dampers, including volume, automatic control, fire and smoke dampers, are installed, accessible and functional. Where applicable, check that the dampers are operating correctly with minimal leakage. Depending on circumstances this may be done visually by reading temperatures or using static pressure drop.
- Terminal boxes are installed, functional and accessible
- All other air distribution devices such as terminal boxes, diffusers, grilles, etc., are installed, accessible and functional

Note any major changes in actual installation. Make corrections on the schematics. Note any changes in capacities. Make corrections on the data and test sheets. Investigate any potential problem areas that you listed while reviewing the contract documents, specifications, catalogs and previous reports.

FIELD TESTING

Record Information on System Components

First, operate a new system for 48 hours before testing and balancing. Then, record information on all areas described in the following sections.

Fan Data

Record fan data as applicable:

Designation	Rotation (CW,CCW)
Location	Capacity (cfm)
Area served	Efficiency (%)
Manufacturer	Type
Serial number	Wheel size
Model number	Tip speed (fpm)
Speed (rpm)	

Fan Rotation

Verify and record fan wheel rotation—clockwise (CW) or counterclockwise (CCW). To determine fan rotation, view a centrifugal fan from the drive side, not the inlet side. The drive side is intended to mean the side that is driven by the motor, but it really varies with the fan configuration. On a single inlet, single wide (SISW) fan, the drive side is the side opposite the inlet. On a double inlet, double wide (DIDW) fan, the drive side is the side that has the drive. On fans with dual drives, the side with the higher horsepower rating is the drive side. Axial fans will normally have an arrow on the housing which indicates direction of rotation.

Observe fan rotation. If any excessive noises or vibrations are observed, stop the fan and investigate. Before continuing the testing phase, determine if the fan can be operated or if it needs repairing. If rotation is backwards, make the necessary corrections. If a centrifugal fan is rotating backwards the air flow can be reduced by 50% or more. An axial fan rotating backwards will produce air flow in the opposite direction.

Reversing Rotation. If the rotation is incorrect, reverse the rotation on a three-phase motor by changing any two of the three power leads at the motor control center or disconnect. In some cases, you may also be able to change rotation in single-phase motors by switching the internal motor leads within the terminal box. Wiring diagrams for single-phase motors are usually found on the motor or inside the motor terminal box. After making changes to the motor, observe the fan rotating in the proper direction.

Fan Pressure Data

Record fan pressure data as applicable: total pressure (*TP*) and static pressure (*SP*).

Other Air Handling Unit Information

Record the following:

Total cfm	Wet bulb (*wb*) temperatures
Outside air cfm	Percent of relative humidity (%RH)
Return air cfm	General condition
Dry bulb (*DB*) temperatures	

Drive Data

Record drive data as applicable:

Shaft sizes	Belt size
Distance between shafts	Number of belts
Sheave sizes	Motor adjustment
Sheave manufacturer	Drive alignment
Adjustable or fixed sheave	Belt tension
Belt manufacturer	

To take drive information, stop the fan and put your personal padlock on the motor disconnect switch so only you have control over starting the fan. Remove the belt guard and read the information off the sheaves and belts. The outside of the sheave generally has a stamped part number which gives the sheave size. If there's no part number on the sheave, measure the outside diameter and refer to the manufacturer's catalog. Most manufacturers list outside diameter and corresponding pitch diameter in their catalogs. Record the number of belts, manufacturer and size of belts.

Measure and record the size of the motor and fan shafts, the distance between the center of the shafts and the adjustment of the motor on its frame. If a sheave has to be changed and adjustment is available on the frame, you may be able to move the motor forwards or backwards to fit the change in sheave size without having to buy a new belt.

Motor Data

Record motor data as applicable:

Manufacturer	Voltage
Frame size	Amperage
Horsepower	Power factor
Phase	Efficiency
Hertz	Brake horsepower
Speed	Starter size
Service factor	Thermal overload protection

If the nameplate is positioned so that it is difficult to read, use a telescoping mirror. In most cases, motors used on HVAC equipment will be either single-phase or three-phase alternating current, single or dual voltage induction motors. A dual voltage motor will also list dual amperage on the nameplate. For example, a three-phase, 50 horsepower, dual voltage motor has the following ratings: 230/460 volts, 120/60 amps. If the motor is wired for 230 volts, the full load amps will be 120. If it's wired for 460 volts, the full load amps will be 60. The volts and amps are inversely proportional. When the voltage doubles, the amperage reduces by one half.

Motor Voltage, Current and Power Factor

Measure and record the motor voltage, amperage and power factor. Voltage and amperage measurements are taken with a portable volt-ammeter. Measure voltage on the line side. Record L1-L2, L1-L3, L2-L3. The most accurate voltage will be read at the motor terminal box, but it's usually much safer to take readings at the motor control center or at the disconnect box. The voltage difference between the two places is generally insignificant. The measured voltage should be plus or minus 10% of the motor nameplate voltage. If it's not within this range, notify the electrical contractor or the utility company. Measure amperage on the terminal side. Record T1-T2, T1-T3, T2-T3. The amperage measured on any phase should not exceed the motor nameplate amperage. If the operating amperage reading is over the nameplate amperage, take one of the following steps to correct the problem: reduce the speed of the fan or close the main discharge damper until the amperage reading is at or below nameplate. If the fan is not serving a critical area, turn the fan off and inform the person responsible for the fan's operation. If desired, measure power factor with a power factor meter. Record L1-L2, L1-L3 and L2-L3.

Motor Speed

The rated rpm on the nameplate is the speed that the motor will turn when it's operating at nameplate horsepower. If the motor is operating at something other than rated horsepower, the rpm may vary a small amount. However, it's the nameplate rpm that's recorded on the report forms and used for drive calculations.

Service Factor

The service factor is the number by which the horsepower or amperage rating is multiplied to determine the maximum safe load that a motor may be expected to carry continuously at its rated voltage and frequency. A service factor of 1.10 for a 50 horsepower motor would allow the motor to operate safely at 55 horsepower (55 × 1.10) and about 132 amps (120 × 1.10) at 230 volts.

Do not leave a motor operating in the service factor area. Under certain circumstances, damage can occur to the windings and shorten the life of the motor. For example, if this motor was operating at 132 amps and the voltage to the motor dropped to 220 volts (because of a reduction in power from the utility company or adding other equipment, etc.), the amperage would go to about 138 amps (230/220 × 132). The motor would overheat and possibly damage the windings.

Motor Overload Protection

Generally, overload protection is installed to protect the motor against current 125% greater than rated amps. In thermal overload selection, it's important to know the ambient temperature at the starter, as compared to the ambient temperature at the motor. Sometimes, these temperatures may vary greatly. In this case, a temperature-compensating thermal overload or a magnetic overload device may be needed. Check with manufacturer or electrical supplier for special cases. Thermal overload protection devices have a letter and/or number on them which is used for proper sizing. A chart listing thermals and their amperage ratings for the specific starters is usually found inside the motor disconnect cover. Thermals must be matched to starter and the rated full load amps of the motor according to the information on the chart or from the manufacturer. If the installed thermal is too large, the motor may not be adequately protected and could overheat. If the thermal is too small, the motor may stop repeatedly. If a new thermal is needed, the following information will be required: motor starter size, full load amps, service factor, class of insulation, motor classification and allowable temperature rise. Refer to the chart or a table provided by the thermal manufacturer to choose the proper size.

Set the Automatic Temperature Controls on Full Cooling

The system must be set for (and balanced for) maximum required air flow. Try to balance in the full cooling mode unless other conditions dictate that balancing must be done in the heating mode. In most HVAC systems the air volume requirement is generally greater for cooling than for heating because the space cooling load is greater than the heating load. Even if the heating and cooling load in a space were the same, more air would be required for the cooling load than the heating load because of the lower temperature difference for cooling.

Example 17.1: Heating

$$Btuh = cfm \times 1.08 \times TD$$

$$cfm = \frac{Btuh}{1.08 \times TD}$$

$$cfm = \frac{50,000}{1.08 \times 105 - 75} \quad \text{room temperature is 75, coil LAT is 105}$$

$$cfm = 1,543$$

Cooling

$$cfm = \frac{50,000}{1.08 \times 75 - 55}$$ room temperature is 75, coil

$$cfm = 2315$$

Dry vs. Wet Cooling Coil

Set the controls for full cooling and check that the cooling coil is dehumidifying and in a "wet" condition. A wet cooling coil has a greater resistance to airflow than a dry coil because of the moisture on the coil. If it is necessary to balance with a dry coil:

- Proportionally balance the system.
- If the fan speed is increased based on measurements made on a dry coil (to get desired airflow at the wet coil conditions), calculate and check the motor current in the dry condition to ensure that the motor isn't overloaded.
- Recheck the system when the coil is wet.
- The fan speed may need to be increased when the system is rechecked with a wet coil.
- Measure motor current.

Set All Dampers and Diverters

Automatic Temperature Control Dampers. Check that automatic temperature controls are properly sequencing and holding dampers in place. If the dampers are not operating properly, disconnect the control linkage and block the dampers as follows:

- For systems with face and bypass dampers, make sure the face dampers are full open and the bypass dampers are full closed.
- For systems with hot and cold deck dampers, make sure the cold deck dampers are full open and the hot deck dampers are full closed. Some systems may have a cooling coil with a diversity factor. Diversity means that the cooling coil is designed for less air than the fan delivers. If this is the case, the total air out of the fan will be divided into the approximate amount required for the cooling coil while the rest of the air will go through the heating coil.
- Check that the return air damper is full open.
- Check that the outside air damper is set approximately at minimum position. After balancing the supply air, the outside air damper will be set to the required minimum position.

Fire and Smoke Control Dampers. Verify that fire and smoke dampers are set to specifications.

Manual Dampers, Straighteners, Pattern Devices and Diverters. Verify the following:

- All extractors, distribution grids and other accessories are set for maximum air flow.
- All supply and return dampers, including dampers at diffusers and registers, are full open.

- All air pattern devices in diffusers and grilles are properly set.
- All splitter devices are set in a nondiverting mode.

Operate All Fans at (or as close as possible to) Design Speeds

Take fan speed with a contact or noncontact tachometer until you have two consecutive, repeatable values. Adjust fan speed as indicated.

Take Static Pressure Measurements

Take the readings with a Pitot tube and a vertical inclined liquid filled manometer, or electronic manometer. The holes will normally be ⅜ inch in diameter to accommodate the standard Pitot tube. For pressure readings see Chap. 1.

In the Unit. Drill test holes in the fan unit to take static pressure readings. Place the test holes:

- Before and after the filter for static pressure drop across the filter.
- Before and after the coil(s) for static pressure drop across the coil(s).
- Before and after the fan for total static pressure rise across the fan.

In the Duct System. Drill test holes in the duct to take static pressure readings at traverse points or across components such as reheat coils, dampers, etc.

At Terminal Boxes. For constant air volume mixing boxes, take measurements at the ends of the system (before the box). Determine if the inlet static pressure is at or above the minimum required for mixing box operation. Increase or decrease fan speed as needed. Check the static pressure drop across the box. Consult box manufacturer for pressure drop requirements. Generally, 0.75 inches wg is required for mechanical regulators. Additional pressure is needed for the low pressure distribution system downstream of the box. Approximately 0.1 inch wg per 100 foot of duct (equivalent length) and 0.05 to 0.1 inch wg for the outlet is needed.

For constant air volume induction boxes (Chap. 4), check the side of the box for a "cfm versus static pressure" chart. If not available, contact the terminal manufacturer for the data. Use a liquid filled or electronic manometer or an air pressure differential gauge to read the primary air pressure at a typical nozzle in the first and last induction terminal on each riser.

Take Total Air Measurements

Measuring Total Air in the Unit

The air in the unit may be read to give an approximate total volume. The reading will be an approximate value because of the probability of unfavorable field conditions existing in the unit. However, the results will serve as a general check of system performance. Adjustments to the fan speed will bring the system into the proper range for balancing.

If a fan is tested and it has more air volume than required, the fan speed should be reduced to bring the air volume to approximately 110% of design. If the fan is tested and it is low on air, the fan speed should be increased to bring the air volume to approximately 110% of design. If the fan volume is 80% of design or less, calculate the fan speed and the horsepower required to bring the unit to 100% capacity. If the present motor can handle the speed, increase or change the drives as indicated. If it's determined that a new motor will be needed to bring the system to design requirements, contact a responsible authority to make a decision to:

- Install a new motor before continuing the balancing.
  Operate the fan at the highest speed possible without overloading the motor and balance the system. Then:
  Install a new motor at a later time to achieve 100% capacity.

- Elect not to install a new motor but leave the system in a balanced condition at less than 100% capacity.

Measuring Air Flow Across Coils and Filters

To determine the approximate total air volume in the unit use an anemometer, capture hood or velocity grid to make a traverse across the filter bank or across the coil. The most common instrument used is the 4-inch rotating vane anemometer, but a capture hood, deflecting vane anemometer, thermal (hot wire) anemometer or other electronic air sensing instrument could also be used. All anemometers are position sensitive, so care must be taken when using them. As with all instruments, follow the manufacturer's instructions for best results.

The airflow measurements should be made on the downstream (leaving air) side of the filter or coil. Any reading will only be an approximation of the total flow. Only experience in using the instruments for this type of application will tell you if the readings are accurate or not.

Taking airflow measurements at the filter or coil is usually not the best way to determine total air volume because it is difficult to get accurate field measurements. Difficulties are caused by:

- The close proximity of the filters to the outside and return air openings and the turbulence of the mixed air.

- The vena contracta (jet velocities) effect as the air passes through the coils.

To determine airflow at the filter or coil:

- Measure the filter or coil and calculate square footage. If the filter or coil is very large, divide it into more manageable sections.

- Measure the velocity in each section and determine the average velocity in feet per minute. Generally, velocities read at the coil will be about 25 to 35% high because of jet velocities.

- For the filters, multiply the area (in square feet) times the average velocity (in feet per minute) of each section to find volume in cubic feet per minute ($cfm = A \times V$). Add the section volumes together to find total air. A capture hood will read volume directly.

- For the coils, multiply the area (in square feet) times the average velocity (in feet per minute) of each section times a 30% correction factor to find volume in cubic feet per minute (cfm = $A \times V \times 0.70$). Add the section volumes together to find total air. The capture hood will read volume directly. Always consult the instrument manufacturer and component manufacturer for advice on how to take proper readings and use correction factors.

Example 17.1: A filter bank is 72 inches $\times$ 72 inches (nine 24 inches $\times$ 24 inches filters). The average velocity at each filter is:

496 fpm	585 fpm	500 fpm
375 fpm	635 fpm	534 fpm
450 fpm	350 fpm	576 fpm

The average velocity of the bank is 500 fpm (4,501/9).
The area of filter bank is 36 square feet (72 inches $\times$ 72 inches/144).
The air volume is: 18,000 cfm (500 $\times$ 36).

Example 17.2: The cooling coil is 78 inches $\times$ 78 inches. The coil readings are taken in quadrants. The velocity in each quadrant is:

651	624
625	662

The average velocity of the coil is 641 fpm (2,562/4).
The area of coil is 42.25 square feet (78 inches $\times$ 78 inches/144).
The air volume is: 18,958 cfm (641 $\times$ 42.25 $\times$ 0.70).

Calculating Coil Pressure Drop

Another method to determine approximate total airflow is to use the coil as an orifice.

• Take the static pressure drop across the coil.

• Get the table or graph of "coil cfm vs. coil pressure drop" from the coil manufacturer to determine airflow.

This is not a recommended practice for finding air volume because of the difficulty of getting reliable rated pressure drop data and accurate field static pressures.

Measuring Total Air in the Duct

Take Pitot tube traverses in the duct. See Chap. 3 for more information.

For Constant air volume multizone systems:
See if the system has diversity (diversity means that the cooling coil is designed for less air than the fan delivers.) Determine the diversity ratio. Keep the proportion of cooled air to total volume constant during the balance by:

• Setting some mixing dampers so the hot deck dampers are partially open at full cooling, or

- Set enough zones to full cooling to equal the design flow through the cooling coil. The remaining zones will be set to heating.

For constant air volume dual duct medium to high pressure systems:

    If all the boxes are constant volume, set the thermostats for full flow through the cold duct. Traverse the cold and hot ducts. If more than 10% of the total rated airflow is measured in the hot duct, check for leaking dampers in the hot duct or mixing boxes with crossed supplies. (The hot and cold duct connections are reversed into the mixing box.) If the system has diversity, determine the diversity ratio and keep the proportion of cooled air to total volume constant during the balance. Set enough boxes to full cooling to equal the design flow through the cooling coil. Set the remaining boxes to heating.

CHAPTER 18 Proportional Balance of the Low Pressure Side of Any System

This chapter outlines the procedure for proportionally balancing the low pressure side of any constant volume, variable air volume, single zone, multizone or dual duct system.

MEASURING AIRFLOW

Using Anemometers to Measure AirFlow

There are 3 types of anemometers: rotating vane anemometer, deflecting vane anemometer and thermal anemometer. Anemometers require a correction factor (Ak or flow factor) from the outlet/inlet manufacturer for each piece of air distribution (AD) in order to convert velocity (fpm) readings to volume (cfm). In addition to the correction factor for the outlets/inlets (AD), the anemometer itself may also have a correction factor.

A correction factor (also called a flow factor, K-factor or Ak factor), is the effective area of the grille or diffuser as determined by the manufacturers' own airflow tests. The flow factors apply only when using the testing instrument specified by the AD manufacturer in the prescribed method. Consult the AD manufacturer's catalog of grilles and diffusers and their corresponding correction factors. These flow factors are used to calculate cfm. The equation is: cfm is equal to the average velocity at the AD times the area of the AD times the correction factor for the AD (cfm = V × Ak). See Chap. 3

If a correction factor is not available or it's producing an unsatisfactory result, a new correction factor can be field determined if it's possible to take a Pitot tube traverse in the duct for the AD in question. The duct must be free from obstructions and air leaks from the point of the traverse to the AD. Take readings at the traverse point and the outlet. Calculate the new flow factor.

Example 18.1: The manufacturer of a 20 inch × 8 inch sidewall supply air register is unknown. A typical supply register of this design has a 0.75 Ak factor using the rotating vane anemometer (RVA). The register is tested and the average air velocity is 645 fpm. Using the 0.75 Ak, the air volume is 484 cfm. (645 × 0.75) = 483.75.

A Pitot tube traverse is taken in the supply duct to the register. The volume of air is 600 cfm. Therefore, the actual correction factor should be 0.93. This Ak will be used when an RVA is used on this model and size AD.

Original Ak	*New Ak*
$Ak = \dfrac{cfm}{V}$	$Ak = \dfrac{cfm}{V}$
$Ak = \dfrac{484}{645}$	$Ak = \dfrac{600}{645}$
$Ak = 0.75$	$Ak = 0.93$

Using Capture Hoods to Measure Airflow

Capture hoods are considered the easiest and most reliable instrument to take outlet or inlet readings. With the hood, airflows are read directly in cfm. Analog capture hoods can be read to the nearest 5 cfm. Electronic capture hoods read to 1 cfm.

Correction factors for the AD are not needed when using captures. However, if a capture hood reading is in question because of very high or very low velocities or an unusual use of the capture hood, take a Pitot tube traverse of the duct and calculate a correction factor for the capture hood. Consult the capture hood manufacturer's information for correction factors for the hood. Capture hoods can also be used to determine A_k factors.

Example 18.2: Readings are taken on several 6 foot linear air diffusers (LAD) located in a drop ceiling. The capture hood has a 24 inch × 24 inch bonnet. It's decided that the procedure will be to read the first two feet, the second two feet and the last two feet of the diffuser and add the three readings together for a total airflow through the LAD. To determine if a field correction factor will be needed, a Pitot tube traverse will be made and the readings will be compared with the capture hood readings. The cfm calculated at the outlet divided by the cfm at the traverse point is the correction factor. Field testing indicates that some capture hoods can read low when measuring linear air diffusers.

Example 18.3: The first of 3 sidewall registers can be read with a capture hood. Because of limited access the other two registers will require the RVA. The Ak factors for the registers is unknown. All three registers are 18 × 6 inches. The registers are full open. All registers are mounted in 18 inch × 6 inch takeoffs.

The first register is read with a capture hood. A piece of cardboard is taped over the top of the bonnet. An 18 inch × 6 inch opening is cut into the cardboard and the hood is placed on the register. The air volume is 350 cfm. The anemometer is then used and the air velocity is 390 fpm.

The calculated Ak is 0.90. This Ak factor will be used on the other two registers.

$$Ak = \frac{cfm}{V}$$

$$Ak = \frac{350}{390}$$

$$Ak = 0.90$$

PROPORTIONAL BALANCING

The principles of proportional balancing require:

- That all the volume dampers in the air distribution system be full open and other dampers are set as outlined in Chap. 17.
- That the AD with the lowest percent of design flow will remain open.
- That the volume damper in the branch with the lowest percent of design flow will remain full open.

The following describes the procedure:

- *Determine which AD in the entire system has the lowest percent of design flow*—Percent of design flow is equal to the measured flow divided by the design flow (%D = measured flow/design flow). Design flow can be defined as either the original design flow per contract specifications or a new calculated design flow based on space conditions. Measured and design flows will be in feet per minute (fpm) when using anemometers. Measured and design flows will be in cubic feet per minute (cfm) when using capture hoods. All the balancing examples in this book will use cfm as the measured and design flow. Typically, the AD with the lowest %D will be on the branch farthest from the fan. This AD will be called the "key" AD.

- *Proportionally balance each AD to within a tolerance of* 10%—The ratio of the percent of design flow between each AD must be within 10% (1.00 to 1.10). Ratio of design flow is equal to the %D of the AD being adjusted divided by %D of the key AD.

$$\text{Ratio} = \frac{\%\,D\ \text{Adjusted AD}}{\%\,D\ \text{Key AD}}$$

Example 18.4: The %D of the adjusted AD is 100%. The %D of the key AD is 80%. The ratio is 1.25 (100/80 = 1.25). These AD are not in tolerance. To be in tolerance the ratio must be between 1.00 and 1.10.

Example 18.5: The %D of the adjusted AD is 88%. The %D of the key AD is 78%. These AD are not in tolerance. The ratio is 1.13 (88/78 = 1.13).

Example 18.6: The %D of the adjusted AD is 108%. The %D of the key AD is 99%. These AD are in tolerance. The ratio is 1.09 (108/99 = 1.09).

- Adjust each AD from the lowest percent of design flow (key AD) to the highest percent of design flow by branches—Start with the key AD. Adjust each AD from the lowest percent of design flow to the highest percent of design flow. To reduce airflow adjust volume dampers in the system at the takeoffs and not at the AD. Dampering at the AD can cause excessive noise and poor air distribution. Proportionally balance all AD on this branch.

- Go to the branch that has the AD with the next lowest percent of design flow as determined from the initial readout.—This "key" AD will usually be on the second farthest branch. Balance all the AD on this branch to the key AD to within plus or minus 10% of design flow.

- Continue balancing until all the AD on all the branches have been balanced to within plus or minus 10% of each other.

- Determine which branch has the lowest percent of design flow (key branch)—Proportionately balance all branch ducts from the lowest %D flow (key branch) to the highest %D flow to within 10% of each other.

- Continue proportionally balancing until all branches have been balanced.

- If needed, adjust the fan speed to bring the system to within 10% of design flow.

- Reread all the terminals and make any final adjustments.

- Complete the report.

Exercise 18.1: Proportional Balancing of the Air Distribution

Balance the constant volume low pressure system illustrated in Fig. 18.1. The system has been inspected and all dampers and air distribution devices are properly set. The fan is operating at the correct speed. Total design airflow is 1,900 cfm. Branch A design airflow is 800 cfm. Branch B design airflow is 1,100 cfm.

FIGURE 18.1.

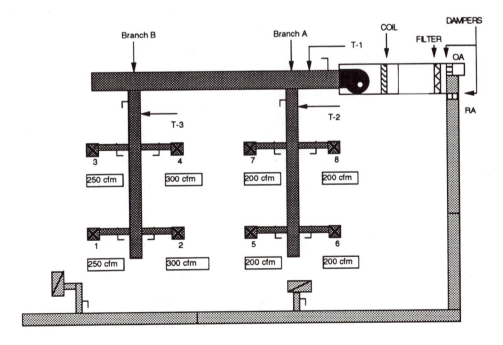

Duct	Traverse Point	Design cfm	Measured cfm	%D
Main	T-1	1,900	2,100	111

Duct	Traverse Point	Design cfm	Measured cfm	%D
Branch A	T-2	800	925	116
Branch B	T-3	1,100	1,120	102
Total		1,900	2,045	108

Initial Reading	AD	Design cfm	Measured cfm
	1	250	235
	2	300	280
	3	250	265
	4	300	340
	5	200	230
	6	200	210
	7	200	260
	8	200	225
Total		1,900	2,045

Determine which AD has the lowest percent of design flow.

AD	Design cfm	Measured cfm	%D
1	250	235	94
2	300	280	93**
3	250	265	106
4	300	340	113
5	200	230	115
6	200	210	105
7	200	260	130
8	200	225	113
Total	1,900	2,045	108

- AD 2 on Branch B is the key AD for the system. The balancing damper at the takeoff to AD 2 remains open.
- AD 1 is the next highest %D at 94%.
- Determine if AD 1 and AD 2 are within 10% of each other.
- The ratio of AD 1 to AD 2 is 1.01 (94%D/93%D).
- The ratio between AD 1 and AD 2 is within 10%. This ratio will remain until the dampers are moved.

Example 18.7: Because the speed of the fan is increased, the airflow in AD 2 is increased to 95%D. The airflow in AD 1 will increase to 96%D (95 × 1.01 = 96).

AD	Design cfm	Measured cfm	%D	Ratio
1	250	235	94	1:2 = 1.01
2	300	280	93	Key

Air Distribution 3

- Go to AD 3 (the AD with the next highest %D).
- Compare AD 3 to AD 2 (the key AD).
- The ratio of AD 3 to AD 2 is 1.14 (106%D/93%D).
- The ratio between AD 3 and AD 2 is greater than 10% (greater than 1.10).
- Balance AD 3 to AD 2.
- Close the volume damper for AD 3.
- Add the %D together and divide by 2 to arbitrarily select a cfm value.
- The value selected is 100%D (106 + 93/2 = 100).
- Arbitrarily close the damper for AD 3 until the AD reads 250 cfm (100%D).
- Measure the air volume in AD 2.
- The airflow at AD 2 is 285 cfm.
- Determine %D for AD 3 and AD 2 and the ratio between them.
- AD 3 is 100%D (250/250). This is the value set by the damper.
- AD 2 is 95%D (285/300).
- The ratio between AD 3 and AD 2 is 1.05 (100%/95%).
- The ratio is within 10%.
- The damper on AD 3 is locked. Once the ratio between two AD being balanced is between 1.00 and 1.10, the damper is locked.

AD	Design cfm	Measured cfm	%D	Ratio
1	250	240*	96*	1:2 = 1.01
2	300	285	95	Key
3	250	250	100	3:2 = 1.05

*Calculated cfm and %D for the purposes of this example only to illustrate what is happening at the AD already set.

Air Distribution 4

- Go to AD 4 (the AD with the next highest %D).
- Measure the air volume in AD 4.
- The airflow at AD 4 is 350 cfm.
- Determine the %D for AD 4, %D for AD 2 and the ratio between them.
- AD 4 is 117%D.
- AD 2 is 95%.
- The ratio of AD 4 to AD 2 is 1.19 (117%D/95%D).
- The ratio between AD 4 and AD 2 is greater than 10% (greater than 1.10).
- Balance AD 4 to AD 2.
- Close the volume damper for AD 4.

- Add the %D together and divide by 2 to arbitrarily select a cfm value.
- The value selected is 106%D (117 + 95/2 = 106).
- Arbitrarily close the damper for AD 4 until the AD reads 318 cfm (106%D).
- Measure the air volume in AD 2.
- The airflow at AD 2 is 290 cfm.
- Determine %D for AD 4, %D for AD 2 and the ratio between them.
- AD 4 is 106%D (318/300). This is the value set by the damper.
- AD 2 is 97%D (290/300).
- The ratio between AD 4 and AD 2 is 1.09 (106%/97%).
- The ratio is within 10%.
- The damper on AD 4 is locked.
- All the AD on Branch B are balanced at within 10% of each other.

AD	Design cfm	Measured cfm	%D	Ratio
1	250	245*	98*	1:2 = 1.01
2	300	290	97	Key
3	250	255*	102*	3:2 = 1.05
4	300	318	106	4:2 = 1.09

*Calculated

Branch A. Go to the key AD on Branch A. This will be AD 6 at 105%D. Balance the other AD on Branch A to AD 6.

AD	Design cfm	Measured cfm	%D
5	200	230	115
6	200	210	105
7	200	260	130
8	200	225	113

Air Distribution 8

- Go to Ad 8 (the AD with the next highest %D).
- Compare AD 8 to AD 6 (the key AD).
- The ratio of AD 8 to AD 6 is 1.08 (113%D/105%D).
- The ratio between AD 8 and AD 6 is less than 10% (less than 1.10).

AD	Design cfm	Measured cfm	%D	Ratio
6	200	210	105	Key
8	200	225	113	8:6 = 1.08

Air Distribution 5

- Go to AD 5 (the AD with the next highest %D).
- Compare AD 5 to AD 6 (the key AD).
- The ratio of AD 5 to AD 6 is 1.10 (115%D/105%D).
- The ratio between AD 5 and AD 6 is 10% (1.10).

AD	Design cfm	Measured cfm	%D	Ratio
5	200	230	115	5:6 = 1.10
6	200	210	105	Key
8	200	225	113	8:6 = 1.08

Air Distribution 7

- Go to AD 7 (the AD with the next highest %D).
- Compare AD 7 to AD 6 (the key AD).
- The ratio of AD 7 to AD 6 is 1.24 (130%D/105%D).
- The ratio between AD 5 and AD 6 is greater than 10% (greater than 1.10).
- Balance AD 7 to AD 6.
- Close the volume damper for AD 7.
- Add the %D together and divide by 2 to arbitrarily select a cfm value.
- The value selected is 118%D (130 + 105/2 = 118).
- Arbitrarily close the damper for AD 7 until the AD reads 236 cfm (118%D).
- Closing the damper results in a value of 230 cfm (115%D).
- Measure the air volume in AD 6.
- The airflow at AD 6 is 218 cfm.
- Determine %D for AD 7 and AD 6. Determine the ratio between them.
- AD 7 is 115%D (230/200). This is the value set by the damper.
- AD 6 is 109%D (218/200).
- The ratio between AD 7 and AD 6 is 1.06 (115%/109%).
- The ratio is within 10%. The damper on AD 7 is locked.
- All the AD on Branch A are balanced at within 10% of each other.

AD	Design cfm	Measured cfm	%D	Ratio
5	200	240*	120*	5:6 = 1.10
6	200	218	109	Key
7	200	230	115	7:6 = 1.06
8	200	236*	118*	8:6 = 1.08

*Calculated

The condition of system after balancing AD:

AD	Design cfm	Measured cfm	%D	Ratio
1	250	245*	98*	1:2 = 1.01
2	300	290	97	Key
3	250	255*	102*	3:2 = 1.05
4	300	318	106	4:2 = 1.09
Subtotal	1,100	1,108*		
5	200	240*	120*	5:6 = 1.10
6	200	218	109	Key
7	200	230	115	7:6 = 1.06
8	200	236*	118*	8:6 = 1.08
Subtotal	800	924*		
Total	1,900	2,032*		

*Calculated

Exercise 18.2: Proportional Balancing of the Branches

Use static pressure to proportionally balance the branches:

- Take static pressure readings in each branch after the branch volume damper and before any AD takeoff.
- Take static pressure readings in each branch at the traverse point.
- Traverse the duct or total the AD to find total cfm in each branch.
- The measured airflow readings will be designated cfm1.
- The measured static pressure reading will be designated SP1.
- Use Fan Law #2 to calculate the required static pressure (SP2) that will result in the design airflow (cfm2).
- Ensure that all the dampers are open before starting the balance. Make sure that no adjustments or changes are made downstream of the branch damper during the branch balancing process.

Example 18.7: After taking the branch traverse and static pressure, an AD damper is found closed. If it is opened now, the cfm would increase and the static pressure would decrease at the traverse point. The fan law can not be used until a new cfm and static pressure are measured.

Example 18.8: After taking the branch traverse and static pressure, an AD damper is closed in one of the takeoffs. Closing the damper will reduce the total cfm at the traverse and the static pressure will be increased. The fan law can not be used until a new cfm and static pressure are measured.

Proportionally balance each branch to within a tolerance of within 10%.

- The ratio of the percent of design flow between each branch must be within 10% (1.00 to

1.10). Ratio of design flow is equal to the %D of the branch being adjusted divided by %D of the key branch.

$$\text{Ratio} = \frac{\% \text{ D adjusted branch}}{\% \text{ D key branch}}$$

Adjust each branch from the lowest percent of design flow (key branch) to the highest percent of design flow as determined after the balance of the outlets.

- Start with the key branch.
- Adjust each branch from the lowest percent of design flow to the highest percent of design flow. To reduce airflow, adjust volume dampers in the branches.
- Proportionally balance all branches.

Go to the branch with the next lowest percent of design flow as determined after the balance of the outlets.

- This "key" will usually be the second farthest branch.
- Balance this branch to the key branch to within 10% of design flow.

Continue balancing until all the branches have been balanced to within plus or minus 10% of each other.

Determine which branch has the lowest percent of design flow.

Branch	Static Press Point	Design cfm2	Measured cfm1	%D	SP1 Measured	Ratio
A	T-2	800	924	116	1.00	
B	T-3	1,100	1,108	101	0.85	A:B 1.18
Total		1,900	2,032	107		

- Branch B is the key branch for the system (101%D). The balancing damper at Branch B remains open.
- Branch A is the next highest %D at 116%.
- Determine if Branch A and Branch B are within 10% of each other.
- The ratio of Branch A and Branch B is 1.15 (116%D/101%D).
- The ratio between Branch A and Branch B is not within 10%.

Branch A

- Balance Branch A to Branch B.
- Close the volume damper for Branch A.
- Add the %D together and divide by 2 to arbitrarily select a cfm value.
- The value selected is 109%D (116 + 101/2 = 109). The cfm value is 872 cfm.

- Use Fan Law #2 to calculate the new static pressure at T-2 (Branch A) which will correspond to 872 cfm (109%D).
- Arbitrarily close the volume damper on Branch A until 0.89 inches wg SP2 is measured.

$$\left[\frac{cfm_2}{cfm_1}\right]^2 = \frac{SP_2}{SP_1}$$

$$SP_2 = SP_1 \left[\frac{cfm_2}{cfm_1}\right]^2$$

$$SP_2 = 1.00 \left[\frac{872}{924}\right]^2$$

$$SP_2 = 0.89 \text{ in. wg}$$

- Measure the static pressure at T-3.
- The static pressure is 0.90 inches wg.
- Calculate the new cfm in Branch B.

$$\left[\frac{cfm_2}{cfm_1}\right]^2 = \frac{SP_2}{SP_1}$$

$$cfm_2 = cfm_1 \sqrt{\frac{SP_2}{SP_1}}$$

$$cfm_2 = 1,108_1 \sqrt{\frac{0.90}{0.85}}$$

$$cfm_2 = 1,140$$

- New airflow in Branch B is 1,140 cfm.
- Determine %D for Branch B and Branch A. Determine the ratio between them.
- Branch B is 104%D (1140/1108 = 104%D).
- Branch A is 109% (872/800).
- The ratio between Branch A and Branch B is 1.05 (109%/104%).
- The ratio is within 10%.
- The branch damper is locked.

Branch	Design cfm	Measured cfm	%D	SP1	Ratio
A	800	872	109	0.89	
B	1,100	1,140	104	0.90	A:B 1.05
Total	1,900	2,012	106		

The branches are now balanced to each other within 10%. Since all the AD have been proportionally balanced to each other by branch, an adjustment at the branch damper will increase or decrease all the AD on each branch proportionally. Since no AD volume dampers have been changed, the ratio of each AD remains the same as when initially set. To determine the cfm at each AD, read AD 2 and 6 and calculate new cfm for the other AD.

The condition of system after balancing branches:

AD	Design cfm	Measured cfm	%D	Ratio
1	250	253*	101*	1:2 = 1.01
2	300	300	100	Key
3	250	263*	105*	3:2 = 1.05
4	300	327*	109*	4:2 = 1.09
Subtotal	1,100	1,143*		
5	200	224*	1.12*	5:6 = 1.10
6	200	204	1.02	Key
7	200	216*	1.08*	7:6 = 1.06
8	200	220*	1.10*	8:6 = 1.08
Subtotal	800	864*		
Total	1,900	2,007*		

*Calculated

- Adjust the fan speed if needed to bring the system to within 10% of design flow. The system is within 10% of total design flow and no speed adjustments are needed.

- Reread all the terminals and make any final adjustments. AD 5 needed minor adjustment to bring it down to 110%D.

- Complete the report.

CHAPTER 19 Testing, Adjusting and Balancing—
Final Conditions

After the air distribution and branches have been proportionally balanced, set the minimum outside air dampers (Fig. 19.1). Outside air volume (cfm) may be determined by:

- Traversing the outside air duct.

- Traversing the supply air duct and the return air duct and subtracting the air quantities.

- Totaling the supply and return AD and subtracting the measured return air volume from the measured supply air volume.

- Reading the outside, return and mixed (or supply) air temperatures and calculating cfm.

The first method is preferred. Generally, temperature readings are difficult to accurately

FIGURE 19.1.

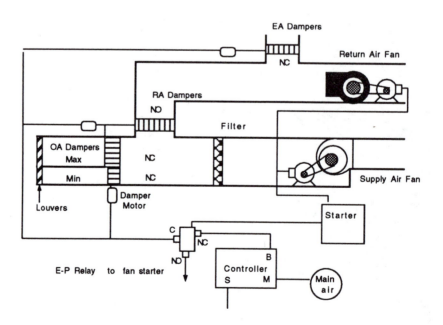

take and are time consuming. If other methods aren't satisfactory, take temperature readings and average them. Then, use the mixed air equations to set the minimum outside air dampers.

Equations

MAT-RAT-OAT Equations

Equation 19.1

$$MAT = (\%RA \times RAT) + (\%OA \times OAT)$$

Equation 19.2

$$\%OA = \frac{(MAT - RAT) \times 100}{(RAT - OAT)}$$

SAT-RAT-OAT Equations

Equation 19.3

$$SAT = (\%RA \times RAT) + (\%OA \times OAT) + 05*(TSP)$$

Equation 19.4

$$\%OA = \frac{RAT - [SAT - 0.5^*(TSP)]}{RAT - OAT} \times 100$$

MAT = mixed air temperature
SAT = supply air temperature
$\%OA$ = percent of outside air, cfm
OAT = outside air temperature
$\%RA$ = percent of return air, cfm
RAT = return air temperature
TSP = total static pressure rise across the fan, in. wg
0.5^* = correction for fan heat of compression. 0.5°F per inch of total static pressure when motor is out of the air stream.
0.6^* = correction for fan heat of compression and motor heat when motor is in the air stream. 0.6°F per inch of total static pressure.

Use a digital single or multiple sensor thermometer to take a temperature traverse across the mixed air plenum. Read temperatures at the center of each filter (Chap. 1). Take temperature readings when the temperature difference between the outside air and return air is greatest.

During the temperature traverse you may find some air stratification (Chap. 1). Air stratification results in uneven heat transfer across the coil and can lead to coil freezeup or the air handling unit shutting down when the freezestat trips (Fig. 19.2). To correct air stratification, install baffles or some other type of air mixing device.

If the mixed air temperature are not satisfactory, the fan discharge air temperature (supply air temperature, SAT) can be used along with the outside air temperature and the return air temperature to determine the mixed air quantities. The fan will add about 0.5°F per inch of measured total static pressure to the air because of mechanical heat of compres-

FIGURE 19.2.

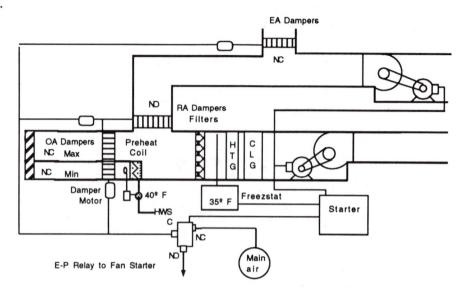

sion. Therefore, an amount equal to this temperature rise across the fan must be subtracted from the discharge air temperature. If the motor is in the airsteam, add an additional 0.1° (total of 0.6°) for the heat from the motor.

VERIFYING THE OPERATION OF THE ECONOMIZER

Verify the operation of the economizer by cycling the return air and the outside air dampers:

- Set the return air dampers to full open.
- Set minimum outside air dampers to minimum.
- Set the economizer (maximum) outside air dampers to full closed.
- Observe the static pressure (SP) in the mixed air plenum.
- Use the economizer controls (mixed air, outside air and morning warmup) to close the return air dampers 10%.
- Check that the economizer dampers begin to open.
- Continue to close the return air dampers in 10% increments.
- Continue to observe the static pressure (SP) in the mixed air plenum (see Fig. 19.3). Any change in static pressure indicates a change in airflow quantity.
- Take a mixed air static pressure for each point of damper operation. This will give a static pressure profile of the economizer damper operation and confirm that the automatic dampers operate simultaneously to close the return air as the outside air is opening. Lagging, or leading damper operation, is a common cause of reduced airflow resulting from the checking effect of one damper closing before the other opens. Correct any problems.

FIGURE 19.3.

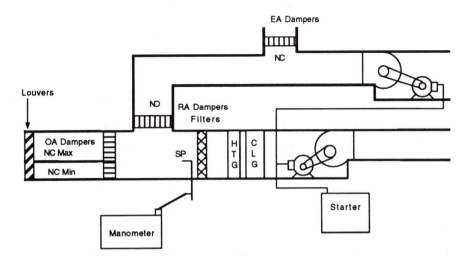

HOW TO CHANGE FAN SPEED

If a speed change is needed to bring the system to within +/− 10% of design airflow:

• Use Fan law #1 to calculate new fan speed:

Equation 19.5: Fan law #1

$$rpm_2 = rpm_1 \times \frac{cfm_2}{cfm_1}$$

    cfm_1 = original volume of airflow in cubic feet per minute
    cfm_2 = new volume of airflow in cubic feet per minute
    rpm_1 = original fan speed in revolutions per minute
    rpm_2 = new fan speed in revolutions per minute

• Use the tip speed equation to calculate new tip speed if there is an increase in fan speed. Fans are built to withstand the increased stress of centrifugal force caused by increases in fan speed to a certain limit designated by the fan class. Before increasing the fan speed, calculate the tip speed and consult the fan performance table or the fan manufacturer for the maximum allowable fan speed.

Equation 19.6: Tip speed equation

$$TS = \frac{3.14 \times D \times rpm}{12}$$

    TS = tip speed, in feet per minute
    3.14 = constant
    D = fan wheel diameter in inches
    rpm = fan speed

• Use Fan law #3 to calculate new horsepower requirements:

Equation 19.7: Fan law #3

$$bhp_2 = bhp_1 \times \left(\frac{rpm_2}{rpm_1}\right)^3$$

bhp_1 = original brake horsepower
bhp_2 = new brake horsepower

- Use the drive equation to determine the size of the sheaves needed:
  —Increasing the size of a fixed pitch motor sheave, or adjusting the belts to ride higher in a variable pitch motor sheave, will result in an increase in fan speed.
  —Decreasing the size of a fixed pitch motor sheave, or adjusting the belts to ride lower in a variable pitch motor sheave, will result in a decrease in fan speed.
  —Increasing the pitch diameter of the fan sheave decreases the fan speed.
  —Decreasing the pitch diameter of the fan sheave increases the fan speed.

Equation 19.8: Drive equation

$$rpm_m \times D_m = rpm_f \times D_f$$

rpm_m = speed at the motor shaft
D_m = pitch diameter of the motor sheave
rpm_f = speed at the fan shaft
D_f = pitch diameter of the fan sheave

Notice that pitch diameter is used in the calculations. For field calculations, use the outside diameter of a fixed sheave for pitch diameter. For variable pitch sheaves, when the belt is riding down in the groove, an approximation of the pitch diameter will be used for calculation purposes.

- Verify motor and fan shaft diameters.
- Verify distance between the centers of the shafts.
- Verify motor and fan bushing sizes.
- Verify the number of belt grooves.
- Verify belt size.
- Verify the allowance for motor movement on the motor slide rail to allow for adjustment of belt tension.
- Use the belt length equation to calculate the new belt length needed. If it's necessary to change any of the belts on a multiple groove sheave, buy a matched set of belts. Otherwise, because belt lengths and tension strengths vary, some belts could end up being too tight and others too loose, resulting in excessive wear.

Equation 19.9: Belt length equation

$$L = 2C + 1.57(D + d) + \frac{(D - d)^2}{4C}$$

L = belt pitch length
C = center to center distance of the shafts

D = pitch diameter of the large sheave
d = pitch diameter of the small sheave
1.57 = constant

- Remove the existing belts—Loosen and slide the motor for easier removal of the belts. Don't force the belts over the sheaves.

- Adjust existing sheave(s) or install new sheave(s).

- Align the motor and fan sheaves.

- Install the original belts or install new belts.

- Adjust the belts for proper tension.

- Secure the motor.

Example Problem 19.1

A fan has been balanced at 25,000 cfm supply air. The measured air temperatures are:

Mixed air temperature (MAT) = 82°F
Return air temperature (RAT) = 75°F
Outside air temperature (OAT) = 98°F

- Determine percent of outside air at these conditions.

- Determine the MAT required for 15% outside air.

- Set the minimum outside air dampers.

$$\%OA = \frac{(MAT - RAT) \times 100}{(RAT - OAT)}$$

$$\%OA = \frac{(82 - 75) \times 100}{(75 - 98)}$$

$$\%OA = \frac{(7) \times 100}{(23)}$$

$$\%OA = 30.4\%$$

$$MAT = (\%RA \times RAT) + (\%OA \times OAT)$$
$$MAT = (0.85 \times 75) + (0.15 \times 98)$$
$$MAT = (63.75) + (14.7)$$
$$MAT = 78.5°F$$

Adjust the minimum outside air dampers until an average temperature of 78.5°F is measured in the mixed air plenum. The outside air dampers are now set for 15%, or 3,750 cfm of outside air.

Reduce the fan output to 23,000 cfm. The fan speed is 900 rpm. The fan sheave is 15 inches Pd. The motor speed is 1,725 rpm. The motor sheave is variable pitch sheave (7.8 inches). The brake horsepower is 14.7 bhp. The center to center distance of the shafts is 36 inches. The belt length is 108 inches. The new conditions are:

$$\text{rpm}_2 = \text{rpm}_1 \times \frac{\text{cfm}_2}{\text{cfm}_1}$$

$$\text{rpm}_2 = 900 \times \frac{23{,}000}{25{,}000}$$

$$\text{rpm}_2 = 828$$

$$\text{bhp}_2 = \text{bhp}_1 \times \left(\frac{\text{rpm}_2}{\text{rpm}_1}\right)^3$$

$$\text{bhp}_2 = 14.7 \times \left(\frac{828}{900}\right)^3$$

$$\text{bhp}_2 = 11.44$$

$$D_m = \text{rpm}_f \times \frac{Df}{\text{rpm}_m}$$

$$D_m = 828 \times \frac{15}{1725}$$

$$D_m = 7.2 \text{ in.}$$

$$L = 2C + 1.57(D + d) + \frac{(D - d)^2}{(4C)}$$

$$L = 2(36) + 1.57(15 + 7.2) + \frac{(15 - 7.2)^2}{144}$$

$$L = 107 \text{ in.}$$

BALANCING THE RETURN AIR SYSTEM

After the supply system has been balanced:

- Gather plans and specifications and prepare report forms.
- Inspect the return air duct system.
- Set all return air dampers to full open positions.
- Set the outside air damper to minimum position.
- Operate all fans associated with the return air system.
- Take total air measurements in the return main.
- Balance and adjust the return air distribution (AD) system:

    —Read the AD.

    —Total the inlets and compare the total air at the fan.

    —Reconcile any differences.

    —Proportionally balance the outlets starting with the inlet with the lowest percent of design flow.

    —Proportionally balance the branches.

- Take AD final readings.

- Open the outside air dampers to maximum.
- Close the return air dampers as applicable.
- Operate the system in full economizer mode.
- Record final data.
- Complete reports.

Balancing Systems with a Return Air Fan

If the return air system has a fan in addition to the standard return air balancing procedure, then also do the following items:

- Test the return air fan:
  —Verify correct rotation.
  —Take electrical measurements.
  —Take speed measurements.
- Take static pressure measurements at the return fan.
- Change fan speed as needed.

TESTING AND ADJUSTING SYSTEM IN FULL ECONOMIZER MODE

The outside air duct normally has less resistance than the return air duct. This means that, with the same fan speed, more airflow will come through the outside air duct than through the return duct (air takes the easiest path). Therefore, the system must be checked in the full outside air mode to ensure that the motor does not over-amp when the maximum outside dampers are full open and the return air dampers are closed.

- Set system to full economizer (100% outside air).
- Take static pressure in the mixed air plenum or where applicable (Fig. 19.4).
- Take motor amperage (Fig. 19.4).

If the static pressure is excessive and/or the motor is over-amping (meaning that the airflow has increased significantly), adjust the manual volume dampers (install dampers as needed if not already in place) to bring static pressure and/or current within specifications.

The supply fan should "see" the same static pressure no matter what configuration the economizer dampers (outside, return, and exhaust dampers) are in. Continue to test the static pressure and adjust the manual dampers until the static pressure remains relatively constant.

Recording Final Data

After all balancing and adjusting is completed, retest the following:

FIGURE 19.4.

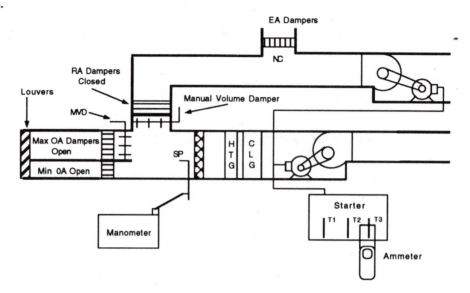

- Fan rpm
- Motor amperage
- Fan static pressure

Make a final pass of the air distribution and record these readings on the final data sheets. Compare final AD readings with traverses. If there are no differences, mark the dampers so they can be reset if tampering or accidental changes occur. Walk the system to determine if there are any drafts. If some are found, they usually can be corrected by adjusting the pattern control device on the air distribution.

Review the final report and make sure that nothing has been omitted. Be certain that all notes on problems, deficiencies and other abnormal conditions are fully explained.

CHAPTER 20 Testing, Adjusting and Balancing—Example System

This chapter shows the TAB forms for a test and balance of a typical constant volume system (Fig. 20.1).

OFFICE PREPARATION

Fig. 20.2—fan data
Fig. 20.3—drive data
Fig. 20.4—motor data
Fig. 20.5—air distribution data

FIGURE 20.1. CONSTANT VOLUME SYSTEM

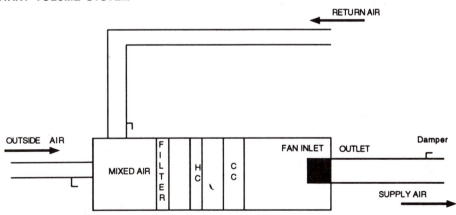

FIGURE 20.2.   AIR HANDLING EQUIPMENT DATA AND TEST SHEET

PROJECT: 92 EASY STREET	ENGINEER: AE	
Fan Information	**Specified**	**Actual**
Designation	AHU-024	
Location	EQUIP. ROOM	
Area Served	LOW BAY	
Manufacturer	AFC	
Serial #	123–456–789	
Model #	AF–30-91	
Class	II	
Speed rpm	1586	
Rotation	CW	
Efficiency	63%	
Wheel Type/Inlet	AIR FOIL/SWSI	
Wheel Size	30″	
Tip Speed	12,450	
Air Volume		
Total Supply Air	21,000	
Total Supply Air Outlets	21,000	
Total Outside Air	4,200	
Total Return Air	16,800	
Fan Pressure		
Inlet Static Pressure	NA	
Outlet Static Pressure	NA	
Total Static Pressure	4.0	
Pressure Differential		
Filter		
Entering Static Pressure	NA	
Leaving Static Pressure	NA	
Static Pressure Differential	NA	
Heating Coil		
Entering Static Pressure	NA	
Leaving Static Pressure	NA	
Static Pressure Differential	0.30	
Cooling Coil		
Wet or Dry	WET	
Entering Static Pressure	NA	
Leaving Static Pressure	NA	
Static Pressure Differential	0.50	

FIGURE 20.3.   **DRIVE DATA SHEET**

PROJECT: 92 EASY STREET	**ENGINEER: AE**
FAN	AHU-024
Drive Information	**Specified**
Fan	
Shaft Size	$2^{7}/_{16}$ in.
Sheave Size	4TB110
Sheave Manufacturer	ASC
Fixed or Adjustable	FIXED
Motor	
Shaft Size	$1^{7}/_{8}$ in.
Sheave Size	4TB94
Sheave Manufacturer	ASC
Fixed or Adjustable	ADJUSTABLE
Shaft Center Distance	NA
Belts	
Manufacturer	ABC
Number	4
Size	NA

Drives	**Condition**	**Problems**
Sheaves		
Shafts		
Bearings		
Belt Tension		
Drive Alignment		

Motor Adjustment	**Inches**
Up	
Down	
Front	
Back	

FIGURE 20.4. MOTOR DATA AND TEST SHEET

PROJECT: 92 EASY STREET	ENGINEER: AE
FAN	AHU-024
Motor Information	**Specified**
Manufacturer	AMC
Frame Size	286T
Horsepower	30
Phase	3
Hertz	60
Speed rpm	1800
Service Factor	1.15
Voltage	230/460
Amperage	72.4/36.2
Power Factor	0.83
Efficiency	93%
Brake Horsepower	24.95
Starter Size	3
Thermal OLP	W65

FIGURE 20.5. AIR DISTRIBUTION TEST SHEET

PROJECT: 92 EASY STREET **ENGINEER: AE**

Area Served	Terminal No. Type	Size	Ak	Design cfm	Initial cfm	Initial %D	Proportioned cfm	Proportioned %D	Final cfm	Final %D
					READINGS					
LOW BAY	1 CD	24 × 24	1.00	700						
LOW BAY	2 CD	24 × 24	1.00	700						
LOW BAY	3 CD	24 × 24	1.00	700						
LOW BAY	4 CD	24 × 24	1.00	700						
LOW BAY	5 CD	24 × 24	1.00	700						
LOW BAY	6 CD	24 × 24	1.00	700						
LOW BAY	7 CD	24 × 24	1.00	700						
LOW BAY	8 CD	24 × 24	1.00	700						
LOW BAY	9 CD	24 × 24	1.00	700						
LOW BAY	10 CD	24 × 24	1.00	700						
LOW BAY	11 CD	24 × 24	1.00	700						
LOW BAY	12 CD	24 × 24	1.00	700						
LOW BAY	13 CD	24 × 24	1.00	600						
LOW BAY	14 CD	24 × 24	1.00	600						
LOW BAY	15 CD	24 × 24	1.00	600						
LOW BAY	16 CD	24 × 24	1.00	600						
LOW BAY	17 CD	24 × 24	1.00	200						
Branch A 32″ × 30″				11,000						
LOW BAY	18 CD	24 × 24	1.00	1,000						
LOW BAY	19 CD	24 × 24	1.00	1,000						
LOW BAY	20 CD	24 × 24	1.00	1,000						
LOW BAY	21 CD	24 × 24	1.00	1,000						
LOW BAY	22 CD	24 × 24	1.00	1,000						
LOW BAY	23 CD	24 × 24	1.00	600						
LOW BAY	24 CD	24 × 24	1.00	400						
LOW BAY	25 CD	24 × 24	1.00	600						
LOW BAY	26 CD	24 × 24	1.00	400						
LOW BAY	27 CD	24 × 24	1.00	1,000						
LOW BAY	28 CD	24 × 24	1.00	1,000						
LOW BAY	29 CD	24 × 24	1.00	1,000						
Branch B 32″ ø				10,000						
System total				21,000						

FIELD TESTING

Fig. 20.6—fan data

Fig. 20.7—total static pressure

Fig. 20.8—static pressure drop across the filters

Fig. 20.9—static pressure drop across the heating coil

Fig. 20.10—static pressure drop across the cooling coil

Fig. 20.11—drive data

Fig. 20.12—motor data

Fig. 20.13—traverse data, main duct

Fig. 20.14—traverse data, branch B duct

Fig. 20.15—traverse summary sheet

Fig. 20.16—traverse locations

Fig. 20.17—layout of main duct traverse

Fig. 20.18—layout of round traverse for branch B duct

FIGURE 20.6.   AIR HANDLING EQUIPMENT DATA AND TEST SHEET

PROJECT: 92 EASY STREET	ENGINEER: AE	
Fan Information	**Specified**	**Actual**
Designation	AHU-024	AHU-024
Location	EQUIP. ROOM	EQUIP. ROOM
Area Served	WEST BAY	WEST BAY
Manufacturer	AFC	AFC
Serial#	123–456–789	123–456–789
Model#	AF–30–91	AF–30–91
Class	II	II
Speed rpm	1586	1545
Rotation	CW	CW
Efficiency	63%	
Wheel Type/Inlet	AIR FOIL/SWSI	AIR FOIL/SWSI
Wheel Size	30 in.	30 in.
Tip Speed	12,450	12,128
Air Volume		
Total Supply Air	21,000	
Total Outside Air	4,200	
Total Return Air	16,800	
Fan Pressure		
Inlet Static Pressure	NA	2.20
Outlet Static Pressure	NA	1.80
Total Static Pressure	4.0	4.00
Pressure Differential		
Filter		
Entering Static Pressure	NA	1.20
Leaving Static Pressure	NA	1.40
Static Pressure Differential	NA	0.20
Heating Coil		
Entering Static Pressure	NA	1.40
Leaving Static Pressure	NA	1.68
Static Pressure Differential	0.30	0.28
Cooling Coil		
Wet or Dry	WET	WET
Entering Static Pressure	NA	1.68
Leaving Static Pressure	NA	2.20
Static Pressure Differential	0.50	0.52

FIGURE 20.7.   TOTAL STATIC PRESSURE

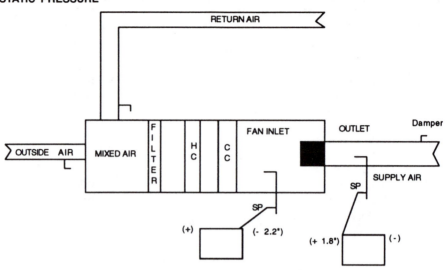

FIGURE 20.8.   STATIC PRESSURE DROP ACROSS THE FILTERS

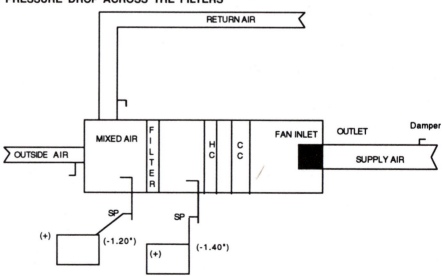

FIGURE 20.9. STATIC PRESSURE DROP ACROSS THE HEATING COIL

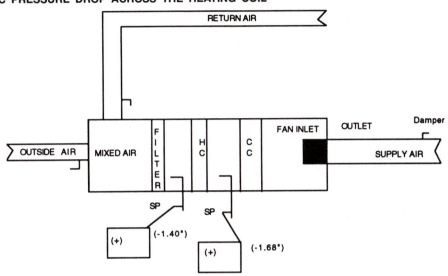

FIGURE 20.10. STATIC PRESSURE DROP ACROSS THE COOLING COIL

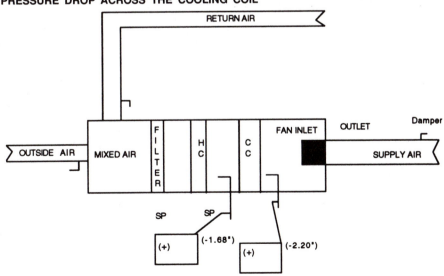

FIGURE 20.11.   DRIVE DATA SHEET

PROJECT: 92 EASY STREET	ENGINEER: AE	
	Specified	*Actual*
FAN	AHU–024	AHU–024
Drive Information		
Fan		
Shaft Size	2⁷⁄₁₆ in.	2⁷⁄₁₆ in.
Sheave Size	4TB110	4TB110
Sheave Manufacturer	ASC	ASC
Fixed or Adjustable	FIXED	FIXED
Motor		
Shaft Size	1⁷⁄₈ in.	1⁷⁄₈ in.
Sheave Size	4TB94	4TB94
Sheave Manufacturer	ASC	ASC
Fixed or Adjustable	ADJUSTABLE	ADJUSTABLE
Shaft Center Distance	NA	34¼ in.
Belts		
Manufacturer	ABC	ABC
Number	4	4
Size	NA	B95
Drives	**Condition**	**Problems**
Sheaves	GOOD	NONE
Shafts	GOOD	NONE
Bearings	GOOD	NONE
Belt Tension	GOOD	NONE
Drive Alignment	GOOD	NONE
Motor Adjustment	**Inches**	
Up	NA	
Down	NA	
Front	3″	
Back	3″	

FIGURE 20.12. **MOTOR DATA SHEET**

PROJECT: 92 EAST STREET	ENGINEER: AE	
	Specified	*Actual*
FAN	AHU–024	AHU–024
Motor Information		
Manufacturer	AMC	AMC
Frame Size	286T	286T
Horsepower	30	30
Phase	3	3
Hertz	60	60
Speed rpm	1,800	1,800
Service Factor	1.15	1.15
Voltage	230/460	478/477/480
Amperage	72.4/36.2	24.6/24.7/24.5
Power Factor	0.83	0..83*
Efficiency	93%	0.93*
Brake Horsepower	24.95	21.05*
Starter Size	3	3
Thermal OLP	W65	W65

*Power factor and efficiency used for calculation of brake horsepower.

FIGURE 20.13.   **RECTANGULAR DUCT TRAVERSE**

PROJECT: 92 EASY STREET		**ENGINEER: AE**
FAN		AHU–024
TRAVERSE		MAIN DUCT

	Specified	*Actual*
Duct Size (Inches)	40″ × 40″	40″ × 40″
Duct Area (Square Feet)	11.1 sq. ft.	11.1 sq. ft.
Volume (Cubic Feet per Minute)	21,000 cfm	20,845 cfm
Average Velocity (Feet Per Minute)	1,892 fpm	1,878 fpm
Center Line Static Pressure	Not Specified	1.76 in. SP
Density (pounds per cubic foot)	0.075 lbs/cf	0.075 lbs/cf
Instrument Correction for Density	None	None

No.	*VP*	*Vel*	*No.*	*VP*	*Vel*	*No.*	*VP*	*Vel*	*No.*	*VP*	*Vel*
1	0.17	1,651	18	0.20	1,791	35	0.19	1,746	52	0.22	1,879
2	0.19	1,746	19	0.21	1,835	36	0.20	1,791	53	0.19	1,746
3	0.20	1,791	20	0.22	1,879	37	0.21	1,835	54	0.17	1,651
4	0.21	1,835	21	0.22	1,879	38	0.21	1,835	55	0.16	1,602
5	0.22	1,879	22	0.23	1,921	39	0.22	1,879	56	0.17	1,651
6	0.19	1,746	23	0.24	1,962	40	0.22	1,879	57	0.18	1,699
7	0.21	1,835	24	0.24	1,962	41	0.20	1,791	58	0.17	1,651
8	0.23	1,921	25	0.26	2,042	42	0.23	1,921	59	0.18	1,699
9	0.24	1,962	26	0.26	2,042	43	0.23	1,921	60	0.19	1,746
10	0.22	1,879	27	0.25	2,003	44	0.24	1,962	61	0.20	1,791
11	0.23	1,921	28	0.24	1,962	45	0.26	2,042	62	0.19	1,746
12	0.24	1,962	29	0.23	1,921	46	0.26	2,042	63	0.19	1,746
13	0.21	1,835	30	0.22	1,879	47	0.27	2,081	64	0.19	1,746
14	0.20	1,791	31	0.21	1,835	48	0.25	2,003			
15	0.19	1,746	32	0.22	1,879	49	0.24	1,962			
16	0.18	1,699	33	0.20	1,791	50	0.23	1,921			
17	0.19	1,746	34	0.21	1,835	51	0.23	1,921			
	Subtotal 30,945			Subtotal 34,380			Subtotal 32,532			Subtotal 22,353	

Total 120,210

Total velocity 120,210 fpm ÷ 64 readings = 1,878 fpm
1,878 fpm × 11.1 SF = 20,845 cfm

FIGURE 20.14.   ROUND DUCT TRAVERSE SHEET—12″ OR LARGER DIAMETER

PROJECT: 92 EASY STREET   **ENGINEER: AE**

FAN   AHU–024

TRAVERSE   BRANCH B

	Specified	Actual
Duct Size (inches)	32″ Diameter	32″ Diameter
Duct Area (square feet)	5.58 sq. ft.	5.58 sq. ft.
Volume (cubic feet per minute)	10,000 cfm	11,215 cfm
Average Velocity (feet per minute)	1,792 fpm	2,010 fpm
Center Line Static Pressure	Not Specified	1.20 in.
Density (pounds per cubic foot)	0.075 lb/cf	0.075 lb/cf
Instrument Correction for Density	None	None

No.	Factor	Inches	VP	Vel	No.	Factor	Inches	VP	Vel
1	0.052	3/4	0.22	1,879	11	0.052	3/4	0.23	1,921
2	0.165	2-3/4	0.23	1,921	12	0.165	2-3/4	0.24	1.962
3	0.293	4-3/4	0.24	1,962	13	0.293	4-3/4	0.25	2,003
4	0.454	7-1/4	0.25	2,003	14	0.454	7-1/4	0.26	2,042
5	0.684	11	0.25	2,003	15	0.684	11	0.27	2,081
6	1.316	21	0.26	2,042	16	1.316	21	0.29	2,157
7	1.547	24-3/4	0.27	2,081	17	1.547	24-3/4	0.28	2,119
8	1.707	27-1/4	0.25	2,003	18	1.707	27-1/4	0.25	2,003
9	1.835	29-1/4	0.25	2,003	19	1.835	29-1/4	0.25	2,003
10	1.948 31-1/4 Subtotal 19,900		0.25	2,003	20	1.948 31-1/4 Subtotal 20,294		0.25	2,003

Total 40,194

Total velocity 40,194 fpm ÷ 20 readings = 2,010 fpm.

FIGURE 20.15.   TRAVERSE SUMMARY SHEET

PROJECT: 92 EASY STREET   **ENGINEER: AE**

FAN   **AHU–024**

No.	Traverse Name	Duct Size	Square Foot	Static Press	Design cfm	Measured cfm	Percent Design
1	Main	40 × 40	11.1	1.76	21,000	20,845	99%
2	Branch A	32 × 30	6.67	***	11,000	***	***
3	Branch B	32″ ø	5.58	1.2	10,000	11,215	112%

***No traverse on branch A.

FIGURE 20.16.   PITOT TUBE TRAVERSE LOCATIONS

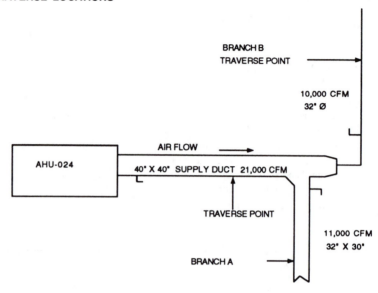

FIGURE 20.17.   AHU–024 LOCATION OF TEST HOLES AND MARKING THE PITOT TUBE (MAIN DUCT)

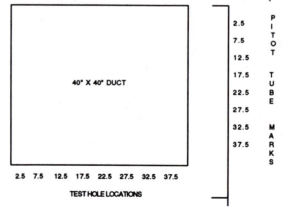

FIGURE 20.18.   AHU–024 PITOT TUBE TRAVERSE FOR 32″ ROUND DUCT

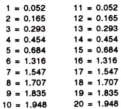

1 = 0.052	11 = 0.052
2 = 0.165	12 = 0.165
3 = 0.293	13 = 0.293
4 = 0.454	14 = 0.454
5 = 0.684	15 = 0.684
6 = 1.316	16 = 1.316
7 = 1.547	17 = 1.547
8 = 1.707	18 = 1.707
9 = 1.835	19 = 1.835
10 = 1.948	20 = 1.948

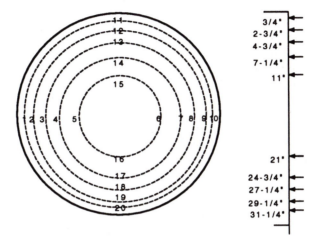

FIELD AIR DISTRIBUTION BALANCING

Fig. 20.19—air distribution, initial reading

Fig. 20.20—air distribution, proportional balance

The air distribution is now proportionally balanced at within 10% of each other by branch.

FIELD BRANCH BALANCING

Branch B is 105% of design with 1.2 inch static pressure at the traverse point.

Branch A is 83% of design with 0.97 inch static pressure at the test point after the volume damper. Close Branch B volume damper to 1.01 inch static pressure (96% of design). Branch A reads 1.15 inch static pressure (90% of design) at the test point. The branches are now balanced to each other. The ratio is 1.07 (96/90). The air distribution is still proportionally balanced at within 10% of each other by branch.

FIELD ADJUSTMENT OF SYSTEM

The system is operating at 93% of design (19,585/21,000).

The following calculations are done to determine the effect of increasing the system to 20,500 cfm at the air distribution: (20,500/19,585 = 1.05). New brake horsepower and amperage using 24.7 as amp_1:

FIGURE 20.19.   AIR DISTRIBUTION TEST SHEET

PROJECT: 92 EASY STREET   **ENGINEER: AE**

| Area Served | Air Distribution | | | Ak | Design cfm | Initial | | Proportioned | | Final | |
	No. Type	Size				cfm	%D	cfm	%D	cfm	%D
LOW BAY	1 CD	24 × 24	1.00		700	515	74%				
LOW BAY	2 CD	24 × 24	1.00		700	500	71%				
LOW BAY	3 CD	24 × 24	1.00		700	540	77%				
LOW BAY	4 CD	24 × 24	1.00		700	600	86%				
LOW BAY	5 CD	24 × 24	1.00		700	490	70%	THE KEY FOR BRANCH A			
LOW BAY	6 CD	24 × 24	1.00		700	700	100%				
LOW BAY	7 CD	24 × 24	1.00		700	600	86%				
LOW BAY	8 CD	24 × 24	1.00		700	660	94%				
LOW BAY	9 CD	24 × 24	1.00		700	630	90%				
LOW BAY	10 CD	24 × 24	1.00		700	570	81%				
LOW BAY	11 CD	24 × 24	1.00		700	580	83%				
LOW BAY	12 CD	24 × 24	1.00		700	625	89%				
LOW BAY	13 CD	24 × 24	1.00		600	700	117%				
LOW BAY	14 CD	24 × 24	1.00		600	550	92%				
LOW BAY	15 CD	24 × 24	1.00		600	580	97%				
LOW BAY	16 CD	24 × 24	1.00		600	480	80%				
LOW BAY	17 CD	24 × 24	1.00		200	200	100%				
Branch A 32″ × 30″					11,000	9,520	86.5%				
LOW BAY	18 CD	24 × 24	1.00		1,000	1,150	115%				
LOW BAY	19 CD	24 × 24	1.00		1,000	965	96%				
LOW BAY	20 CD	24 × 24	1.00		1,000	1,200	120%				
LOW BAY	21 CD	24 × 24	1.00		1,000	1,145	115%				
LOW BAY	22 CD	24 × 24	1.00		1,000	1,080	108%				
LOW BAY	23 CD	24 × 24	1.00		600	675	113%				
LOW BAY	24 CD	24 × 24	1.00		400	450	113%				
LOW BAY	25 CD	24 × 24	1.00		600	580	97%				
LOW BAY	26 CD	24 × 24	1.00		400	350	88%	THE KEY FOR BRANCH B			
LOW BAY	27 CD	24 × 24	1.00		1,000	1,125	113%				
LOW BAY	28 CD	24 × 24	1.00		1,000	900	90%				
LOW BAY	29 CD	24 × 24	1.00		1,000	1,120	112%				
Branch B 32″ ø					10,000	10,740	107%				
System total					21,000	20,260	96.5%				

Note: #5 is the key for the system.

FIGURE 20.20. **AIR DISTRIBUTION TEST SHEET**

PROJECT: 92 EASY STREET   **ENGINEER: AE**

Area Served		Air Distribution			Design	Initial		Proportioned		Final	
	No. Type		Size	Ak	cfm	cfm	%D	cfm	%D	cfm	%D
LOW BAY	1 CD		24 × 24	1.00	700	515	74%	595	85%		
LOW BAY	2 CD		24 × 24	1.00	700	500	71%	567	81%		
LOW BAY	3 CD		24 × 24	1.00	700	540	77%	616	88%		
LOW BAY	4 CD		24 × 24	1.00	700	600	86%	588	84%		
LOW BAY	5 CD		24 × 24	1.00	700	490	70%	**560**	**80%**		
LOW BAY	6 CD		24 × 24	1.00	700	700	100%	602	86%		
LOW BAY	7 CD		24 × 24	1.00	700	600	86%	588	84%		
LOW BAY	8 CD		24 × 24	1.00	700	660	94%	574	82%		
LOW BAY	9 CD		24 × 24	1.00	700	630	90%	567	81%		
LOW BAY	10 CD		24 × 24	1.00	700	570	81%	560	80%		
LOW BAY	11 CD		24 × 24	1.00	700	580	83%	560	80%		
LOW BAY	12 CD		24 × 24	1.00	700	625	89%	574	82%		
LOW BAY	13 CD		24 × 24	1.00	600	700	117%	492	82%		
LOW BAY	14 CD		24 × 24	1.00	600	550	92%	498	83%		
LOW BAY	15 CD		24 × 24	1.00	600	580	97%	516	86%		
LOW BAY	16 CD		24 × 24	1.00	600	480	80%	486	81%		
LOW BAY	17 CD		24 × 24	1.00	200	200	100%	162	81%		
Branch A 32″ × 30″					11,000	9,520	86.5%	9,105	83%		
LOW BAY	18 CD		24 × 24	1.00	1,000	1,150	115%	1,080	108%		
LOW BAY	19 CD		24 × 24	1.00	1,000	965	96%	1,070	107%		
LOW BAY	20 CD		24 × 24	1.00	1,000	1,200	120%	1,050	105%		
LOW BAY	21 CD		24 × 24	1.00	1,000	1,145	115%	1,030	103%		
LOW BAY	22 CD		24 × 24	1.00	1,000	1,080	108%	1,030	103%		
LOW BAY	23 CD		24 × 24	1.00	600	675	113%	635	106%		
LOW BAY	24 CD		24 × 24	1.00	400	450	113%	425	106%		
LOW BAY	25 CD		24 × 24	1.00	600	580	97%	650	108%		
LOW BAY	26 CD		24 × 24	1.00	400	350	88%	**390**	**98%**		
LOW BAY	27 CD		24 × 24	1.00	1,000	1,125	113%	1,070	107%		
LOW BAY	28 CD		24 × 24	1.00	1,000	900	90%	1,000	100%		
LOW BAY	29 CD		24 × 24	1.00	1,000	1,120	112%	1,050	105%		
Branch B 32″ ø					10,000	10,740	107%	10,480	105%		
System total					21,000	20,260	96.5%	19,585	93%		

$$(1.05)^3 \times 21.05 = 24.37 \text{ bhp}_2$$
$$(1.05)^3 \times 24.7 = 28.6 \text{ amp}_2$$

The increase in fan volume can be accommodated by the existing motor (30 hp, design 24.95 brake horsepower).

Increase in Fan Speed

$$1,545 \times 1.05 = 1,622 \text{ rpm}$$

Motor Sheave Size Needed

$$1.05 \times 9.4 = 9.87''$$

After consulting ASM drive catalog, a new fixed motor sheave and new fixed fan sheave are selected.

New motor sheave is 4TB80 (8 inch pitch diameter with a "B belt").

New fan sheave is 4TB90:

$$\frac{1,800 \text{ rpm}_m \times 8.0''}{9.0''} = 1,600 \text{ rpm}_f$$

New belt length (pitch diameter):

$$2(34.25) + 1.57 (9 + 8) + \frac{(9 - 8)^2}{137} = 95.2''$$

New belt size is B94.

Calculated new tip speed:

$$\frac{3.14 \times 30 \times 1,622}{12} = 12,733 \text{ fpm}$$

The changes are made and the system is tested.

FIELD ADJUSTMENT OF RETURN AND OUTSIDE AIR SYSTEM

The design for the outside air (Fig. 20.21) is 4,200 cfm, or 20% of the total (4,200/21,000). A traverse (Fig. 20.22) shows 5,020 cfm outside air, or 24% (5,020/21,000). The layout of the holes in the duct and the marks on the Pitot tube for the traverse of the outside air duct is:

Drill holes in 24 inch side at 3 inches, 9 inches, 15 inches and 21 inches.

Mark the Pitot tube for the 20 inch side at 2.5 inches, 7.5 inches, 12.5 inches and 17.5 inches.

Temperatures are also measured with the following results.

$$OA = 98°F$$
$$RA = 78°F$$
$$MA = 83°F$$

The static pressure at the traverse is 0.85 inches.

Closing the volume damper in the outside air duct to 0.59 inch static pressure reduces the airflow to 4,200 cfm. The mixed air temperature is 82 F, verifying the 20% outside air.

After setting the outside air, a traverse of the return duct (Fig. 20.23) indicates 16,295 cfm, or 97% of design (16,295/16,800).

FIGURE 20.21.   RETURN AND OUTSIDE AIR DUCT

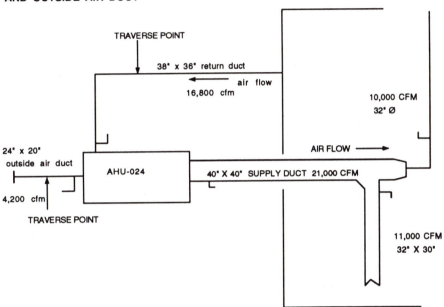

FIGURE 20.22.   RECTANGULAR DUCT TRAVERSE SHEET

PROJECT: 92 EASY STREET	ENGINEER: AE
FAN	AHU–024
TRAVERSE	OUTSIDE AIR DUCT

	Specified	*Actual*
Duct Size (inches)	24 in. × 20 in.	24 in. × 20 in.
Duct Area (square feet)	3.33 sq. ft.	3.33 sq. ft.
Volume (cubic feet per minute)	4,200 cfm	5,020 cfm
Average Velocity (feet per minute)	1,261 fpm	1,507 fpm
Center Line Static Pressure	Not Specified	0.85 in. SP
Density (pounds per cubic foot)	0.075 lb/cf	0.075 lb/cf
Instrument Correction for Density	None	None

No	VP	Vel
1	See	1,387
2	note 1	1,397
3	below	1,403
4		1,450
5		1,475
6		1,486
7		1,404
8		1,497
9		1,455
10		1,589
11		1,607
12		1,656
13		1,540
14		1,567
15		1,590
16		1,610
17		
	Subtotal 24,113	

Total velocity 24,113 fpm ÷ 16 readings = 1,507 fpm
1,507 fpm × 3.33 = 5,020 cfm

Note: Readings taken with an electronic manometer which reads directly in fpm.

FIGURE 20.23. **RECTANGULAR DUCT TRAVERSE SHEET**

PROJECT: 92 EASY STREET	ENGINEER: AE
FAN	AHU–024
TRAVERSE	RETURN AIR DUCT

	Specified	Actual
Duct Size (inches)	38 × 36 in.	38 × 36 in.
Duct Area (square feet)	9.5 sq. ft.	9.5 sq. ft.
Volume (cubic feet per minute)	16,800 cfm	16,295 cfm
Average Velocity (feet per minute)	1,768 fpm	1,715 fpm
Center Line Static Pressure	Not Specified	0.95 in. SP
Density (pounds per cubic foot)	0.075 lb/cf	0.075 lb/cf
Instrument Correction for Density	None	None

No.	VP	Vel	No.	VP	Vel	No.	VP	Vel
1	See	1,700	18	See	1,690	35	See	1,690
2	note 1	1,715	19	note 1	1,700	36	note 1	1,695
3	below	1,725	20	below	1,710	37	below	1,690
4		1,750	21		1,709	38		1,600
5		1,748	22		1,675	39		1,654
6		1,797	23		1,798	40		1,663
7		1,703	24		1,701	41		1,674
8		1,724	25		1,778	42		1,600
9		1,788	26		1,756			
10		1,730	27		1,768			
11		1,727	28		1,779			
12		1,700	29		1,790			
13		1,777	30		1,789			
14		1,750	31		1,680			
15		1,727	32		1,690			
16		1,700	33		1,659			
17		1,727	34		1,600			
	Subtotal 29,488			Subtotal 29,272			Subtotal 13,266	
							TOTAL 72,026	

72,026 fpm ÷ 42 readings = 1,715 fpm
1,715 fpm × 9.5 = 16,295

Note: Readings taken with an electronic manometer which reads directly in fpm.

FINAL FIELD DATA

Fig. 20.24—final fan data

Fig. 20.25—final drive data

Fig. 20.26—final motor data

Fig. 20.27—final air distribution data

FIGURE 20.24.   AIR HANDLING EQUIPMENT FINAL DATA AND TEST SHEET

PROJECT: 92 EASY STREET	ENGINEER: AE	
Fan Information	*Specified*	*Actual*
Designation	AHU-024	AHU–024
Location	EQUIP. ROOM	EQUIP. ROOM
Area Served	LOW BAY	LOW BAY
Manufacturer	AFC	AFC
Serial #	123–456–789	123–456–789
Model #	AF–30–91	AF–30–91
Class	II	II
Speed rpm	1586	1605
Rotation	CW	CW
Efficiency	63%	58%
Wheel Type/Inlet	AIR FOIL/SWSI	AIR FOIL/SWSI
Wheel Size	30 in.	30 in.
Tip Speed	12,450	12,599
Air Volume		
Total Supply Air	21,000	20,428
Total Outside Air	4,200	4,240
Total Return Air	16,800	16,295
Fan Pressure		
Inlet Static Pressure	NA	2.43
Outlet Static Pressure	NA	1.98
Total Static Pressure	4.0	4.41
Pressure Differential		
Filter		
Entering Static Pressure	NA	1.32
Leaving Static Pressure	NA	1.54
Static Pressure Differential	NA	0.22
Heating Coil		
Entering Static Pressure	NA	1.54
Leaving Static Pressure	NA	1.85
Static Pressure Differential	0.30	0.31
Cooling Coil		
Wet or Dry	WET	WET
Entering Static Pressure	NA	1.85
Leaving Static Pressure	NA	2.42
Static Pressure Differential	0.50	0.57

FIGURE 20.25. DRIVE FINAL DATA SHEET

PROJECT: 92 EASY STREET	**ENGINEER: AE**	
FAN		
Drive Information	*Specified*	*Actual*
Fan	AHU-024	AHU–024
Shaft Size	2⁷⁄₁₆ in.	2⁷⁄₁₆ in.
Sheave Size	4TB110	4TB90
Sheave Manufacturer	ASC	ASC
Fixed or Adjustable	FIXED	FIXED
Motor		
Shaft Size	1⅞ in.	1⅞ in.
Sheave Size	4TB94	4TB80
Sheave Manufacturer	ASC	ASC
Fixed or Adjustable	ADJUSTABLE	ADJUSTABLE
Shaft Center Distance	NA	34¼ in.
Belts		
Manufacturer	ABC	ABC
Number	4	4
Size	NA	B94
Drives	*Condition*	Problems
Sheaves	GOOD	NONE
Shafts	GOOD	NONE
Bearings	GOOD	NONE
Belt Tension	GOOD	NONE
Drive Alignment	GOOD	NONE
Motor Adjustment	*Inches*	
Up	NA	
Down	NA	
Front	3 in.	
Back	3 in.	

FIGURE 20.26. MOTOR FINAL DATA TEST SHEET

PROJECT: 92 EASY STREET	ENGINEER: AE	
	Specified	Actual
FAN	AHU-024	AHU–024
Motor Information		
Manufacturer	AMC	AMC
Frame Size	286T	286T
Horsepower	30	30
Phase	3	3
Hertz	60	60
Speed rpm	1800	1800
Service Factor	1.15	1.15
Voltage	230/460	478/477/480
Amperage	72.4/36.2	28.5/28.6/28.4
Power Factor	0.83	0.83*
Efficiency	93%	0.93*
Brake Horsepower	24.95	24.37
Starter Size	3	3
Thermal OLP	W65	W65

*Power factor and efficiency used for calculation of brake horsepower.

FIGURE 20.27. AIR DISTRIBUTION TEST SHEET

PROJECT: 92 EASY STREET					ENGINEER: AE					
					READINGS					
Area Served	Air Distribution			Design cfm	Initial		Proportioned		Final	
	No. Type	Size	Ak		cfm	%D	cfm	%D	cfm	%D
LOW BAY	1 CD	24 × 24	1.00	700	515	74%	595	85%	679	97%
LOW BAY	2 CD	24 × 24	1.00	700	500	71%	567	81%	644	92%
LOW BAY	3 CD	24 × 24	1.00	700	540	77%	616	88%	700	100%
LOW BAY	4 CD	24 × 24	1.00	700	600	86%	588	84%	672	96%
LOW BAY	5 CD	24 × 24	1.00	700	490	70%	560	80%	637	**91%**
LOW BAY	6 CD	24 × 24	1.00	700	700	100%	602	86%	679	97%
LOW BAY	7 CD	24 × 24	1.00	700	600	86%	588	84%	672	96%
LOW BAY	8 CD	24 × 24	1.00	700	660	94%	574	82%	658	94%
LOW BAY	9 CD	24 × 24	1.00	700	630	90%	567	81%	644	92%
LOW BAY	10 CD	24 × 24	1.00	700	570	81%	560	80%	637	91%
LOW BAY	11 CD	24 × 24	1.00	700	580	83%	560	80%	637	91%
LOW BAY	12 CD	24 × 24	1.00·	700	625	89%	574	82%	665	95%
LOW BAY	13 CD	24 × 24	1.00	600	700	117%	492	82%	570	95%
LOW BAY	14 CD	24 × 24	1.00	600	550	92%	498	83%	576	96%
LOW BAY	15 CD	24 × 24	1.00	600	580	97%	516	86%	588	98%
LOW BAY	16 CD	24 × 24	1.00	600	480	80%	486	81%	552	92%
LOW BAY	17 CD	24 × 24	1.00	200	200	100%	162	81%	184	92%
Branch A 32″ × 30″				11,000	9,520	87%	9,105	83%	10,394	95%
LOW BAY	18 CD	24 × 24	1.00	1,000	1,150	115%	1,080	108%	1,030	103%
LOW BAY	19 CD	24 × 24	1.00	1,000	965	96%	1,070	107%	1,020	102%
LOW BAY	20 CD	24 × 24	1.00	1,000	1,200	120%	1,050	105%	1,000	100%
LOW BAY	21 CD	24 × 24	1.00	1,000	1,145	115%	1,030	103%	990	99%
LOW BAY	22 CD	24 × 24	1.00	1,000	1,080	108%	1,030	103%	990	99%
LOW BAY	23 CD	24 × 24	1.00	600	675	113%	635	106%	612	102%
LOW BAY	24 CD	24 × 24	1.00	400	450	113%	425	106%	408	102%
LOW BAY	25 CD	24 × 24	1.00	600	580	97%	650	108%	618	103%
LOW BAY	26 CD	24 × 24	1.00	400	350	88%	390	98%	376	**94%**
LOW BAY	27 CD	24 × 24	1.00	1,000	1,125	113%	1,070	107%	1,020	102%
LOW BAY	28 CD	24 × 24	1.00	1,000	900	90%	1,000	100%	960	96%
LOW BAY	29 CD	24 × 24	1.00	1,000	1,120	112%	1,050	105%	1,010	101%
Branch B 32″ ø				10,000	10,740	107%	10,480	105%	10,034	100%
System total				21,000					20,428	97%

Testing, Adjusting and Balancing Constant Air Volume—Multizone, Dual Duct and Induction Systems

This chapter provides balancing procedures for multizone systems, dual duct systems, induction unit systems, and return or exhaust systems.

MIXING DAMPER AIR HANDLING UNIT AND MULTIZONE (MZ) SYSTEM

Multizones are dual path units. They normally have a cooling coil and heating coil in parallel in the unit. The heating coil is in the hot deck. The cooling coil is in the cold deck. Mixing dampers are located directly after the coils. Air passes through the heating and cooling coils, through the mixing dampers and into zone ducts to the various conditioned spaces.

Some systems do not have a heating coil. When the system is in the heating mode, air from the mixed air plenum goes through a resistance apparatus in the hot deck and is bypassed around the cooling coil.

Multizone systems are designed as constant volume systems, but the actual volume may vary somewhat during normal operation because of the changes in resistance between the smaller heating coil and the larger cooling coil. Multizone systems generally have between 5 and 12 zones.

Balancing Procedure

- Prepare report forms.
- Conduct field inspection.
  - —Inspect the job site.
  - —Inspect the equipment.
  - —Inspect the air distribution system.
- Set all dampers at the balancing position (see Chap. 17).
- Check mixing dampers for proper operation and leakage.
- Make initial tests of all the equipment applicable to the system being balanced.

324

—Verify motor and fan rotation.

—Take electrical measurements. Voltage, current, power factor.

—Take speed measurements.

—Measure static pressure at the fan.

• Take total air measurements.

—Normal Systems:

   • Traverse the zone ducts or read all the air distribution and total.

—Systems with diversity:

   • Determine the diversity ratio.

   • Keep the proportion of cooled air to total volume constant.

   • Set enough zones to full cooling to equal the design flow through the cooling coil.

   • Set the remaining zones to full heating.

   • Set some mixing dampers so the hot deck mixing dampers remain partially open at full cooling.

• Balance the distribution system.

—Set the zones to be tested on full cooling.

—Proportionally balance the air distribution.

—Proportionally balance the branches.

—Proportionally balance the zones.

   • Take a centerline static pressure reading after the zone volume damper.

   • Using the air volume from a Pitot tube traverse of the zone duct or the total of the zone outlets, calculate the percent of design flow for each zone.

   • Starting with the zone with the lowest percent of design flow, proportionally balance the zones using static pressure readings and Fan Law #2 (see branch balancing in Chap. 18).

• Change fan speed as needed.

—Change drives as needed.

• As applicable, place system in the maximum outside air mode and test.

—If the motor overloads or the airflow is excessive, adjust the system as needed.

• Place the system in the heating mode and verify proper operation.

—Measure the airflow through the heating coil or the resistance apparatus in the hot deck.

• Place the system in the cooling mode and record final test data.

• Complete the balance report.

AIR HANDLING UNIT WITH MIXING BOXES AND DUAL DUCT (DD) SYSTEM

Dual duct or double duct units are dual path. They normally have a cooling coil and heating coil in parallel in the unit. The heating coil is in the hot deck. The cooling coil is in the cold deck. Air passes through the coils into hot and cold ducts.

Most constant volume dual duct systems are designed for medium to high pressure. They are pressure independent systems. The hot and cold ducts terminate at mixing boxes. The mixing boxes contain dampers which mix the airflows and supply the conditioned space with heated, cooled or mixed temperature air.

Some constant volume dual duct systems are low pressure and pressure dependent. These systems use mixing dampers in the hot and cold duct to supply heated, cooled or mixed temperature air to the space through a common secondary duct. Mixing boxes are not used. Some systems have manual balancing dampers in both the hot and cold duct before they combine into the common supply duct to the conditioned space. Other systems have only one manual balancing damper located in the common duct.

Some systems do not have a heating coil. When the system is in the heating mode, air from the mixed air plenum goes through a resistance apparatus in the hot deck and is by-passed around the cooling coil.

Dual duct systems are designed as constant volume systems, but the actual volume may vary somewhat during normal operation because of the changes in resistance between the smaller heating coil and the larger cooling coil.

Balancing Procedure

- Prepare report forms.
- Conduct field inspection.
  —Inspect the job site.
  —Inspect the equipment.
  —Inspect the air distribution system.
- Set all dampers at the balancing position (see Chap. 17).
- Check mixing dampers for proper operation and leakage.
- Make initial tests of all the equipment applicable to the system being balanced.
  —Verify motor and fan rotation.
  —Take electrical measurements (voltage, current, power factor).
  —Take speed measurements.
  —Measure static pressure at the fan.
- Take static pressure measurements at the end of the system.
  —The static pressure must be at or above the minimum required for mixing box operation.
  —Increase or decrease fan speed as needed to get required static pressure.
  —Measure the static pressure drop across the box. Consult box manufacturer for pressure drop requirements. Generally, 0.75 inches wg is required for mechanical regulators. Additional pressure is needed for the low pressure air distribution system downstream of the box.
- Take total air measurements.
  —Normal Systems:
    - Traverse the cold and hot ducts. If more than 10% of the total rated airflow is measured in the hot duct, check for leaking hot air dampers or boxes with crossed supplies.
  —Systems with diversity:

- Determine the diversity ratio.
- Keep the proportion of cooled air to total volume constant.
- Set the thermostats for full flow through the cold duct.
- Set enough boxes to full cooling to equal the design flow through the cooling coil.
- Set the remaining boxes to heating.

- Balance the distribution system.

  —Pressure Independent Systems:

  - Set the mixing box regulators according to manufacturer's recommendations for airflow through the boxes.
  - Take Pitot tube traverses of the low pressure duct off the boxes and/or a total of the outlets to confirm box setting and determine duct leakage.
  - Proportionally balance the outlets.

  —Pressure Dependent Systems:

  - Proportionally balance the air distribution.
  - Proportionally balance the common ducts.
  - Adjust manual volume dampers or automatic static pressure dampers for correct airflow.

- Change fan speed as needed.

  —Change drives as needed.

- As applicable, place system in the maximum outside air mode and test.

  —If the motor overloads or the airflow is excessive, adjust the system as needed.

- Place the system in the heating mode and verify proper operation.

  —Pressure Independent Systems:

  - Measure the airflow through the heating coil or the resistance apparatus in the hot deck.

  —Pressure Dependent Systems:

  - If separate manual volume dampers are provided in the hot and cold ducts, adjust the hot duct damper for correct airflow. If only one damper is provided in the common supply duct, do not change this damper from the cooling setting.
  - Read the air distribution. The airflow should be close to the volume read when the system was balanced in the cooling position.

- Place the system in the cooling mode and record final test data.
- Complete the balance report.

AIR HANDLING UNIT AND INDUCTION UNIT SYSTEMS

Balancing Procedure

- Prepare report forms.
- Conduct field inspection.

—Inspect the job site.

—Inspect the equipment.

—Inspect the air distribution system.

- Set all dampers at the balancing position (see Chap. 17).
- Make initial tests of all the equipment applicable to the system being balanced.

  —Verify motor and fan rotation.

  —Take electrical measurements (voltage, current, power factor).

  —Take speed measurements.

  —Measure static pressure at the fan.

- Take static pressure measurements at the induction units at the end of the system.

  —The static pressure must be at or above the minimum required for induction unit operation.

  —Increase or decrease fan speed as needed to get required static pressure.

- Take total air measurements.

  —Traverse the main ducts.

  —Traverse the risers.

- Balance the distribution system.

  —Proportionally balance induction units to within +/− 10% of design airflow. Determine the airflow being delivered to each unit from a manufacturer-supplied "static pressure versus airflow" chart.

  —Balance all risers at +/− 10%.

- Change fan speed as needed.

  —Change drives as needed.

- As applicable, place system in the maximum outside air mode and test.

  —If the motor overloads or the airflow is excessive, adjust the system as needed.

- Place the system in the heating mode and verify proper operation.

  —Pressure Independent Systems:

  - Measure the airflow through the heating coil or the resistance apparatus in the hot deck.

  —Pressure Dependent Systems:

  - If separate manual volume dampers are provided in the hot and cold ducts, adjust the hot duct damper for correct airflow. If only one damper is provided in the common supply duct, do not change this damper from the cooling setting.
  - Read the air distribution. The airflow should be close to volume read when the system was balanced in the cooling position.

- Place the system in the cooling mode and record final test data.
- Complete the balance report.

RETURN OR EXHAUST SYSTEMS

Balancing Procedure

- Prepare report forms.
- Conduct field inspection.
  - —Inspect the job site.
  - —Inspect the equipment.
  - —Inspect the air distribution system.
- Set all dampers at the balancing position (see Chap. 17).
- Make initial tests of all the equipment applicable to the system being balanced.
  - —Verify motor and fan rotation.
  - —Take electrical measurements (voltage, current, power factor).
  - —Take speed measurements.
  - —Measure static pressure at the fan.
- Take total air measurements.
- Balance the distribution system.
  - —Read the system.
  - —Total the return or exhaust air inlets and compare the total air at the return or exhaust fan. Reconcile any differences.
  - —Proportionally balance the inlets starting with the return or exhaust inlet with the lowest percent of design flow.
  - —Proportionally balance the branches.
- Change fan speed as needed.
  - —Change drives as needed.
- Complete the balance report.

CHAPTER 22

Testing, Adjusting and Balancing Pressure Independent and Pressure Dependent Variable Air Volume Systems

This chapter provides balancing procedures for various types of variable air volume systems. Also included are examples of such procedures.

GENERAL BALANCING PROCEDURE

- Do the preliminary office work. Gather the plans and specifications and prepare the report forms.

- Do the preliminary field inspection. Inspect the job site, the equipment and the distribution system. Determine if the duct has been leak tested.

- Set all dampers at the full open position except the outside air. Set the outside air damper to simulate minimum outside air. If the fan has inlet (vortex) dampers, set them for minimum position.

- Set the VAV terminals for proper operation. Verify that all VAV terminal dampers are operating properly and controlled by the correct space thermostat. Verify the action of the thermostat (direct acting or reverse acting) and the volume damper position (normally closed or normally open). As applicable, verify the spring range of the damper motor as it responds to the velocity controller. Refer to the terminal manufacturer's instructions. Discontinue the balancing process if the terminals do not operate correctly and repairs are needed.

- Set space thermostats to heating or cooling to satisfy the diversity. Select the thermostats to simulate, as closely as practical, the manner in which the system will respond to the building's cooling load shift.

- Bump the supply and return fan motors to check fan rotation.

- Start the supply fan. If there is a return fan, it should be interlocked with the supply fan. Operate all associated fans.

- Take electrical measurements on the supply and return fan with the volume control set for minimum airflow. Gradually increase airflow to maximum. Take electrical measurements and observe the system for any adverse effects caused by the higher pressures.

- Adjust the static pressure control device for the supply fan to provide design total volume. Some systems may have a static pressure controller with both a high and low limit.

- Take speed measurements.

- Read the static pressure at the unit.

- Take the static pressure at the sensor for the static pressure control device.

- Take a static pressure in the inlet duct to each end run terminal box. Determine if the inlet static pressure is at or above the minimum required by the manufacturer for VAV box operation. Increase or decrease the fan speed as needed. Consult the manufacturer's data for the required pressure drop range across the box. Add the pressure losses needed for the low pressure distribution system downstream of the box [approximately 0.1 inch wg per 100 foot of duct (equivalent length) and 0.05 to 0.1 inch wg for the outlet]. This is the total required inlet static pressure. If the system serves multiple floors, record the static pressure at each floor takeoff and each end run.

- Take total airflow using a Pitot tube traverse. If the traverse indicates excessive leakage, record the problem on the report. When the total air volume cannot be determined by the Pitot tube traverse, use the summation of the outlet quantities. If the total fan capacity is less than 95% of design, notify the appropriate person before proceeding with the balancing.

- Set the terminal boxes. Consider each terminal box and associated downstream low pressure ductwork as a separate, independent system. Connect a differential pressure gauge, such as a manometer, to the controller's pressure taps and read differential pressure. Use the manufacturer's published data (usually found on the side of the box) to convert the pressure readings to cfm. Adjust the box's minimum and maximum settings following manufacturer's recommendations.

- Set the terminals being tested on full cooling or for diversity as applicable. Verify that the box is at maximum flow. Take the static pressure drop across the box and the inlet static pressure. Set the box's controller for correct cfm using:

    —the box's pressure tap readings

    —a Pitot tube traverse of the low pressure duct

    —the total of the outlets, or

    —a proportioned outlet

Note: Field conditions may be such that the inlet duct configuration to the box may give an erroneous reading at the sensor.

   To verify the box's pressure tap reading, take a Pitot tube traverse of the low pressure duct off the boxes and a total of the outlets to confirm the box setting. This will provide the actual cfm delivered by the box and help to determine amount and location of any low pressure duct leakage.

- Set the terminal box to maximum flow by setting the box's space thermostat below the space temperature. Proportionally balance the air distribution.

- Set the terminal box to minimum flow by setting the box's space thermostat above the conditioned space temperature. Read the cfm at the individual outlets. Most of the outlets should still be in proportion. However, it's possible that some outlets will be out of tolerance in the minimum setting. Do not rebalance. Leave the system balanced for maximum flow. Include both maximum and minimum airflow quantities on the report.

- If the system pressure is low, put enough boxes adjacent to the test box in the minimum flow position to bring the inlet static pressure at the test box to the required amount. Following the manufacturer's directions, set the test box for maximum and minimum airflow. Set the test box for maximum airflow and proportionally balance the outlets. Set the test box for minimum flow. Read and record the cfm at the outlets.

- After all the boxes have been set, change fan speed as needed to get required system static pressure.

- Set the system for diversity if applicable.

- Take volume readings on the entire system.

- After completing the proportional balance, change the system from minimum outside air to maximum outside air. Take motor amperage. Take plenum static pressure. If the motor overloads or the airflow is excessive, adjust outside air dampers or fan speed as needed.

- Place the system in a logical nonmodulating mode. Make a final pass of the outlets, recording all readings on data sheets. Compare final outlet readings with the final traverses.

- Mark the dampers in case they need to be reset due to tampering or accidental changing. Walk the system and be alert for drafts. Adjust outlet pattern control devices to eliminate drafts.

- For fan powered or induction terminal boxes, take discharge air temperature and airflow on each side of all secondary ductwork where tee fittings are installed to determine if there is stratification. Set the terminal in a mode where there will be a mixture of cold primary air and warm secondary air. Record any evidence of stratification on the report.

- Verify that all the automatic control dampers are sequencing properly and are not leaking.

- Record the following final conditions: motor currents and voltages, fan speeds, static pressures and temperatures.

- Complete the report. Review the report and be sure that nothing has been omitted. Be certain that all notes on problems, deficiencies and other abnormal conditions are explained fully.

FIELD SETTING OF THE VAV BOX

Example 22.1: A single inlet terminal. The following procedure is for a particular single inlet VAV box. The steps outlined *do not* apply to all boxes. Consult the box manufacturer for specific instructions.

Procedure

1. Verify that the damper action (normally open or normally closed), set on the face of the controller, matches the settings printed on the controller label.

2. Verify that the thermostat setting (direct acting or reverse acting), on the face of the controller, matches the actual installed thermostat.

3. Refer to the calibration curve on the side of the terminal. From the curve, read the differential pressure across the sensor for the required airflow.

4. Connect an air differential pressure gauge (normally 0 to 2 in. wc) to the airflow sensor in the cold duct inlet. The controller should have main air applied to it.

5. Set the minimum. Read the gauge. If the reading matches the required differential from the calibration chart, no adjustment is necessary. If the reading is different from what is required, rotate the minimum adjustment device up or down as indicated until the gauge shows the required differential pressure.

6. Set the maximum. Read the gauge. If the reading matches the required differential from the calibration chart, no adjustment is necessary. If the reading is different from what is required, rotate the maximum adjustment device up or down as indicated until the gauge shows the required differential pressure.

Example 22.2: A dual inlet terminal. The following procedure is for a particular dual inlet VAV box. The steps outlined *do not* apply to all boxes. Consult the box manufacturer for specific instructions.

Procedure

1. Verify that the damper action (normally open or normally closed), set on the face of the controller, matches the settings printed on the controller label.

2. Verify that the thermostat setting (direct acting or reverse acting), on the face of the controller, matches the actual installed thermostat.

3. Refer to the calibration curve on the side of the terminal. From the curve, read the differential pressure across the sensor for the required airflow.

4. Set the minimum to zero.

5. Connect an air differential pressure gauge (normally 0 to 2 in. wc) to the airflow sensor in the cold duct inlet. The controller should have main air applied to it.

6. Read the gauge. If the reading matches the required differential from the calibration chart, no adjustment is necessary. If the reading is different from what is required, rotate the maximum adjustment device up or down until the gauge shows the required differential pressure.

7. To set the hot duct inlet for maximum flow, remove the main air (0 psi applied to the hot duct controller) and repeat step 6.

8. To set the hot duct for minimum flow, place a 0-30 psi pressure gauge in the line between the controller and the damper actuator on the cold duct.

9. Hook the differential pressure gauge to the hot duct flow sensor.

10. Reduce the pressure to the actuator until the cold duct damper just closes. The pressure gauge should read about 5 psi.

11. Read the differential pressure gauge. If the reading matches the required differential from the calibration chart, no adjustment is necessary. If the reading is different from

what is required, rotate the maximum adjustment device up or down until the gauge shows the required differential pressure.

BALANCING PROCEDURE FOR PRESSURE DEPENDENT SYSTEMS WITHOUT DIVERSITY (Fig. 22.1)

- Do the preliminary office work.
- Do the preliminary field inspection.
- Set all dampers at the full open position except the outside air (which is set at minimum position).
- Set the VAV terminals for proper operation. Discontinue the balancing process if the terminals do not operate correctly and repairs are needed.
- Set the system for full cooling.
- Bump the supply and return fan motors to check fan rotation.
- Operate all associated fans (supply, return and exhaust) at or near design speeds.
- Take electrical measurements.

FIGURE 22.1.   PRESSURE DEPENDENT VAV SYSTEMS

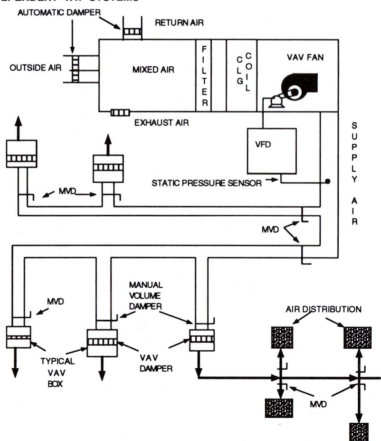

- Adjust the static pressure control device for the supply fan to provide design total volume.

- Take speed measurements.

- Take static pressure measurements at the unit.

- Take total air readings.

- Balance and adjust the distribution system. Nondiversity systems are balanced similar to constant volume systems. Start with the box that has the lowest percent of design flow. Take a Pitot tube traverse of the low pressure duct off the box and the total of the outlets to determine duct leakage. Proportionally balance the outlets. Proportionally balance the VAV terminals using the inlet volume damper. Proportionally balance the branches.

- Change fan speed as needed.

- Take final readings and check the system in the heating/minimum mode.

- Check the system in the maximum outside air mode. If the motor overloads or the flow is excessive, adjust outside air dampers or fan speed as needed.

- Complete report. Review the report to be sure that nothing has been omitted. Make sure that all notes on problems, deficiencies and other abnormal conditions are explained fully.

BALANCING PROCEDURE FOR PRESSURE DEPENDENT SYSTEMS WITH DIVERSITY (Fig. 22.1)

- Do the preliminary office work.

- Do the preliminary field inspection.

- Set all dampers at the full open position except the outside air (which is set at minimum position).

- Set the VAV terminals for proper operation. Discontinue the balancing process if the terminals do not operate correctly and repairs are needed.

- Set the system for full cooling and diversity.

- Bump the supply and return fan motors to check fan rotation.

- Operate all associated fans (supply, return and exhaust) at or near design speeds.

- Take electrical measurements.

- Adjust the static pressure control device for the supply fan to provide design total volume.

- Take speed measurements.

- Take static pressure measurements at the unit.

- Take total air readings. Take a Pitot tube traverse to get total air volume. Compare the Pitot tube reading with a total of the outlets to determine air leakage.

- Balance and adjust the distribution system. At the completion of balancing, the inlet manual damper will be fully open to at least one VAV terminal box in the system. At least one damper in each branch duct will be fully open and at least one outlet on each branch duct will be fully open. Take a Pitot tube traverse of the low pressure duct off the box and the total of the outlets to determine duct leakage. Proportionally balance the outlets. Proportionally balance the terminals using the inlet volume damper. Proportionally balance the branches.

- Change fan speed as needed.

- Take final readings and check the system in the heating and minimum air modes.

- Check the system in the maximum outside air mode. If the motor overloads or the flow is excessive, adjust outside air dampers or fan speed as needed.

- Complete the report. Review the report to be sure that nothing has been omitted. Be certain that all notes on problems, deficiencies and other abnormal conditions are explained fully.

BALANCING PROCEDURE FOR PRESSURE DEPENDENT FAN POWERED SYSTEMS WITH DIVERSITY (Fig. 22.2)

- Do the preliminary office work.

- Do the preliminary field inspection.

FIGURE 22.2.   PRESSURE DEPENDENT FAN POWERED VAV BOXES

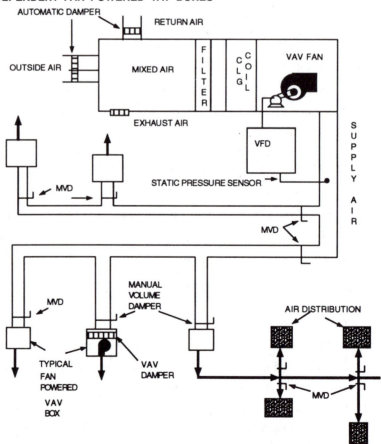

- Set all dampers at the full open position except the outside air (which is set at minimum position).
- Set the VAV terminals for proper operation. Discontinue the balancing process if the terminals do not operate correctly and repairs are needed.
- Set the system for full cooling and diversity.
- Bump the supply and return fan motors to check fan rotation.
- Operate all associated fans (supply, return and exhaust) at or near design speeds. Set the secondary fan at design maximum speed with the terminal operating on full return. Consult the terminal manufacturer for the proper operation and setting of the secondary fan.
- Take electrical measurements.
- Adjust the static pressure control device for the supply fan to provide design total volume.
- Take speed measurements.
- Take static pressure measurements at the unit.
- Take total air readings. Take a Pitot tube traverse to get total air volume. Compare the Pitot tube reading with a total of the outlets to determine air leakage.
- Take a Pitot tube traverse of the low pressure duct off the box and the total of the outlets to determine duct leakage.
- Proportionally balance the outlets.
- Proportionally balance the terminals using the inlet volume damper. Start with the terminal with the lowest percent of design airflow. Proportionally balance all terminal boxes to receive the same ratio of required quantities of primary air. Set the terminal to full cooling (maximum cfm) and adjust the manual volume damper in the inlet to the terminal. There must be adequate static pressure at all times in the primary air duct.
- Change fan speed as needed.
- Take final readings and check the system in the heating and minimum air modes.
- Check the system in the maximum outside air mode. If the motor overloads or the flow is excessive, adjust outside air dampers or fan speed as needed.
- Complete the report. Review the report to be sure that nothing has been omitted. Make sure that all notes on problems, deficiencies and other abnormal conditions are explained.

BALANCING PROCEDURE FOR PRESSURE INDEPENDENT SINGLE DUCT (Fig. 22.3)

- Do the preliminary office work.
- Do the preliminary field inspection.
- Set all dampers at the full open position except the outside air. Set the outside air damper to simulate minimum outside air. If the fan has inlet (vortex) dampers, set them for minimum position.
- Set the VAV terminals for proper operation. Discontinue the balancing process if the terminals do not operate correctly and repairs are needed.

FIGURE 22.3.   PRESSURE INDEPENDENT VAV BOXES

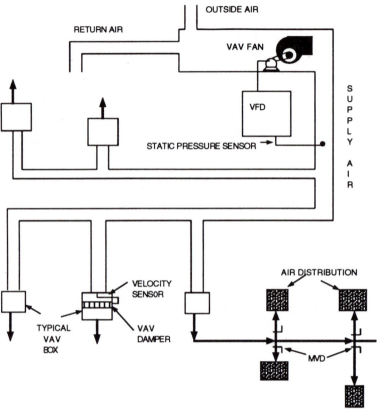

- Set space thermostats to heating or cooling to satisfy the diversity. Select the thermostats to simulate, as closely as practical, the manner in which the system will respond to the building's cooling load shift.

- Bump the fan motors to check fan rotation.

- Operate all associated fans.

- Take electrical measurements on the supply and return fan with the volume control set for minimum airflow. Gradually increase airflow to maximum. Take electrical measurements and observe the system for any adverse effects caused by the higher pressures.

- Adjust the static pressure control device to provide design total volume.

- Take speed measurements.

- Read the static pressure at the unit.

- Take the static pressure at the sensor for the static pressure control device.

- Take a static pressure in the inlet duct to each end run terminal box. Increase or decrease the fan speed as needed. Record the static pressure at each floor takeoff and each end run.

- Take total airflow using a Pitot tube traverse.

- Set the terminal boxes.

- Set the terminals being tested on full cooling or diversity as applicable.
- Set the terminal box to maximum flow. Proportionally balance the outlets.
- Set the terminal box to minimum flow. Read the cfm at the individual outlets.
- Change the system from minimum outside air to maximum outside air. Take motor amperage. Take plenum static pressure. If the motor overloads or the airflow is excessive, adjust outside air volume dampers or fan speed as needed.
- Place the system in a logical nonmodulating mode. Make a final pass of the outlets, recording all readings on data sheets.
- Mark the dampers.
- Adjust outlet pattern control devices to eliminate drafts.
- Verify that all the automatic control dampers are sequencing properly and are not leaking.
- Record final currents, voltages, fan speeds, static pressures and temperatures.
- Complete the report.

BALANCING PROCEDURE FOR PRESSURE INDEPENDENT DUAL DUCT (Fig. 22.4)

- Do the preliminary office work.
- Do the preliminary field inspection.
- Set all dampers at the full open position, except the outside air. Set the outside air damper to simulate minimum outside air.
- Set the VAV terminals for proper operation. Discontinue the balancing process if the terminals do not operate correctly and repairs are needed.
- Set space thermostats to either neutral (deadband with no heating or cooling) or full cooling as required to satisfy the design diversity factor. Select thermostats to simulate as near as practical the manner in which the system will respond to the building's cooling load shift.
- Bump the fan motors to check fan rotation.
- Operate all associated fans.
- Take electrical measurements on the supply and return fan with the volume control set for minimum airflow. Gradually increase airflow to maximum. Take electrical measurements and observe the system for any adverse effects caused by the higher pressures.
- Adjust the static pressure control device to provide design total volume.
- Take speed measurements.
- Read the static pressure at the unit.
- Take the static pressure at the sensor for the static pressure control device.
- Take a static pressure in the inlet duct to each end run terminal box. Increase or decrease the fan speed as needed. Record the static pressure at each floor takeoff and each end run.

FIGURE 22.4.   PRESSURE INDEPENDENT DUAL DUCT VAV BOXES

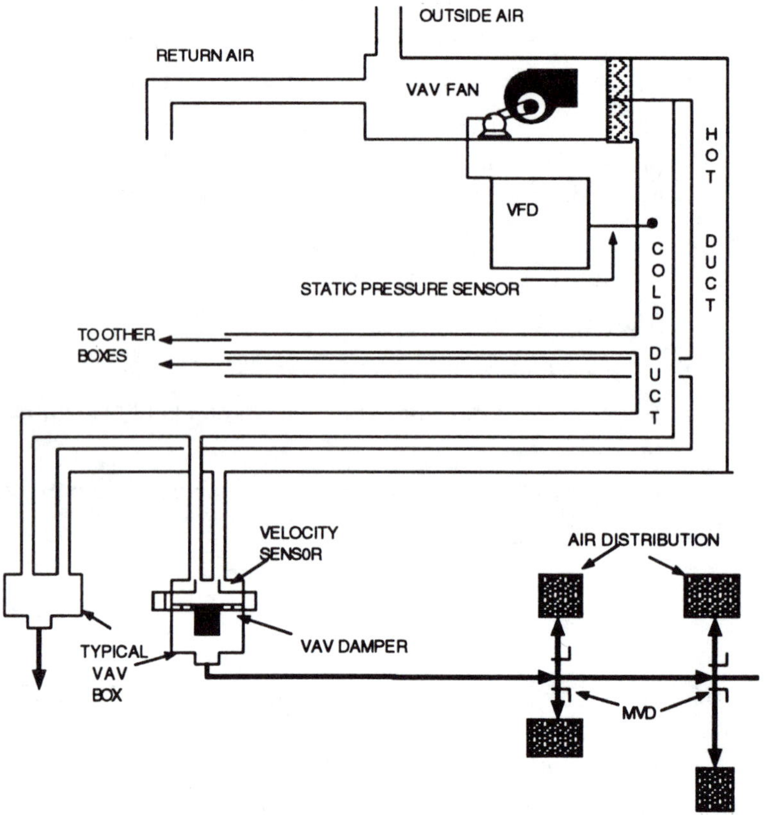

- Take total airflow using a Pitot tube traverse.
- Set the terminal boxes.
- Set the terminals being tested on full cooling or diversity as applicable.
- Set the terminal box to maximum flow. Proportionally balance the outlets.
- Set the terminal box to minimum flow. Read the cfm at the individual outlets.
- Change the system from minimum outside air to maximum outside air. Take motor amperage. Take plenum static pressure. If the motor overloads or the airflow is excessive, adjust outside air dampers or fan speed as needed.
- Place the system in a logical nonmodulating mode. Make a final pass of the outlets, recording all readings on data sheets.
- Mark the dampers.
- Adjust outlet pattern control devices to eliminate drafts.
- Verify that all the automatic control dampers are sequencing properly and are not leaking.
- Record final currents, voltages, fan speeds, static pressures and temperatures.
- Complete the report.

BALANCING PROCEDURE FOR VARIABLE PRIMARY/CONSTANT SECONDARY PRESSURE INDEPENDENT SERIES FAN POWERED VAV BOXES (Fig. 22.5)

- Do the preliminary office work.
- Do the preliminary field inspection.
- Set all dampers at the full open position except the outside air. Set the outside air damper to simulate minimum outside air. If the fan has inlet (vortex) dampers, set them for minimum position.
- Set the VAV terminals for proper operation. Discontinue the balancing process if the terminals do not operate correctly and repairs are needed.
- Set space thermostats to heating or cooling to satisfy the diversity. Select the thermostats to simulate as closely as practical, the manner in which the system will respond to the building's cooling load shift.
- Bump the fan motors to check fan rotation.
- Operate all associated fans. Set the secondary fan at design maximum speed with the

FIGURE 22.5.   PRESSURE INDEPENDENT SERIES FAN POWERED VAV BOXES

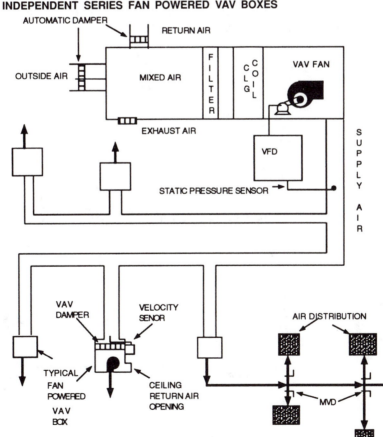

terminal operating on full return. Consult the terminal manufacturer for the proper operation and setting of the secondary fan.

- Take electrical measurements on the supply and return fan with the volume control set for minimum airflow. Gradually increase airflow to maximum. Take electrical measurements and observe the system for any adverse effects caused by the higher pressures.

- Adjust the static pressure control device to provide design total volume.

- Take speed measurements.

- Read the static pressure at the unit.

- Take the static pressure at the sensor for the static pressure control device.

- Take a static pressure in the inlet duct to each end run terminal box. Increase or decrease the fan speed as needed. Record the static pressure at each floor takeoff and each end run.

- Take total airflow using a Pitot tube traverse.

- Set the terminal boxes.

- Set the terminals being tested on full cooling or for diversity as applicable.

- Set the terminal box to maximum flow. Proportionally balance the outlets.

- Set the terminal box to minimum flow. Read the cfm at the individual outlets.

- Change the system from minimum outside air to maximum outside air. Take motor amperage. Take plenum static pressure. Change fan speed as needed. Take final readings and check the system in the heating and minimum air modes. Check the system in the maximum outside air mode. If the motor overloads or the flow is excessive, adjust outside air dampers or fan speed as needed.

- Complete the report. Review the report to be sure that nothing has been omitted. Make sure that all notes on problems, deficiencies and other abnormal conditions are explained fully.

FIGURE 22.6.   PRESSURE INDEPENDENT PARALLEL FAN POWERED VAV BOXES

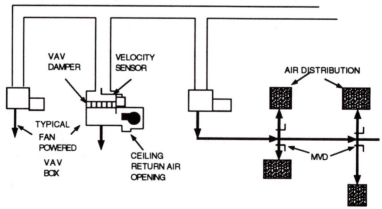

BALANCING PROCEDURE FOR VARIABLE PRIMARY/VARIABLE SECONDARY PRESSURE INDEPENDENT PARALLEL FAN POWERED VAV BOXES (Fig. 22.6)

- Do the preliminary office work.
- Do the preliminary field inspection.
- Set all dampers at the full open position except the outside air. Set the outside air damper to simulate minimum outside air. If the fan has inlet (vortex) dampers, set them for minimum position.
- Set the VAV terminals for proper operation. Discontinue the balancing process if the terminals do not operate correctly and repairs are needed.
- Set space thermostats to heating or cooling to satisfy the diversity. Select the thermostats to simulate, as closely as practical, the manner in which the system will respond to the building's cooling load shift.
- Bump the fan motors to check fan rotation.
- Operate all associated fans. Set the secondary fan at design maximum speed with the terminal operating on full return. Consult the terminal manufacturer for the proper operation and setting of the secondary fan.
- Take electrical measurements on the supply and return fan with the volume control set for minimum airflow. Gradually increase airflow to maximum. Take electrical measurements and observe the system for any adverse effects caused by the higher pressures.
- Adjust the static pressure control device to provide design total volume.
- Take speed measurements.
- Read the static pressure at the unit.
- Take the static pressure at the sensor for the static pressure control device.
- Take a static pressure in the inlet duct to each end run terminal box. Increase or decrease the fan speed as needed. Record the static pressure at each floor takeoff and each end run.
- Take total airflow using a Pitot tube traverse.
- Set the terminal boxes.
- Set the terminals being tested on full cooling or for diversity as applicable.
- Set the terminal box to maximum flow. Proportionally balance the outlets.
- Set the terminal box to minimum flow. Read the cfm at the individual outlets.
- Change the system from minimum outside air to maximum outside air. Take motor amperage. Take plenum static pressure. If the motor overloads or the airflow is excessive, adjust outside air volume dampers or fan speed as needed.
- Place the system in a logical nonmodulating mode. Make a final pass of the outlets, recording all readings on data sheets.
- Mark the dampers. Adjust outlet pattern control devices to eliminate drafts.
- Take discharge air temperature and airflow on each side of all secondary ductwork where tee fittings are installed to determine stratification.
- Verify that all automatic control dampers are sequencing properly and are not leaking.

• Record final currents, voltages, fan speeds, static pressures and temperatures.
• Complete the report.

BALANCING PROCEDURE FOR VARIABLE PRIMARY/INDUCTION SECONDARY PRESSURE INDEPENDENT VAV BOXES (Fig. 22.7)

• Do the preliminary office work.
• Do the preliminary field inspection.
• Set all dampers at the full open position except the outside air. Set the outside air damper to simulate minimum outside air. If the fan has inlet (vortex) dampers, set them for minimum position.
• Set the VAV terminals for proper operation. Discontinue the balancing process if the terminals do not operate correctly and repairs are needed.
• Set space thermostats to heating or cooling to satisfy the diversity. Select the thermostats to simulate, as closely as practical, the manner in which the system will respond to the building's cooling load shift.
• Bump the fan motors to check fan rotation.
• Operate all associated fans.
• Take electrical measurements on the supply and return fan with the volume control set for minimum airflow. Gradually increase airflow to maximum. Take electrical measurements and observe the system for any adverse effects caused by the higher pressures.
• Adjust the static pressure control device to provide design total volume.
• Take speed measurements.
• Read the static pressure at the unit.

FIGURE 22.7.　PRESSURE INDEPENDENT INDUCTION VAV BOXES

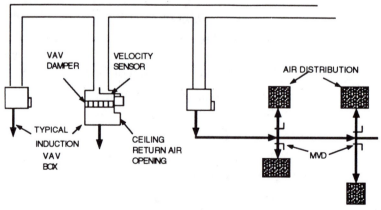

- Take the static pressure at the sensor for the static pressure control device.

- Take a static pressure in the inlet duct to each end run terminal box. Increase or decrease the fan speed as needed. Set the static pressure in the primary duct high enough to overcome the resistance in the secondary low pressure ductwork, plus be able to induce return air. Record the static pressure at each floor takeoff and each end run.

- Take total airflow using a Pitot tube traverse.

- Set the terminal boxes.

- Set the terminals being tested on full cooling or for diversity as applicable.

- Set the terminal box to maximum flow. Proportionally balance the outlets.

- Set the terminal box to minimum flow. With the thermostat set for full heating, measure the quantity of primary air using the velocity controller's pressure taps and the velocity pressure vs. cfm chart on the side of the terminal. If different from design, make the necessary adjustments. On the report, include the quantity of primary air supplied each terminal in both the minimum and maximum primary modes. Read the cfm at the individual outlets.

- Change the system from minimum outside air to maximum outside air. Take motor amperage. Take plenum static pressure. If the motor overloads or the airflow is excessive, adjust outside air dampers or fan speed as needed.

- Place the system in a logical nonmodulating mode. Make a final pass of the outlets, recording all readings on data sheets.

- Mark the dampers. Adjust outlet pattern control devices to eliminate drafts.

- Take discharge air temperature and airflow on each side of all secondary ductwork where tee fittings are installed to determine stratification.

- Verify that all the automatic control dampers are sequencing properly and are not leaking.

- Record final currents, voltages, fan speeds, static pressures and temperatures.

- Complete the report.

BALANCING PROCEDURE FOR VARIABLE PRIMARY/VARIABLE SECONDARY SYSTEM POWERED VAV BOXES (Fig. 22.8)

- Do the preliminary office work.

- Do the preliminary field inspection.

- Set all dampers at the full open position, except the outside air (which is set at minimum position).

- Set the VAV terminals for proper operation. Discontinue the balancing process if the terminals do not operate correctly and repairs are needed.

- Set the system for full cooling and diversity.

FIGURE 22.8.   SYSTEM POWERED VAV BOXES

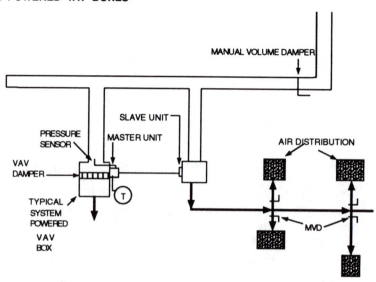

- Bump the supply and return fan motors to check fan rotation.
- Operate all associated fans (supply, return and exhaust) at or near design speeds.
- Take electrical measurements.
- Adjust the static pressure control device for the supply fan to provide design total volume.
- Take speed measurements.
- Take static pressure measurements at the unit.
- Take total air readings. Take a Pitot tube traverse to get total air volume. Compare the Pitot tube reading with a total of the outlets to determine air leakage.
- Balance and adjust the distribution system. Take a Pitot tube traverse of the low pressure duct off the box and the total of the outlets to determine duct leakage. Proportionally balance the outlets. Proportionally balance the terminals using the inlet volume damper. Proportionally balance the branches.
- Change fan speed as needed.
- Take final readings and check the system in the heating and minimum air modes.
- Check the system in the maximum outside air mode. If the motor overloads or the flow is excessive, adjust outside air dampers or fan speed as needed.
- Complete the report. Review the report to be sure that nothing has been omitted. Make sure that all notes on problems, deficiencies and other abnormal conditions are explained fully.

BALANCING PROCEDURE FOR CONSTANT PRIMARY/VARIABLE SECONDARY BYPASS BOX VAV SYSTEM (Fig. 22.9)

- Do the preliminary office work.
- Do the preliminary field inspection.
- Set all dampers at the full open position, except the outside air (which is set at minimum position).
- Set the VAV terminals for proper operation. Discontinue the balancing process if the terminals do not operate correctly and repairs are needed.
- Set the system for full cooling.
- Bump the supply and return fan motors to check fan rotation.
- Operate all associated fans (supply, return and exhaust) at or near design speeds.
- Take electrical measurements.
- Take speed measurements.
- Take total air readings. Take a Pitot tube traverse to get total air volume. Compare the Pitot tube reading with a total of the outlets to determine air leakage.

FIGURE 22.9.   BYPASS BOX VAV SYSTEM

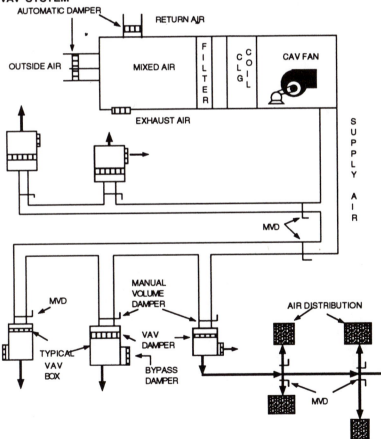

- Balance and adjust the distribution system. At the completion of balancing, the inlet manual damper will be fully open to at least one VAV terminal box in the system. At least one damper in each branch duct will be fully open and at least one outlet on each branch duct will be fully open. Take a Pitot tube traverse of the low pressure duct off the box and the total of the outlets to determine duct leakage. Proportionally balance the outlets. Proportionally balance the terminals using the inlet and bypass volume dampers. Set the dampers so that the bypass air quantity is equal to the supply air quantity. Proportionally balance the branches.

CHAPTER 23 Testing, Adjusting and Balancing Water Systems

This chapter describes procedures for balancing water systems using both flow meters for direct flow measurements, and pressure drops or temperature changes across system components for indirect flow measurements. Direct flow measurements using flow meters is the preferred method because it eliminates the compounding errors introduced by balancing using pressure drops across components or using temperature measurements. See Chap. 5 for review of verification of water system performance, including verifying impeller size.

OFFICE PREPARATION

First, assemble all applicable mechanical contract documents, specifications, catalogs and previous reports. These should include:

- Shop drawings
- "As-built" drawings
- Schematics
- Automatic temperature control drawings
- Manufacturers' catalogs and performance curves
    —Pump description and capacity
    —Pump performance curve
    —Terminal box description and capacities
- Manufacturers' data and recommendations
    —Testing of pumps and heat exchangers
- Equipment operation and maintenance instructions
- Water balance reports

Next, review the drawings, specifications, catalogs and reports. Note any equipment, system component or condition that might change the system or component water volume, shut down the system, change the sequence of balancing the system, or in any way affect the balance procedure.

Evaluate and make plans for the inspection or arrangement of any of the following conditions:

- Devices, controls or system features which might shut down the system or contribute to an unbalanced condition.
- General accessibility

  —Equipment that is difficult to get to

  —Limited access areas such as security areas, clean rooms, hotel rooms, meeting rooms, etc.
- Time delays
- Sequence of balancing systems and means of generating an "out of season" heating or cooling load.
- Marking the final settings of balancing valves.
- Arrangements for witnessing of the balancing.

Then, check your instruments:

- Verify that the instruments selected for the balancing project meet the calibration criteria stated in the balancing specifications.
- Verify that the instruments have been calibrated within the last six months, or to manufacturer's recommendations.
- Include a list of the various instruments used in the test and balance report.
- Determine what instruments will be needed.
- Determine what measurements will be taken and where.

Finally, gather the data and test sheets needed for each system being tested. The test and balance report may include any or all of the following:

  —Pump data and test sheets (see Fig. 23.1)

  —Motor data and test sheets (see Fig. 23.2)

  —Flow meter data and test sheets (see Fig. 23.3)

  —Water coil data and test sheets (see Fig. 23.4)

Enter design information on the appropriate data and test sheet. Design data will include water quantities, pump information, motor information and water distribution information. The test and balance report is a complete record of design, preliminary and final test data. It reflects the actual tested and observed conditions of all systems and components during the balance and lists any discrepancies between the design and the test data and possible reasons for the differences.

FIGURE 23.1.   PUMP DATA AND TEST SHEET

Project: _____   Engineer/Contact: _____

Pump Information	*Specified*	*Actual*
Designation		
Location		
Service		
Manufacturer		
Serial #		
Model #		
Impeller Diameter		
Gallons per Minute		
Total Dynamic Head		
Pump Speed		
Rotation		

	Shutoff	*Operating*
Gallons per Minute		
Discharge Pressure (psig)		
Suction Pressure (psig)		
PSI Rise		
Static Head		
Velocity Head		
Total Dynamic Head		
Impeller Diameter		
Brake Horsepower		
Efficiency		

System Condition

Pump _____

Pipe _____

Notes: _____

FIGURE 23.2.   MOTOR DATA AND TEST SHEET

Project: _____   Engineer/Contact: _____

	Specified	*Actual*
Pump Designation	_____	_____
Motor Information	_____	_____
Manufacturer	_____	_____
Frame Size	_____	_____
Horsepower	_____	_____
Phase	_____	_____
Hertz	_____	_____
Speed rpm	_____	_____
Service Factor	_____	_____
Voltage	_____	_____
Amperage	_____	_____
Power Factor	_____	_____
Efficiency	_____	_____
Brake Horsepower	_____	_____
Starter Size	_____	_____
Thermal OLP	_____	_____

Make a schematic drawing of each system. Show the central system with pressure and temperature drops (rises) across coils and pumps. The schematic should show the general location of the water distribution devices (terminals, automatic valves, manual valves, balancing valves, etc.)

FIELD INSPECTION

Note any major changes in actual installation. Make corrections on the schematics. Note any changes in capacities. Make corrections on the data and test sheets. Check out any potential problem areas that you listed while reviewing the contract documents, specifications, catalogs and previous reports. Note any problems on the report forms. Check to ensure that:

- Strainers are clean and correctly installed
- Construction strainer baskets have been replaced with permanent baskets
- Pumps are properly aligned, grouted and anchored
- Pump bearings are lubricated
- Vibration isolators are properly installed and adjusted
- Flexible connections are properly installed

FIGURE 23.3. FLOW METER DATA AND TEST SHEET

Project: _____   **Engineer/Contact:** _____

	Location	Model	Size	Design		Test		Final
				gpm	*PD*	*gpm*	*PD*	*gpm*
No. 1	___	___	___	___	___	___	___	___
2	___	___	___	___	___	___	___	___
3	___	___	___	___	___	___	___	___
4	___	___	___	___	___	___	___	___
5	___	___	___	___	___	___	___	___
6	___	___	___	___	___	___	___	___
7	___	___	___	___	___	___	___	___
8	___	___	___	___	___	___	___	___
9	___	___	___	___	___	___	___	___
10	___	___	___	___	___	___	___	___
11	___	___	___	___	___	___	___	___
12	___	___	___	___	___	___	___	___
13	___	___	___	___	___	___	___	___
14	___	___	___	___	___	___	___	___
15	___	___	___	___	___	___	___	___
16	___	___	___	___	___	___	___	___
17	___	___	___	___	___	___	___	___
18	___	___	___	___	___	___	___	___
19	___	___	___	___	___	___	___	___
20	___	___	___	___	___	___	___	___
21	___	___	___	___	___	___	___	___
22	___	___	___	___	___	___	___	___
23	___	___	___	___	___	___	___	___
24	___	___	___	___	___	___	___	___
25	___	___	___	___	___	___	___	___
26	___	___	___	___	___	___	___	___
27	___	___	___	___	___	___	___	___
28	___	___	___	___	___	___	___	___
29	___	___	___	___	___	___	___	___
30	___	___	___	___	___	___	___	___

Notes: If temperature readings are used, change PD to Temp.

_____

_____

_____

FIGURE 23.4.   WATER COIL DATA AND TEST SHEET

Project: _____   **Engineer/Contact:** _____

	Specified	*Actual*
Designation		
Manufacturer		
Model #		
Size		
Service		
Location		
Gallons per Minute		
Water PD, feet		
Entering Water Temp		
Leaving Water Temp		

	Specified	*Actual*
Designation		
Manufacturer		
Model #		
Size		
Service		
Location		
Gallons per Minute		
Water PD, feet		
Entering Water Temp		
Leaving Water Temp		

	Specified	*Actual*
Designation		
Manufacturer		
Model #		
Size		
Service		
Location		
Gallons per Minute		
Water PD, feet		
Entering Water Temp		
Leaving Water Temp		

Notes:

_____

_____

_____

- Pumps are mechanically ready
- Motors are electrically ready
- Motors and pump rotation is correct
- Proper motor starter and overload protection is installed
- Correct fuses have been installed
- Motors are properly secured on their frames
- Motor bearings are lubricated
- System is filled to the proper level
- The pressure reducing valve is set
- Expansion/compression tanks are at the proper level
- Piping is cleaned and flushed
- Air vents (automatic and manual) are properly installed and functional
- Air has been purged from the system
- Valves (automatic and manual) have been installed correctly, and are accessible and functional
- Flow meters have been installed correctly and are accessible and functional
- Temperature/pressure ports have been installed correctly and are accessible and functional
- Coils/heat exchangers have been installed and piped correctly and are accessible
- Coils are clean and have been properly installed and sealed
- All operating and safety settings for temperature and pressure are correct
- Pressure relief valve is functional
- Boiler is started and operates properly
- Chiller and condenser are started and operate properly
- Automatic controls are complete and functional

Field Testing

First, operate a new system for 48 hours before testing and balancing. Then, record information on all system components while they are being tested.

Recording Pump Data

Record the following pump data, as applicable:

- designation
- location
- area served
- manufacturer
- serial number
- model number
- impeller diameter—verify from pump curve

- gpm (capacity)
- total dynamic head (pressure)
- pump speed
- rotation (CW, CCW)
- shutoff conditions
- operating conditions
- general system conditions

Calculate operating brake horsepower and pump efficiency

Verifying Pump Rotation

Verify and record pump rotation. Note whether it is in a clockwise (CW) or counter-clockwise (CCW) rotation. Centrifugal pumps will normally have an arrow on the housing which indicates direction of rotation.

Observe pump rotation. If any excessive noises or vibrations are observed, stop the pump and investigate. Before continuing the test phase, determine if the pump can be operated or if it needs repairing. If rotation is backwards, make the necessary corrections. If a centrifugal pump is rotating backwards, the water flow can be reduced by 50% or more.

If the rotation is incorrect, reverse the rotation on a three-phase motor by changing any two of the three power leads at the motor control center or disconnect. In some cases, you may also be able to change rotation in single-phase motors by switching the internal motor leads within the terminal box. Wiring diagrams for single-phase motors are usually found on the motor or inside the motor terminal box. After making changes to the motor, observe the pump rotating in the proper direction.

Recording Pump Head (Pressure) Data

Read and record pump pressure data—both discharge static pressure (psig) and suction static pressure (psig).

Calculate and record pump pressure and head:

- Pressure rise (psi), shutoff and operating
- Static head (feet of water)
- Velocity head (feet of water) (not normally calculated, see Chap. 5)
- Total dynamic head (feet of water)

Recording Motor Data

Record the following motor data, as applicable:

manufacturer	Hertz	amperage	starter size
frame size	speed	power factor	thermal overload protection
horsepower	service factor	efficiency	
phase	voltage	brake horsepower	

If the nameplate is positioned so that it is difficult to read, use a telescoping mirror. In most cases, motors used on HVAC equipment will be either single-phase or three-phase alternating current, single or dual voltage induction motors. A dual voltage motor will also list dual amperage on the nameplate. For example, a three-phase, 50 horsepower, dual voltage motor has the following ratings: 230/460 volts, 120/60 amps. If the motor is wired for 230 volts the full load amps will be 120. If it's wired for 460 volts, the full load amps will be 60. The volts and amps are inversely proportional. When the voltage doubles, the amperage reduces by one half.

Measuring Motor Voltage, Current and Power Factor

Measure and record the motor voltage, amperage and power factor. Voltage and amperage measurements are taken with a portable volt-ammeter. Measure voltage on the line side. Record L1-L2, L1-L3, L2-L3. The most accurate voltage will be read at the motor terminal box. However, it's usually much safer to take readings at the motor control center or at the disconnect box. The voltage difference between the two places is generally insignificant. The measured voltage should be plus or minus 10% of the motor nameplate voltage. If it's not within this range, notify the electrical contractor or the utility company.

Measure amperage on the terminal side. Record T1-T2, T1-T3, T2-T3. The amperage measured on any phase should not exceed the motor nameplate amperage. If the operating amperage reading is over the nameplate amperage, take one of the following steps to correct the problem: reduce the impeller size or close the main discharge valve until the amperage reading is at or below nameplate. If the pump is not serving a critical area, turn it off and inform the person responsible for the pump's operation. If desired, measure power factor with a power factor meter. Record L1-L2, L1-L3, L2-L3.

Checking Motor Speed

The rated rpm on the nameplate is the speed that the motor will turn when it's operating at nameplate horsepower. If the motor is operating at something other than rated horsepower, the rpm may vary a small amount. However, it's the nameplate rpm that's recorded on the report forms.

Checking the Motor Service Factor

The service factor is the number by which the horsepower or amperage rating is multiplied to determine the maximum safe load that a motor may be expected to carry continuously at its rated voltage and frequency. A service factor of 1.10 for a 50 horsepower motor would allow the motor to operate safely at 55 horsepower (55×1.10) and about 132 amps (120×1.10) at 230 volts. Do not leave a motor operating in the service factor area because, under certain circumstances, damage can occur to the windings and shorten the life of the motor. For example, if this motor was operating at 132 amps and the voltage to the motor dropped to 220 volts because of a reduction in power from the utility company or adding other equipment, etc., the amperage would go to about 138 amps ($230/220 \times 132$). The motor would overheat and possibly damage the windings.

Verifying Motor Overload Protection

Generally, overload protection is installed to protect the motor against current 125% greater than rated amps. However, in thermal overload selection, it's important to know the ambient temperature at the starter as compared to the ambient temperature at the motor. Sometimes, these temperatures may vary greatly and a temperature-compensating thermal overload or a magnetic overload device may be needed. Check with the manufacturer or electrical supplier for special cases. Thermal overload protection devices have a letter and/or number on them which is used for proper sizing. A chart listing thermals and their amperage ratings for the specific starters is usually found inside the motor disconnect cover. Thermals must be matched to starter and the rated full load amps of the motor according to the information on the chart or from the manufacturer. If the installed thermal is too large, the motor may not be adequately protected and could overheat. If the thermal is too small, the motor may stop repeatedly. If a new thermal is needed, the following information will be required: motor starter size, full load amps, service factor, class of insulation, motor classification and allowable temperature rise. Refer to the chart or table provided by the thermal manufacturer to choose the proper size.

Recording Flow Meter Data

Record flow meter data as applicable: location, model, size, design gpm and pressure drop (PD), tested gpm and pressure drop (PD) and final gpm. If flow meters are not installed, use the flow meter data and test sheet to record information on temperature or pressure test ports.

Recording Water Coil Data

Record water coil/heat exchanger data as applicable: manufacturer, model, size, service, location, design gpm and pressure drop (PD) and entering/leaving water temperatures.

GENERAL PROCEDURE FOR BALANCING WATER SYSTEMS

Determine Total Flow

Determine total flow at the pump (or in the main if possible). If the flow isn't within +/– 10% of design, try to determine the reasons for the difference. For balancing purposes, the water volume should be 10 to 15% above design. This is because during the balancing process, there generally will be between 5 and 10% loss of total water.

If the flow is well below design (80% of design or less), a decision must be made to determine if the system will be proportionally balanced low, or if a new impeller is needed. Use the pump laws to calculate the new impeller size needed to bring the water flow as close to 100% of design as possible. Increasing the impeller size may also require a motor and/or pump change. Therefore, after calculating the impeller size, the new required brake horsepower must be calculated to determine if there's adequate horsepower presently available. Never increase the pump flow to a point where the motor becomes overloaded.

To determine the total volume, use one or more of the following methods (as applicable):

- Pitot tube traverse of the main
- Differential pressure gauge reading of the main flow meter
- Bourdon tube test gauge reading of the main flow meter
- Differential pressure gauge reading of the pump
- Bourdon tube test gauge reading of the pump

If the total flow is greater than 20% above design, do one or more of the following (as applicable):

- Close the pump discharge valve.
- Adjust the main balancing valve.
- Trim the impeller.
- Install a smaller impeller.

If the total flow is below design, do one or more of the following (as applicable):

- Verify that the discharge valve is open.
- Verify that the main balancing valve is full open.
- Observe for poor inlet or outlet connections at the pump.
- Observe for restrictive fittings in the piping system (consult with the mechanical contractor if piping changes are necessary).

For Systems with Two-way Valves

- Set the temperature controls for full flow through the water coils.
- Set the pressure reducing valve (PRV) to maintain the proper pressure in the system. The PRV should be adjusted so there's a minimum of 5 psi additional pressure at the highest terminal.
- Set all manual balancing valves full open.
- Proportionally balance the system.
- If necessary, change impeller size to increase or decrease water flow.
- Complete report.

For Systems with Three-way Valves

Systems that use three-way automatic control valves (Fig. 23.5) will mean (in most cases) that the bypass manual balancing valve will have to be throttled. Normally, the coil loop will have a higher pressure drop than the bypass loop. When this is the case and the bypass balancing valve is not adjusted, more water than is required will flow through the bypass when the valve is in a modulated position. The coil then receives less than design flow. For example, if the bypass balancing valve has not been set and the valve is modulated at 50% bypass and 50% flow through, the coil may receive only 25% of the total flow, while the bypass gets 75% of the flow. To adjust the bypass balancing valve:

- Set all manual balancing valves full open.
- Set the temperature controls so the three-way valves are open to full flow through the coil and closed to flow through the bypasses.
- Proportionally balance the water flow through the coils.
- For each coil, change the controls so there is full flow through the bypass and no flow through the coil.
    —Determine water flow through the bypass.
    —Set the manual balancing valve in the bypass pipe so full flow gallons per minute (gpm) through the bypass is equal to full flow gpm through the coil.

FIGURE 23.5.

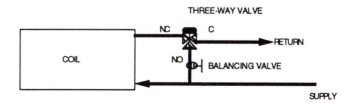

- If necessary, change impeller size to increase or decrease water flow.
- Complete report.

PROPORTIONALLY BALANCING THE WATER DISTRIBUTION SYSTEM

General Procedure

To proportionally balance the system, (if the system has a primary and secondary loop the primary circuit is balanced first) the entire water distribution system is measured. Then manual balancing valves are adjusted to proportionally balance the system. The balance will be accomplished in a logical sequence from the coil (branch, riser, and header) with the lowest percent of design flow to the one with the highest percent of design flow.

When a valve (manual or automatic) is closed the static head upstream of the valve is increased. This means that the pump is working against a greater system static head. An increase in static head results in a decrease in total water flow out of the pump. Therefore it is general practice, when possible, to set the pump at approximately 110% of design as some loss of flow can be expected when the balance is completed. Test and balance in the following order (as applicable):

- Each terminal (coil)
- Each branch
- Each riser
- Each header

The principles of proportional balancing require:

- That all balancing valves in the water distribution system being balanced are full open.
- That the balancing valve to the terminal with the lowest percent of design flow will remain open.
- That the balancing valve in the branch with the lowest percent of design flow will remain full open.
- That the balancing valve in the riser with the lowest percent of design flow will remain full open.
- That the balancing valve in the header with the lowest percent of design flow will remain full open.

Determine which terminal in the entire system has the lowest percentage of design flow. Percent of design flow is equal to the measured flow divided by the design flow (%D = measured flow/design flow). Design flow can be defined as either the original design flow per the contract specifications or a new calculated design flow based on space conditions. Measured and design flows will be in gallons per minute (gpm). Typically, the terminal with the lowest percent of design flow will be on the branch farthest from the pump. This terminal will be called the "key" terminal.

If a part of the distribution system is extremely low and the problem can't be corrected before starting the balancing—for instance, because the problem is poor piping design—

don't sacrifice the entire system by cutting the major portion of the system in an attempt to force water to the problem area. Simply ignore the problem area and proportionally balance the rest of the system. If the balance can be delayed, consult with the design engineer for solutions to correct the problem. The engineer might suggest redesigning restrictive piping or design and install a new separate secondary system.

Proportionally balance each terminal to within a tolerance of 10%. The ratio of the percent of design flow between each terminal must be within 10% (1.00 to 1.10). Ratio of design flow is equal to the percent of design flow of the terminal being adjusted, divided by the percent of design flow of the key terminal.

$$\text{Ratio} = \frac{\%\text{D adjusted terminal}}{\%\text{D key terminal}}$$

Adjust each terminal from the lowest percent of design flow (key terminal) to the highest percent of design flow by branches. Start with the key terminal. Adjust each terminal from the lowest percent of design flow to the highest percent of design flow. Proportionally balance all terminals on this branch.

Go to the branch that has the terminal with the next lowest percent of design flow as determined from the initial readout. This "key" terminal will usually be on the second farthest branch. Balance all the terminals on this branch to the key terminal to within 10% of each other.

Continue balancing until all the terminals on all the branches have been balanced to within +/– 10% of each other. Determine which branch has the lowest percent of design flow (key branch). Proportionately balance all branches from the lowest percent of design flow (key branch) to the highest percent of design flow to within 10% of each other.

Continue proportionally balancing until all branches have been balanced. Determine which riser has the lowest percent of design flow (key riser). Proportionately balance all risers from the lowest percent of design flow (key riser) to the highest percent of design flow to within 10% of each other.

Continue proportionally balancing until all risers have been balanced. Determine which header has the lowest percent of design flow (key header). Proportionately balance all headers from the lowest percent of design flow (key header) to the highest percent of design flow to within 10% of each other.

Continue proportionally balancing until all headers have been balanced. Adjust the pump impeller (or pump speed in the case of belt driven pumps) if needed to bring the system to within +/– 10% of design flow. Reread all the terminals and make any final adjustments. Complete the report.

Proportional Balancing Using Direct Measurement Flow Meters

The use of direct reading flow meters is considered the most reliable and accurate method of determining flow in a water system. Use the general balancing procedure as outlined above. Proportionally balance the terminals using flows determined from the flow meters.

Proportional Balancing Using Pressure Drops

Use the general balancing procedure as outlined above. Proportionally balance the terminals using flows determined from pressure drop across terminals (or other heat exchangers) or valves.

Specific Procedure

Use a differential pressure gauge or a Bourdon tube test gauge to measure the pressure drop across a terminal or other primary heat exchanger. When using a Bourdon tube gauge, ensure that the gauge is at the same height for both the entering and leaving reading, or that a correction is made for a difference in height. To eliminate the need for correction and the possibility of error resulting from the correction, use one test gauge hooked up to a manifold (Fig. 23.6). With a single gauge connected in this manner, the gauge is alternately valved to the high pressure side and then the low pressure side of the terminal to determine the pressure differential.

Measuring Flow Through a Terminal

To use this method of flow measurement, the coil or heat exchanger must be a new coil or in "like new" condition. The coil is acting like an orifice plate or any other flow meter with a known pressure drop. The tubing in the coils has a known resistance to a certain flow rate. If the flow rate is increased, the resistance to flow is also increased (by the square of the flow). This can only happen if the coil is new. If the coil is old and scaled up, the inside diameter of the pipe is reduced and the resistance is increased at rated flow. The original rated pressure drop for rated flow is no longer valid and Eq. 23.1 can't be used.

The coil or heat exchanger must also have certified data from the equipment manufacturer showing actual tested water flow and pressure drop at rated flow. If the data was derived by calculation instead of from actual tests, the information may not be accurate.

Finally, the coil or heat exchanger must also have properly located pressure ports for measuring differential pressure. Some installations have the pressure ports at poor locations. This may lead to erroneous readings because of added losses from piping and fittings.

FIGURE 23.6.

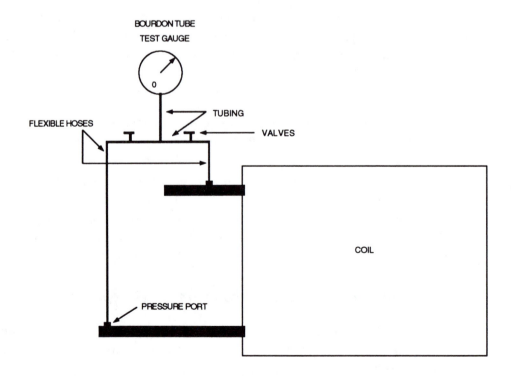

When the rated flow and pressure drop are known, use Eq. 23.1 to determine actual flow (see Chap. 5 for examples):

Equation 23.1

$$\text{gpm}_C = \text{gpm}_R \sqrt{\frac{\Delta P_M}{\Delta P_R}}$$

gpm_C = calculated flow rate
gpm_R = rated flow rate
ΔP_M = measured pressure drop
ΔP_R = rated pressure drop

Measuring Flow Through a Valve

Another way to determine flow is to use the drop through a control valve. As with taking pressures across a terminal, the valve must be in new condition and the pressure ports must be installed as close to the inlet and outlet of the valve as possible. Control valve manufacturers rate their valves based on the relationship between pressure drop and flow. The term for this relationship is "flow coefficient" (C_V) and is the flow rate in gpm of 60° water which will cause a pressure drop of 1 psi (2.31') across a wide open valve. Where the Cv is known, the flow in gpm may be determined by the following equation (see Chap. 5 for examples):

Equation 23.2

$$\text{gpm} - C_V \sqrt{\Delta P}$$

gpm = water flow rate in gallons per minute
C_V = flow coefficient or valve constant
ΔP = measured pressure drop across the valve in psi

Proportional Balancing Using Temperatures

Low temperature hot water heating systems (to 200°F) can, if care is taken, be balanced using temperatures. Its best if the system has the same design temperature differentials (40°F or less) across the coils. Although not a recommended procedure, pipe surface temperature measurements can be used for approximating the temperature of low temperature hot water heating systems. The surface temperature of water pipes should be above 150°F and the pipe surface must be cleaned to a bright finish. To help prevent errors in measurement:

• Make sure temperature wells or other temperature measuring stations are properly installed at good locations.

• Use the same thermometer for all readings.

• Use care to prevent reading errors in using the thermometers.

• Allow ample time for the temperatures to stabilize.

• Stabilize load conditions.

• Eliminate air stratification across the coil.

- Use care to avoid error caused by the thermometer sensing ambient air temperature when taking pipe surface temperatures.

Use the general balancing procedure as outlined above. Proportionally balance the terminals using flows calculated from entering and leaving water temperatures.

Calculating Coil gpm.

Equation 23.3

$$\text{gpm} = \frac{\text{Btuh}}{500 \times TD}$$

gpm = flow rate through the coil in gallons per minute
500 = constant
TD = temperature difference between entering and leaving water
Btuh = For heating coils and chilled water coils with no latent load (dry coil):
Btuh = cfm × 1.08 × ΔT
(ΔT = temperature difference between the entering and leaving air)
For chilled water coils with latent load (wet coil):
Btuh = cfm × 4.5 × Δh
(Δh = enthalpy difference, in Btu/lb, between entering and leaving air)

ADDITIONAL INFORMATION ON BALANCING WATER SYSTEMS

Constant Flow Systems

If the system uses three-way valves, set the coils for full flow. Balance on full flow through the coil. Change the three-way valve to full bypass and set the bypass balancing valve. If system uses two-way valves with a bypass loop at the end of the system or at the pump, set the system to full flow and balance with the bypass closed. Measure the differential pressure between the supply and return mains and set the bypass balancing valve to maintain the differential pressure.

Constant Flow Systems with Primary and Secondary Loops

Balance the primary loop first. To balance, the system must be set for full flow in both the primary and the secondary loop.

Constant Flow Primary and Variable Flow Secondary Loops

As the two way valves at the terminals in the variable flow loop start to close down, the secondary pumps will use either a variable frequency, variable speed drive on the motor (which is based on the sensing system differential pressure), or the pump will simply backup on its curve as if the pump discharge valve was being throttled. Some systems will use multiple pumps and stage them to come on or go off to increase or decrease the volume of flow.

Variable Flow Systems

Variable flow hydronic systems are used to achieve either full- or part-load heating or cooling conditions while reducing the energy consumed by the pump. Variable flow systems may use a constant speed pump with two-way automatic control valves or a variable speed pump with either two- or three-way automatic control valves. Both types of systems reduce flow to maintain a constant water temperature difference across the terminal. The equation is $Q = \text{gpm} \times 500 \times \Delta T$. As the heat load ($Q$) in the conditioned space changes, the flow (gpm) changes to maintain constant temperature (ΔT) across the terminal.

Variable flow systems that use a constant speed pump and two-way valves do not reduce the energy consumed by the pump nearly as much as variable speed systems. In a typical constant speed system, a temperature control device in the conditioned space sends a signal to a terminal's two-way valve to reduce or increase flow. If the valve closes to reduce flow, there is an increase in resistance in the system. This increased resistance "backs the pump up on its curve," reducing water flow through the pump and decreasing the horsepower and the energy required by the pump. For example, a pump that uses 33.2 brake horsepower when circulating 1,250 gallons per minute of water at 82 feet of head will require only 29.9 bhp when pumping 1,000 gpm. The flow and the horsepower are reduced because the two-way valves in the system partially close to increase the resistance to 100 feet of head.

Systems that use variable speed pumping to control the water flow rate use an electronic device called a "variable frequency drive" to vary the speed of the pump motor and the pump. This type of system will reduce pumping horsepower approximately by the cube of the pump speed change. In the example above, if the system had variable speed pumping, the horsepower would be reduced to approximately 17 bhp with a reduction in flow from 1,250 to 1,000 gpm.

A differential pressure (DP) device is installed in the piping to control the speed of the pump. In a typical commercial system, as the temperature controls are satisfied, the automatic control valves for the terminals begin to close. The differential pressure devices senses the increase in system pressure and the pump speed is reduced to meet actual system demand.

To set the differential pressure on the DP device, the pump and the system are first set for full flow. The DP is then set to maintain the required differential pressure at that point in the system. For instance, if the device is placed in the ends of the supply and return mains, it is set for the required pressure drop across the last terminal in the system. This pressure drops includes the terminal's associated valves and piping. When the differential pressure device is located at the end of the system, the pump will operate at maximum energy savings.

If the DP device is placed closer to the pump, some energy savings will be lost because the pump will have to operate at a higher speed to overcome the pressure losses in the piping and terminals that are past the device. In some systems, the DP may have to be installed closer to the pump to maintain control of the system. The DP is installed in the correct location when it will control the variable frequency drive and operate the pump at the lowest speed to maintain required demand to all terminals.

All hydronic systems need some means of determining and balancing water flow. Some designers believe that variable flow systems are self-balancing and balancing valves do not need to be installed. This is incorrect. Constant flow systems are not self-balancing, and neither are variable flow systems. Even when special attention is given to the selection of automatic control valves and the design of the piping of variable flow systems, flow meters and manual balancing valves are needed to balance the system and prevent flow problems. Properly designed systems that include manual balancing valves and flow meters are always preferable to systems that rely solely on automatic control valves or varying the pump speed for system balance and control.

PROPORTIONAL BALANCING PROCEDURE FOR VARIABLE FLOW SYSTEMS

Balancing the water distribution system will consist of reading the flow meters and setting the water flow by manually adjusting balancing valves at the mains, headers, risers, branches and terminals as is appropriate for each system. To balance the water distribution system, you will need to:

- Set the pump for full flow. If the system has diversity, set the terminals accordingly.

- Read all the terminals. Start with the terminal with the lowest percent of design flow.

- Proportionally balance all the terminals by branch, all the branches by riser, all the risers by header and then the headers. Balance within 10% of desired flow.

- If the system uses three-way valves, set the terminals for full flow. Balance on full flow through the terminal. Change the three-way valve to full bypass and set the manual balancing valve in the bypass to match terminal flow.

- If the piping has a bypass loop at the end of the system or at the pump, close the two-way valve in the bypass and set the system to full flow. Balance the terminals with the bypass valve closed. After the terminals are balanced, measure the differential pressure between the supply and return lines. Set the manual balancing valve in the bypass to maintain this differential pressure through the bypass.

- If the system is variable speed with a DP device, measure the differential pressure across the DP. Set the DP to maintain the design differential pressure across the device.

- After the system has been balanced it may be necessary to make some slight adjustments at the balancing valves in order to achieve the final balance. Since the system has been proportionally balanced, an adjustment at a branch valve, for example, will increase or decrease all the terminals on that branch proportionally.

TROUBLESHOOTING VARIABLE FLOW SYSTEMS

If a variable flow system is not operating properly, conduct a survey and analysis of the system pressures and components, including the controls. The survey should be conducted at various pump speeds such as 25%, 50%, 75% and 100%. Some typical problem areas and general recommendations for correcting the problems are listed below. Depending on the system, some recommendations may need to be combined. The purpose of implementing these recommendations is to achieve control of the system as close to design intent as possible. With this objective in mind, it is important to note that some of these recommendations may increase operating costs.

System A: The differential pressure (*DP*) sensor is located across the ends of the mains to ensure required DP at the remote terminals.

System B: The DP is located midway in the distribution system.

System C: The DP is located at a critical terminal that has an unusually high pressure drop.

Problem 1: At part loads, some terminals do not receive required flow.

Problem 2: The terminal nearest the pump short-circuits, robbing the remote terminals of required flow.

Recommendations to the individual systems after analysis as applicable:

- Add balancing valves.
- Balance the system.
- Relocate the DP sensor:
    —Closer to the pump if the terminals nearest the pump are not getting required flow.
    —Farther from the pump if the remote terminals are not being satisfied.
    —Closer to the critical terminal.
- Add a separate DP sensor:
    —At the terminal nearest the pump.
    —At a critical terminal.
- Change the setting on the DP to a higher value.
- Install a DP that shifts the differential pressure readout point to one of several different locations along the supply and return mains as the need arises. This DP changes the differential pressure setting to match the requirements at the new location, allowing the variable speed pump to meet whatever requirements the system demands.
- Use variable orifice, self-limiting balancing valves. These pressure-activated valves sense a reduced or increased pressure differential across the valve and open or close to a set maximum.
- Replace existing automatic control valves with valves that have a higher pressure drop.
- Convert a direct return system to a reverse return.

CHAPTER 24 Testing, Adjusting and Balancing—Example Water System

This chapter gives examples of proportionally balancing water systems.

Example 24.1: (Fig. 24.1)

After the system is set up, read the terminals.

Calculate the percent of design for each terminal.

Determine the key terminal.

Terminal #	D	M	%D	Ratio to Key
1	50	72	144	1.36
2	50	70	140	1.32
3	50	68	136	1.28
4	50	53	106	key
5	50	59	118	1.11

Proportional Balancing Procedure

Close #5 to 55 gpm.

Read #4. It reads 54 gpm.
Calculate %D for each:
Calculate ratio to key:
Read #3. It now reads 69 gpm.

Arbitrarily selected (106 + 118/2 = 112%D).
Elect to use 110%.

#4 = 108%   #5 = 110%
#5/#4 = 110/108 = 1.02   Balanced

FIGURE 24.1.

TYPICAL FLOW METER AND BALANCING VALVE

COIL

1   2   3   4   5

PUMP

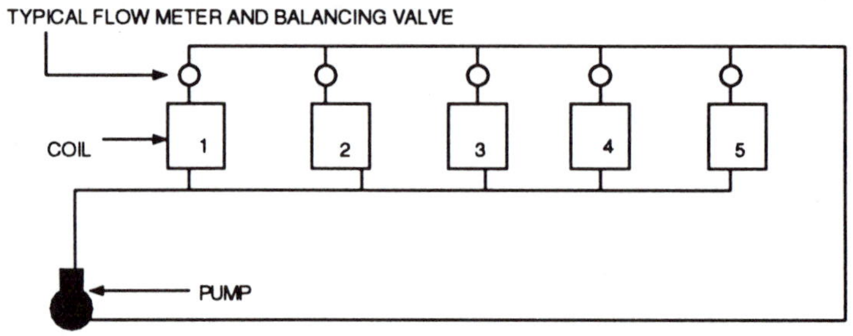

Close #3 to 63 gpm.	Arbitrarily selected.
Read #4. It reads 56 gpm.	
Calculate %D for each:	#4 = 112% #3 = 126%
Calculate ratio to key:	126/112 = 1.13 Not balanced
Close #3 to 59 gpm.	Arbitrarily selected.
Read #4. It reads 57 gpm.	
Calculate %D for each:	#4 = 114% #3 = 118%
Calculate ratio to key:	#3/#4 = 118/114 = 1.04 Balanced
Read #2. It now reads 72 gpm.	
Close #2 to 64 gpm.	Arbitrarily selected.
Read #4. It reads 59 gpm.	
Calculate %D for each:	#4 = 118% #2 = 128%
Calculate ratio to key:	#2/#4 = 128/118 = 1.08 Balanced
Read #1. It now reads 76 gpm.	
Close #2 to 66 gpm.	Arbitrarily selected.
Read #4. It reads 61 gpm.	
Calculate %D for each:	#4 = 122% #1 = 132%
Calculate ratio to key:	#1/#4 = 132/122 = 1.08 Balanced

After proportional balancing, the system now reads:

Terminal #	D	M	%D	Ratio to Key
1	50	66	132	1.08
2	50			1.08
3	50			1.04
4	50	61	122	key
5	50			1.02

Terminal #	D	M	%D	Ratio to Key
1	50	66	132	1.08
2	50	66	*132*	$1.08 \times 122\%D = 132\% \times 50 = 66$ gpm
3	50			1.04
4	50	61	122	key
5	50			1.02

Terminal #	D	M	%D	Ratio to Key
1	50	66	132	1.08
2	50			1.08
3	50	*63*	*127*	$1.04 \times 122\%D = 127\% \times 50 = 63$ gpm
4	50	61	122	key
5	50			1.02

Terminal #	D	M	%D	Ratio to Key
1	50	66	132	1.08
2	50			1.08
3	50			1.04
4	50	61	122	key
5	50	62	*124*	$1.02 \times 122\%D = 124\% \times 50 = 62$ gpm

Terminal #	D	M	%D	Ratio to Key
1	50	66	132	1.08
2	50	66	132	1.08
3	50	63	127	1.04
4	50	61	122	key
5	50	62	124	1.02

The system now reads 318 gpm. The design is 250 gpm. The impeller size is 8 inches. The operating bhp is 11.5. The operating TDH is 100 feet. The pump efficiency is .070.

To predict the impeller size, head and brake horsepower required to bring this system to deisgn, use the pump laws.

$$\frac{gpm_2}{gpm_1} = \frac{d_2}{d_1}$$

$$\frac{TDH_2}{TDH_1} = \left(\frac{d_2}{d_1}\right)^2$$

$$\frac{bhp_2}{bhp_1} = \left(\frac{d_2}{d_1}\right)^3$$

The new impeller size is 6.3 inches.
The new operating TDH is 62.
The new operating bhp is 5.6.

Example 2 (Fig. 24.2)

After the system is set up, read the terminals.

Calculate the percent of design for each terminal.

Determine the key terminal.

Coil #	Design gpm	Measured gpm	%D	Ratio
1	6	4.7	78	Lowest %D
2	6	5.2	87	2:1 = 1.12
3	5	5.0	100	3:2 = 1.15
4	5	6.1	122	4:3 = 1.22
5	6	7.5	125	5:4 = 1.02
	28	28.5		

Starting with coil #1, adjust each coil on branch #1 until the ratio of the percent of design flow between each coil is within 10%. Instead of balancing back to the key, balance to the previous terminal, starting with the lowest percent of design. The balancing sequence will be:

a. Adjust coil #2 to coil #1.

b. Adjust coil #3 to coil #2.

c. Adjust coil #4 to coil #3.

d. Adjust coil #5 to coil #4.

FIGURE 24.2.

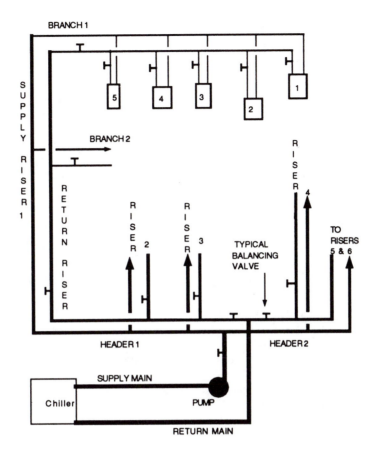

Proportional Balancing Procedure

Close #2 to 5.0 gpm   Arbitrarily selected (87 + 78/2 = 82.5%D).
                                       Elect to close to 83%.

Read #1. It reads 4.7 gpm.
Calculate %D for each: #2 = 83%   #1 = 78%
Calculate ratio: #2/#1 = 83/78 = 1.06 Balanced

Coil #	Design gpm	Measured gpm	%D	Ratio
1	6	4.7	78	Lowest %D
2	6	5.0	83	2:1 = 1.06

Read #3. It still reads 5.0 gpm.
Calculate %D: #3 = 100%   #2 = 83%
Calculate ratio: #3/#2 = 100/83 = 1.20   Not balanced
Close #3 to 4.6 gpm.   Arbitrarily selected. (100 + 83/2 = 91.5%D)
                                       Elect to close to 92%.

Read #2. It reads 5.1 gpm.
Calculate %D: #3 = 92%   #2 = 85%
Calculate ratio: #3/#2 = 92/85 = 1.08   Balanced

Coil #	Design gpm	Measured gpm	%D	Ratio
2	6	5.1	85	2:1 = 1.06
3	5	4.6	92	3:2 = 1.08

Read #4. It still reads 6.1 gpm.
Calculate %D: #4 = 122%   #3 = 92%
Calculate ratio: #4/#3 = 122/92 = 1.33   Not balanced
Close #4 to 5.2 gpm.   Arbitrarily selected.

Read #3. It reads 4.9 gpm.
Calculate %D: #4 = 104%   #3 = 98%
Calculate ratio: #4/#3 = 104/98 = 1.06   Balanced

Coil #	Design gpm	Measured gpm	%D	Ratio
3	5	4.9	98	3:2 = 1.08
4	5	5.2	104	4:3 = 1.06

Read #5. It reads 7.5 gpm.
Calculate %D: #5 = 125%   #4 = 104%
Calculate ratio: #5/#4 = 125/104 = 1.20   Not balanced
Close #5 to 7.0 gpm.   Arbitrarily selected.

Read #4. It reads 5.3 gpm.
Calculate %D for each: #5 = 117%   #4 = 106%
Calculate ratio to key: #5/#4 = 117/106 = 1.10   Balanced

Coil #	Design gpm	Measured gpm	%D	Ratio
4	5	5.3	106	4:3 = 1.09
5	6	7.0	117	5:4 = 1.10

If you were to read the coils on branch #1, the quantities would be approximately:

Coil #	Design gpm	Measured gpm	%D	Ratio
1	6	5.2	87	Lowest %D
2	6	5.5	92	2:1 = 1.06
3	5	5.0	100	3:2 = 1.08
4	5	5.3	106	4:3 = 1.06
5	6	7.0	117	5:4 = 1.10
	28	28.0		

- Balance the coils on branch #2.

- Proportionately balance the branches on riser #1.

Branch #	Design gpm	Measured gpm	%D	Ratio
1	28	28	100	Lowest %D
2	30	36	120	2:1 = 1.20

Close branch #2 to 33 gpm.

Arbitrarily selected (120 + 99/2 = 110%D). Elect to close to 100%.

Read branch #1. It reads 30 gpm.
Calculate %D for each: #2 = 110%   #1 = 107%
Calculate ratio: #2/#1 = 110/107 = 1.03 Balanced

Branch #	Design gpm	Measured gpm	%D	Ratio
1	28	30	107	Lowest %D
2	30	33	110	2:1 = 1.03

After the branches on riser #1 have been proportionally balanced, go to the next riser and start with its lowest branch. When all the branches on all the risers have been proportionally balanced, start with the riser that has the lowest percent of design flow and proportionally balance each riser.

Header #1

Riser #	Design gpm	Measured gpm	%D	Ratio
1	58	63	109	Lowest %D
2	80	89	110	2:1 = 1.01
3	40	48	120	3:2 = 1.09
	178	200		

The branches on header #1 are balanced.

Header #2

Riser #	Design gpm	Measured gpm	%D	Ratio
4	100	138	138	4:5 = 1.13
5	100	122	122	5:6 = 1.11
6	100	110	110	Lowest %D
	300	370		

Balance riser #5 to riser #6.
Balance riser #4 to riser #5.

Close riser #5 to 116 gpm.

Arbitrarily selected (122 + 110/2 = 116%D). Elect to close to 116%.

Read riser #6. It reads 112 gpm.
Calculate %D for each: #5 = 116%   #6 = 112%
Calculate ratio: #5/#6 = 116/112 = 1.04   Balanced

Riser #	Design gpm	Measured gpm	%D	Ratio
5	100	116	116	5:6 = 1.04
6	100	112	112	Lowest %D

Read riser #4. It reads 139 gpm.
Calculate %D for each: #4 = 139%   #5 = 116%
Calculate ratio: #4/#5 = 139/116 = 1.20 Not Balanced

Close riser #4 to 128 gpm.   Arbitrarily selected (139 + 116/2 = 128%D).
                                              Elect to close to 128%.

Read riser #5. It reads 120 gpm.
Calculate %D for each: #4 = 128%   #6 = 120%
Calculate ratio: #4/#5 = 128/120 = 1.07   Balanced

Riser #	Design gpm	Measured gpm	%D	Ratio
4	100	128	128	4:5 = 1.07
5	100	120	120	5:6 = 1.04

Riser #	Design gpm	Measured gpm	%D	Ratio
4	100	128	128	4:5 = 1.07
5	100	120	120	5:6 = 1.04
6	100	115	115	Lowest %D
	300	363		

After the risers have been balanced, proportionally balance header #2 to header #1.

Header	Design gpm	Measured gpm	%D	Ratio
1	178	200	112	Lowest %D
2	300	363	121	2:1 = 1.08
	478	563		

The headers are in balance. The system is 118%D (563/478). Adjust the impeller diameter to bring the system to within 10% of design flow. After the new impeller is installed, read the entire system and make any final adjustments.

CHAPTER 25 Verification of System Performance: Design and Testing of Special Systems and Laboratory Fume Hoods

This chapter outlines the types and basic operating principles of laboratory fume hoods and exhaust systems; the advantages and disadvantages of the various systems; design principles for laboratory supply and exhaust system components; smoke, air volume and velocity testing of laboratory fume hoods; and troubleshooting fume hoods for reverse air flow or incorrect face velocity. The primary concern in the designing and testing of laboratory air conditioning systems and fume hoods is human safety. We want to make the work place safe for all the occupants of the laboratory facility including laboratory personnel, maintenance people, support personnel, visitors and anyone else within the fume hood or exhaust system environment.

LABORATORY FUME HOOD

A laboratory fume hood (Fig. 25.1) is a ventilated, box-like structure enclosing a work space which captures, contains and exhausts contaminated fumes, vapors and particulate matter generated inside the enclosure. Laboratory fume hoods are made of various materials such as transite (a cement and asbestos material), epoxy coated steel, stainless steel, fiberglass, epoxy resin, polypropylene and PVC.

The basic laboratory fume hood is mounted on a bench or table and has two side panels, a front, a back and a top panel. It also has an exhaust plenum with a baffle and an exhaust collar. The front of the hood is called the **face** and is usually equipped with a movable, transparent sash. Sashes may be vertical moving only, or a combination sash which has horizontally sliding panels set in a vertical moving sash. For either type of sash, the vertical sash is placed in the upmost position for easier setup or removal of laboratory apparatus in the hood. For normal hood use, other than setup or removal of apparatus, the vertical sash should be closed whenever someone is not using the hood. In any circumstance, the sash should only be opened high enough to allow for proper operation. With the combination sash, the vertical sash is closed completely and the horizontal sash is opened only wide enough for proper operation.

The laboratory fume hood has a baffle across the back which helps control the pattern

FIGURE 25.1.

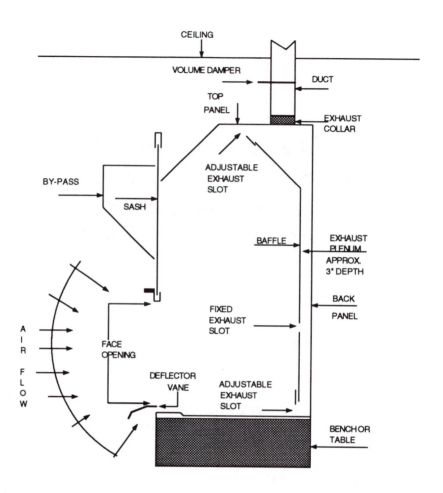

of the air moving into and through the fume hood. The baffle is usually adjustable so the air flow through the hood can be directed up for "lighter than air" fumes, or down for "heavier than air" fumes. The baffle is normally built so it is impossible, by routine adjustment, to restrict air flow through the fume hood more than 20%.

The top panel of the fume hood has an exhaust collar to connect the exhaust duct to the fume hood. The exhaust duct may have a manual or automatic damper for control of the total volume of air through the hood. Total volume of air may also be controlled by changing the speed of the exhaust fan or by moving volume dampers at the exhaust fan.

Most fume hoods have an airfoil, called a **deflector vane,** at the entrance to the work surface. The design of the vane smooths the flow of air across the work surface and deflects it to the lower baffle opening. The vane, which is about 6 inches deep, also functions as a standoff to help keep the hood user at least that distance away from the hood face. When a deflector vane is installed, there is a fixed opening between the work surface and the vane. Therefore, even when the vertical sash is fully closed and the exhaust fan is on, there is still air flow into the hood.

Operating Principles

The laboratory space (Fig. 25.2) is supplied with filtered, conditioned air for temperature and humidity control by the supply air handling equipment. In the conditioned air, there is enough outside air to meet ventilation code requirements and maintain proper space pres-

FIGURE 25.2.

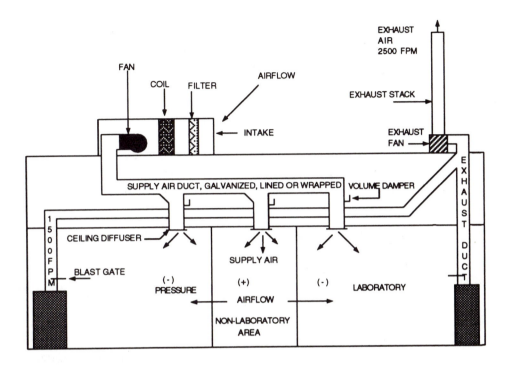

surization. The supply air system may be either constant air volume (CAV), or variable air volume (VAV). The exhaust system may be either constant air volume, variable air volume, or a combination of the two. When the fume hood exhaust fan is energized, conditioned air from the laboratory space is brought into the hood to contain and exhaust the contaminants generated inside the hood. The contaminants are ducted to the outside where they are released into the atmosphere for dilution. To maintain laboratory temperature and pressurization, the exhausted air must be replaced entirely by conditioned air from the supply system. Laboratory fume hoods are generally grouped into two basic classifications: conventional and bypass.

Conventional Fume hood, Constant Air Volume (CAV)

Airflow through the standard conventional fume hood in a constant air volume exhaust system is variable, both in total air volume and face velocity. The conventional hood has a movable, vertical or combination horizontal/vertical sash. At the sash's full open vertical position, the free area of the hood face is generally about 10 to 13 square feet. Therefore, a Class B laboratory fume hood with a minimum average face velocity of 100 feet per minute (fpm) would have about 1,000 to 1,300 cubic feet per minute (cfm) of air exhausted through the hood. The equation for this calculation is Q = AV, where Q is air volume in cfm; A is face area in square feet; and V is face velocity in fpm. The operation of this type of fume hood is such that as the sash is lowered, the face area is reduced and the velocity of the air through the face opening begins to increase as it tries to maintain a constant air volume. However, at some point in the closing of the sash, the total volume of air is also reduced—even with the increase in face velocity. Closing the sash on a conventional hood in a constant air volume system disrupts the air flow pattern and results in high velocities and unwanted turbulence at the hood face. The air turbulence can induce contaminants out of the fume hood into the laboratory space.

Bypass Fume Hood, Constant Air Volume

The airflow through a bypass fume hood in a constant air volume exhaust system is constant in total air volume and face velocity. The standard bypass hood has a movable, vertical or combination horizontal/vertical sash. The construction of the bypass hood is similar to the standard conventional hood described before, with the addition of the bypass. The bypass provides a constant volume of air flow through the fume hood as the sash is closed. The operation of a bypass fume hood is such that as the sash is pulled down, the air volume through the hood face is reduced. However, simultaneously, as the sash is being closed, the bypass is being opened, and more air is then drawn through the bypass. This keeps the total air flow through the hood constant. It also means that the hood face velocities stay constant. This is a great improvement over the standard constant air volume conventional fume hood.

Conventional Fume Hood, Variable Air Volume (VAV)

The air flow through a conventional fume hood in a variable air volume exhaust system is variable in total air volume, but has a constant face velocity. The conventional hood has a movable, vertical or combination horizontal/vertical sash. This type of hood is also equipped with special controls to allow the volume of exhaust air to vary while still maintaining a constant velocity across the hood face. The operation of a variable air volume conventional fume hood is such that as the sash is lowered, the face area is reduced. The face velocity begins to increase to maintain constant volume, as described with the standard conventional hood in the constant air volume system. However, unlike the standard CAV conventional hood, as the velocity of the air through the face increases, a controller in the hood senses the rise in velocity and sends a signal to an air valve or to the fan to decrease air volume through the hood to maintain a preselected face velocity. When the sash is raised, the controller senses a face velocity below the set point and sends a signal to increase air volume to maintain the correct face velocity. The advantages of the variable air volume fume hood and exhaust system are that the volume of conditioned air exhausted is reduced as the sash is closed. This results in energy savings, and the face velocities stay constant.

VERIFYING SYSTEM PERFORMANCE: TESTING LABORATORY FUME HOODS

Study the following mechanical plans, equipment specifications and catalogs to become familiar with the air handling system and its design intent:

- Engineering drawings
- Shop drawings
- "As-built" drawings
- Schematics
- Previous air balance reports and hood certification

- Equipment catalogs:
  —Fan description and capacities
  —Fan performance curves
  —Recommendations for testing hood and fan equipment
  —Operation and maintenance instructions

Ensure that all fans and fume hoods are designated correctly on the drawings, reports and equipment. While reviewing the documents, make note of any piece of equipment, system component or any other condition that should be specifically investigated during the field verification of performance.

As applicable, prepare the following reporting forms for each supply and exhaust system:

- Motor data and test sheet
- Drive data and test sheet

- Air handling equipment data and test sheet
- Fume hood test and data sheet (Fig. 25.3)

A performance test to determine the operating condition of a fume hood must be conducted periodically as required by local code. Performance tests are conducted for volume of flow, face velocity and reverse airflow. The performance test gives a relative and quantitative determination of the efficiency of the fume hood. The test procedure outlined here presumes a bench type laboratory fume hood in an air conditioned environment. If other types of hoods are used, some modification of the test procedure may be required. The performance test does not constitute an engineering investigation of what the causes may be for poor performance of the fume hood, or of ways to improve performance. However, the performance test could be used as an aid to such an investigation.

Generally, the toxicity level of the work done within the hood will determine the hood face velocities and the total amount of the exhaust air. Materials of little or no toxicity need face velocities only sufficient to maintain control under normal operating conditions. As toxicity levels rise, the face velocity should be increased to assure control. The recommended velocities for different toxic levels range from 80 fpm to 150 fpm. Generally, 100 fpm is satisfactory for most applications. The fume hood user, the facility hygienist or the applications engineer must specify what level of performance is required for each hood. The tests must be in agreement with the toxicity level of the work to be performed in the fume hood and the safety standards established by the facility.

The building and laboratory air conditioning systems should be operating normally. General activity in the laboratory should be normal. Air currents in front of the hood should be reduced or eliminated if possible. The velocity of the air currents should never exceed 20% of the average face velocity. Conduct the test with the normal hood apparatus in place and operating, except where clearance must be provided for the test instruments or equipment. Make a sketch of the room showing the general layout of the laboratory, the location of all significant laboratory equipment (including all fume hoods), and the layout of the supply, return and exhaust air systems with types of air devices. The following list is typical of the test instruments and equipment needed to perform the various air flow and smoke tests.

Air Flow	Smoke
Pitot tube	Smoke candles, 1/2 minute
Static pressure tip	Smoke tubes
Manometer	Dry ice
Micromanometer	Titanium tetrachloride
Anemometer	Cotton swabs
	Masking or duct tape

Procedure for Testing the Hood Air Volume and Face Velocity

- Test the operation of the sash. Use one hand to grip the sash at the extreme right side and raise and lower the sash. Repeat at the extreme left side. The sash should glide smoothly and freely and hold at any height without creeping. If the hood has a vertical moving

FIGURE 25.3. FUME HOOD DATA AND TEST SHEET

Project: _____   **Contact:** _____

	Specified	*Actual*
Hood Class	_____	_____
Average Velocity (fpm)	_____	_____
Opening Size ___ ″× ___ ″	_____	_____
Exh. Duct Traverse (cfm)	_____	_____
Aux. Duct Traverse (cfm)	_____	_____
Aux. Air is ___% of Exh. Air	_____	_____

	Yes	*No*	
HVAC sysem has been balanced	_____	_____	
All systems are in normal operation	_____	_____	Date:
Cross drafts are acceptable	_____	_____	
Hood sash is open	_____	_____	fpm:
Hood sash operation is acceptable	_____	_____	% open:
Baffle operation is acceptable	_____	_____	
Baffle position	_____	_____	
Lowest face velocity reading	*****	*****	
Smoke flow into hood is acceptable	_____	_____	fpm:
Smoke flow over work surface is acceptable	_____	_____	

Hood Face Velocity Readings

	A	B	C	D	E	F	G	H
1	____	____	____	____	____	____	____	____
2	____	____	____	____	____	____	____	____
3	____	____	____	____	____	____	____	____
4	____	____	____	____	____	____	____	____
5	____	____	____	____	____	____	____	____
6	____	____	____	____	____	____	____	____
7	____	____	____	____	____	____	____	____

sash, set it in the full open or in the "as used" position. If the hood has a combination sash, close the vertical sash and position the horizontal sash to get the maximum face opening, or set it in the "as used" position.

- Traverse the exhaust duct using a manometer or anemometer to determine total volume of air through the hood. Calculate the total air volume in cubic feet per minute by multiplying the area of the duct by the average duct velocity ($Q = AV$). Take static pressures in the exhaust duct at the collar and compare with design static pressure loss through the fume hood. Record the readings. Any holes drilled for testing must be sealed as specified by the engineer or equipment manufacturer.

- Divide the hood face into a grid using the vertical and horizontal dimensions to get equal areas over the cross section of the sash opening. Establish the center of each area. The maximum distance between the centers should not exceed 6 inches. Make a sketch showing area centers.

- Traverse the sash opening using an anemometer to take a velocity reading at each center point. Record the readings. The hood fails the test if the minimum reading at any point on the traverse is less than 80% of the average face velocity.

- Calculate the average face velocity in feet per minute. Compare the average face velocity with design specifications.

- Calculate the volume of exhaust air in cubic feet per minute at the hood face by multiplying the square feet of the sash opening by the calculated average face velocity ($Q = AV$). Compare with the air volume taken at the exhaust duct traverse point.

- Place a certificate showing test results on the fume hood.

Smoke Test Procedure

- Make a complete traverse of the hood face with either titanium tetrachloride on a cotton swab, a smoke candle or a smoke tube. Determine that a positive air flow is entering the hood over the entire face. Use a smoke tube, a smoke candle or swab a stripe of titanium tetrachloride on pieces of tape to produce the necessary amount of smoke. Direct the smoke at the following test locations:

  —Along both sides, the top and the work surface of the hood about 6 inches behind and in parallel to the hood face

  —Along the back panel and the baffle of the hood, and

  —Around any equipment in the hood.

- Verify that all smoke is carried to the back of the hood and exhausted. The hood fails the test if visible smoke flows out the front of the hood. Reverse air flows or dead air spaces are not permitted.

- With the sash open, ignite a smoke candle within the hood enclosure to observe the exhaust capacity of the hood. All smoke should flow quickly and directly to the back of the hood and be exhausted. Set the candle on the work surface and close the sash. With the sash closed, the hood must have enough air to dilute and exhaust the smoke. The hood fails the test if visible smoke flows out the front of the hood. Reverse air flows or dead air spaces are not permitted.

- Place a pan of hot water in the center of the work area and add enough chunks of dry ice to the hot water to form a large volume of heavy white smoke. All smoke should flow directly to the back of the hood and be exhausted. The hood fails the test if visible smoke flows out the front of the hood. Reverse airflows or dead air spaces are not permitted.

Laboratory Fume Hood and Exhaust System Design

General design principles dictate that outside air intakes should be located to avoid drawing in hazardous fumes coming from either the laboratory building itself or from other structures and devices. Air exhausted from laboratory fume hoods must be ducted to the outside. Laboratories in which there are hazardous materials must be maintained at an air pressure that is negative to the corridors and adjacent nonlaboratory areas. Care should be exercised in the selection and placement of supply air diffusers, grilles or registers to avoid air patterns that would adversely affect the performance of the laboratory fume hoods. Laboratory fume hood face velocities must be sufficient to prevent the escape of contaminants. All laboratory fume hoods should be equipped with visual and audible alarms to warn users of unsafe air flows. The generally accepted design face velocities are:

Hood Class	Threshold Limit Value	Velocity
Class A	<10 PPM	125–150 fpm
Class B	10–500 PPM	100 fpm
Class C	>500 PPM	80 fpm

Caution: Higher face velocities do not necessarily result in improved user protection. In fact, with higher face velocities, the eddy currents in the hood become greater. These eddy currents can drag the contaminants in the hood back into the face of the user. Higher face velocities also mean greater volume of supply air and increased energy usage.

Classifications of exhaust systems are based on the arrangement of the major system components such as fans, plenums, mains, branches, etc., or on the method of operation and control. The general classifications are individual exhaust system, central exhaust system, combination system, constant air volume system and variable air volume system.

Individual Fume Hood Exhaust Systems

Individual fume hood exhaust systems (Fig. 25.4) have a separate exhaust connection, separate duct and a separate fan for each fume hood. These types of exhaust systems have been common but are increasingly used only on single story buildings that have a need for a few hoods. The general operating sequence of this type of system is such that the exhaust fan is always on and is interlocked electrically with the supply fan so that, if the exhaust does go off, the supply fan is shut down.

When installed with constant air volume fans, individual fume hood exhaust systems provide a stable and easily controlled system and are simple to air balance. Each hood is operated by simply starting or stopping the fan motor and the operation of any one fume hood does not affect any other hood. Since each hood has its own ductwork, there is no mixing of exhausted air between the fume hoods in the laboratory spaces and shutdowns for repairs or maintenance are localized. Individual fume hood systems are used in selective applications such as systems that need special exhaust filtration or require special duct and fan construction for corrosive elements, or systems with fumes that contain very hazardous elements. Typical applications are perchloric acid or radioisotope hoods.

FIGURE 25.4.

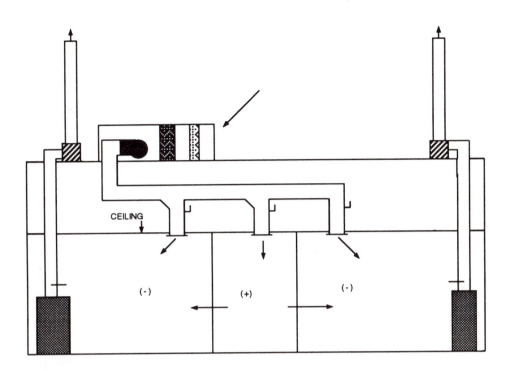

Individual fume hood exhaust systems are inexpensive for small systems with only a few fans. However, if the system is large, the initial investment and the operating costs are high because of the greater number of fans, ductwork, roof penetrations, controls, and wiring that must be installed and maintained. Installation of air locks (an anteroom with airtight doors between a controlled and uncontrolled space) may also be needed to prevent reverse airflows. For instance, the shutdown of one individual exhaust system while others are on could cause the airflow from the corridor into the laboratory to reverse direction. This could cause contaminants and odors to flow out of the laboratory and into the corridor and adjacent labs.

Central Exhaust Systems

Facilities which use many fume hoods will use a central system (Fig. 25.2). Central systems have one normal fan, one standby fan, a common suction plenum and branch connections to multiple exhaust terminals. Central systems are most economical when exhaust devices are grouped by type, proximity and fire, pressurization or contamination zones.

A central exhaust system, as compared to an individual system of equivalent size, has a smaller initial investment and less operating costs. Also, the air is more diluted before being exhausted into the atmosphere. In addition, the central system has a standby fan for safety, and generally has greater flexibility for future expansion.

The main disadvantage to central exhaust systems is that they are more difficult to balance and they also require more periodic rebalancing to ensure proper airflow. Balancing the central system is made more difficult when different types of exhaust devices such as fume hoods, safety cabinets and other special filtered equipment are installed on common duct runs. The different pressure losses associated with each type of device are what cause the balancing problems.

VAV Exhaust Systems

The main advantage of variable air volume exhaust systems over constant volume systems is reduced energy usage while still maintaining constant face velocity. A sensor and controller in the hood responds to changes in air velocity or sash position and sends a signal to a volume control device. The volume control device, either a variable speed drive on the exhaust fan or an air valve in the exhaust duct, modulates to maintain the face velocity within the desired range. For large central systems, static pressure regulators are also installed in the exhaust duct to control fan air volume and pressure. When total air flow through the fume hood is reduced below design, a relief system is needed to maintain the proper supply air volume in the laboratory. Without proper air volume, the laboratory air balance, air flow patterns, temperature and humidity cannot be controlled. The laboratory relief system, consisting of inlets, ductwork, dampers, etc., provides an exhaust path for the supply air out of the laboratory when the air is restricted through the hood.

Variable air volume systems are very flexible and can easily exhaust from different types of devices such as fume hoods, safety cabinets, etc. Since the airflow and pressure in a VAV system is constantly changing, the different pressure losses associated with each type of device is not a problem with VAV systems (as is the case with CAV systems). A disadvantage to VAV exhaust systems is that they can become very complex and difficult to air balance and control.

EXHAUST DUCTWORK AND STACKS

When designing or specifying ductwork, it is important to know the types of effluents or combinations of effluents that will be in the hood and exhaust ducts. Also important is the highest possible dry bulb and dewpoint temperature of the effluents. Specify the ambient temperature of the space in which the ductwork is installed, since the temperature affects the condensation of the vapors in the exhaust system. Condensation contributes to the corrosion of ductwork metals with or without the presence of chemicals. Designate if the fan will have continuous or intermittent operation. Intermittent operation can unbalance air flows and cause unsafe conditions in the laboratory. It may also mean longer periods of wetness due to condensation than would continuous operation. If intermittent operation is used, a one hour time delay should be specified to dry/wet surfaces before shutting the fan off. Determine the length and arrangement of duct runs. The longer the duct run, the longer the exposure to effluents and therefore, more condensation. If condensation is probable, provide sloped ductwork and condensate drains. Use caution—the condensate drains may accumulate hazardous materials.

Exhaust ductwork and exhaust stacks must be of adequate strength. Do not use flexible duct connectors in hidden spaces or with corrosive materials. Seal the duct on both the discharge and the suction sides of the fan. All penetrations in the duct for testing, cleaning, dampers, etc., must be sealed according to codes and specifications. Each hood from a separate laboratory must be separately ducted to the outside or to a shaft or mechanical space where the ducts may be connected to a common exhaust duct. Hoods within the same laboratory may be connected by ducting within that laboratory. Exhaust duct velocities should be maintained between 1,000 and 2,000 feet per minute.

Exhaust stacks should be placed on the highest roof of the building. If architectural enclosures are used, the stack should extend over the enclosure to prevent exhaust contamination of equipment within the enclosure. The stack should extend above any obstruction. The effective stack height is that part that extends above the local recirculation region and any upwind and downwind obstacles. Check local codes for stack height requirements.

When using smoke to adjust stack height, the height is changed until the lower edge of the smoke plume avoids contact with all recirculation, wake and high turbulence regions. The largest recirculation, wake and high turbulence regions occur when the wind is normal to the upwind wall of the building. Observe the smoke patterns on the roof, around air intakes, architectural screens, penthouses and the wake behind the building.

The stack should have a terminal velocity of about 2,500 feet per minute (fpm). If this stack velocity is not possible, increase the stack height. If short stacks must be used, velocities up to 8,000 fpm are generally needed to get the exhaust plume above the roof recirculation region. Although high discharge velocity increases plume rise and reduces the possibility of recirculating contaminated air back into the building, it is a costly substitute for increased stack height when viewed from an energy standpoint.

A stack velocity of 2,500 fpm prevents condensed moisture from draining down the stack and keeps rain from entering the stack. For intermittently operating exhausts, the stack needs a drain. Even with drains in the stack, velocities should still be maintained at 2,500 fpm to provide adequate plume rise and exhaust dilution. However, if the stack condensate is corrosive, the velocities should be about 1,000 fpm. A drain should be installed and the stack should have a converging cone to increase the discharge velocity to 2,500 fpm.

Do not use rain caps on the exhaust stack. They tend to deflect air downward, increasing the chances that contaminated air will be recirculated into the building. Rain caps have high friction losses and provide less rain protection than a properly designed stack. Direct all exhaust discharge away from any present or future air intake. Discharge exhaust air above the building envelope, not on the side of the building. Before discharging exhaust air, contamination should be reduced by filters, collectors or scrubbers as required by code.

To increase dilution of the exhaust effluent, use a central exhaust system and combine exhaust ducts into a single stack (Fig. 25.5). You could also use a central exhaust with variable air volume subsystems (shown in Fig. 25.6). Combining the ducts into a single large stack increases the momentum of the effluent and the exhaust plume will rise higher into the atmosphere. If ducts cannot be combined into a single stack, group individual exhaust stacks on the roof in a tight cluster. Clustering the stacks will also increase the plume rise. However, stacks lined up in a row will not act as a cluster. Combining ducts in a single stack (or clustering stacks) allows building air intakes to be placed as far away as possible from the exhausts.

AIRFLOW PATTERNS AROUND BUILDINGS AND INSIDE BUILDINGS

As the wind moves around a building, it accelerates as it curves upward over the roof and decelerates as it curves downward into the wake on the lee side of the building. The effectiveness of the wind around a building to remove exhaust effluents from the roof depends on a combination of the stack location, the stack height, the stack velocity, the concentration of the contaminants and the momentum of the exhaust effluent. This effectiveness also depends on the air turbulence created by the building itself and the atmospheric turbulence in

FIGURE 25.5.   COMBINING STACKS

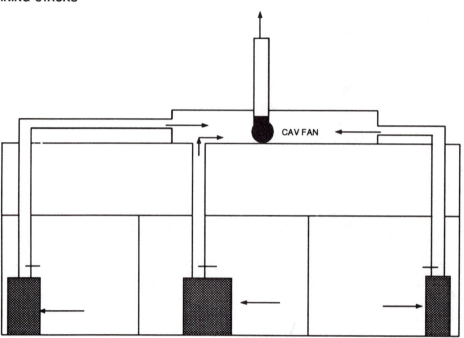

the wind approaching the building. Generally, the air flow above the roof recirculation area is affected by the presence of the building to a distance about 1.5 times the building height. Tests show that atmospheric turbulence is more intense in and near the recirculation areas that occur on the upwind edges of the building. Other tests show that atmospheric turbulence has a significant effect on exhaust removal when the building is less than 300 feet high, or is less than twice as high as the surrounding buildings. Atmospheric turbulence is

FIGURE 25.6.   COMBINING VAV FUME HOODS WITH A CONSTANT AIR VOLUME DISCHARGE

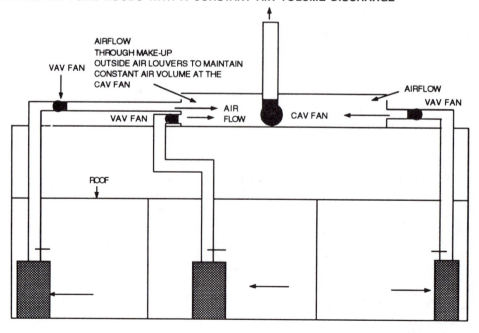

insignificant for buildings higher than 300 feet high, or when they are twice as high as the surrounding buildings.

Control of the direction and pattern of the air flow inside buildings is essential to controlling the spread of airborne contaminants. Airflow patterns are based on the type of materials handled in the laboratory and the permissibility of air transfer from one space to another. All laboratories should be designed to provide directional flow of air from areas of least contamination to areas of greatest contamination. Generally, this means that laboratory areas will be under negative pressure and nonlaboratory areas will be under a positive pressure. This is known as zone pressurization.

LABORATORY FUME HOOD TROUBLESHOOTING GUIDE

To help with troubleshooting the exhaust system, review the previous sections on system and component design and operation.

REVERSE AIRFLOW

Air Currents in the Room

Reverse airflow in a laboratory hood can be caused by air currents in the room. These currents can be produced by people walking past the hood, by opening and closing doors to the laboratory and by the air velocity and pattern from the supply air outlets. To reduce unwanted air currents, consider the following items:

- Move fume hoods at least 6 feet away from doors and active aisles.
- Install supply air devices to avoid air velocities and patterns that would adversely affect the performance of the hood.
- Use several small ceiling diffusers instead of one large diffuser to supply an area.
- Reduce supply outlet terminal velocity to 50 fpm or less.
- Move supply air outlets at least 4 feet from the hood face.
- Place sidewall outlets on adjacent walls instead of the wall opposite the fume hood face.
- Reduce the air flow from supply air outlets discharging directly into the fume hood.
- Change the pattern deflector in supply outlets to discharge away from the hood.
- Block part of the supply air outlet so it cannot discharge into the hood.
- Fix sidewall outlets to discharge air over the top of the fume hood.

Eddy Currents

Reverse airflow in a laboratory fume hood can be caused by eddy currents in the hood. These currents are caused by sharp edges, depressions, sinks, posts and service fittings in the hood. For instance, a plain, sharp edge at the entrance of a fume hood will produce turbulence about 1 inch above the work surface and about 6 inches into the hood. Fumes generated in this area will be disturbed and could possibly escape the hood. Eddy currents can also be created by some research operations using blenders, mixers, etc. To reduce eddy currents, consider the following measures:

- Install an airfoil on hoods with a plain sharp edged entrance.
- Sinks, service fittings, etc. should be placed at least 6 inches inside the hood face.
- Conduct all research operations at least 6 inches into the hood.

Reverse airflow in a laboratory hood could be caused by the *movements of the user* at the hood face. To check for this, do a smoke test of the hood while simulating user movement at the hood face. Correct conditions causing reverse flows. Reverse airflow in a laboratory hood could also be caused by *high face velocities*. To correct this, reduce laboratory fume hood face velocities to 150 fpm or less.

FACE VELOCITIES

Changes in the size of the face opening, the position of the baffle, the position of the volume damper, the ductwork, the exhaust fan speed, the filters or blockages of the air path all affect hood face velocity. Review the methods that will make a permanent change in the hood face velocity. If changes are made, take electrical power measurements to determine the fan motor efficiency and load. If needed, change the motor to increase efficiency or correct an overload condition.

To reduce face velocities, use any of the following methods that are appropriate:

- Slow the fan speed by decreasing motor sheave pitch diameter or increasing fan sheave pitch diameter. Change belts if needed. If the system is VAV and has a speed or static pressure controller, set the controller for a slower speed or a lower static pressure.
- Close the volume damper. If the system is VAV, reduce the air volume at the air valve.
- Open the sash on a conventional hood. If the hood is a bypass type or the system is VAV, the face velocity should not change.

To increase face velocities, use any of the following methods that are appropriate.

- Increase the fan speed by increasing motor sheave pitch diameter or decreasing fan sheave pitch diameter. Change belts if needed. If the system is VAV and has a speed or static pressure controller, set the controller for a faster speed or a higher static pressure.
- Open the volume damper. If the system is VAV, increase the air volume at the air valve.
- Close the sash on a conventional hood. If the hood is a bypass type or the system is VAV, the face velocity should not change.
- Remove any blockages in the air path. Visually inspect the baffle openings and behind the baffle. Take static pressures in the duct and across the filters and the fan. Visually inspect where required.
- Change ductwork to relieve restrictions and reduce duct friction and dynamic losses.

HVAC CONTROLS

General

Individual laboratories should have separate thermostats. Cooling and heating loads in laboratories are highly variable due to the operation of laboratory equipment. Air Volume Dampers must fail to the open position to assure continuous draft through the lab. Controls must be placed outside of any location where flammable or combustible materials are conveyed or created.

Testing Controls

When possible, simulate all control modes to ensure that room pressures and airflows are as specified. Verify the proper operation of all controls and monitoring devices. Verify that readings between sensor, readout and control devices are correct and properly calibrated. Verify that all controls operate in all modes in a minimum amount of time between sequences and without erratic behavior.

Controls to Maintain Room Pressurization and Temperature with Variable Air Volume Systems

The two basic methods to control room pressure and temperature in laboratories with variable air volume systems are *flow tracking* or *flow differential* (Fig. 25.7) and *differential space pressurization* (Fig. 25.8). The flow tracking method sets an air volume differential (e.g., 200 cfm) between the amount of air supplied to the lab and the amount exhausted from the lab. In other words, the lab room supply air is controlled to track at a value less than total room (hood and general) exhaust. With the differential space pressurization method, a room to corridor differential pressure controller maintains a set differential value (inches of water) between the lab and the corridor (or other appropriate space). The amount of lab room supply air is controlled by the differential pressure.

With either control scheme the fume hood exhaust is controlled by a hood face velocity sensor or a sash position sensor. The supply air temperature to the lab is usually between 55 and 60°F. The supply air VAV box has a reheat coil which is controlled by the room thermostat. A typical control sequence is:

- The hood sash is closed.
- The exhaust air velocity through the hood increases.
- The velocity sensor in the hood senses the increased face velocity and closes the fume hood exhaust VAV to reduce airflow and maintain the correct hood face velocity. (To this point the sequence has been the same for either control method. The differences in the control schemes follows.)

Flow Tracking

- The room exhaust VAV controller signals the supply VAV controller to reduce the volume of supply air into the room to maintain the correct volume differential (which in turn maintains the negative pressure in the room).

FIGURE 25.7.   FLOW TRACKING

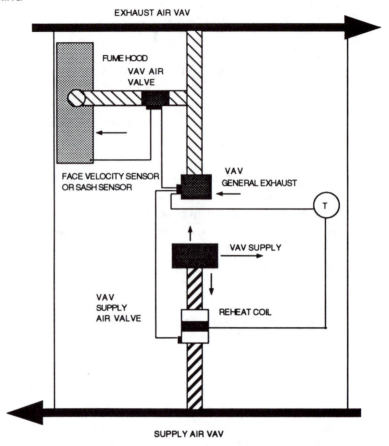

- The room temperature rises above thermostat setpoint because of he reduction in supply air.

- The room thermostat opens the general exhaust VAV which increases the total amount of air being exhausted.

- The room exhaust VAV controller signals the supply VAV controller to increase the volume of supply air into the room to maintain the correct volume differential (which in turn cools the room). If the room overcools, the general exhaust closes and the supply VAV closes to maintain the correct volume differential. If the room is still too cool, the reheat coil is energized.

Differential Space Pressurization

- The room pressurization sensor senses a decrease in differential pressure (there is an increase in room pressure because the hood is exhausting less air).

- The room pressurization controller signals the supply VAV to reduce the volume of air into the room to maintain the correct pressure differential.

- The room temperature rises above thermostat setpoint because of the reduction in supply air.

- The room thermostat opens the general exhaust VAV.

- The room pressurization sensor senses an increase in differential pressure (the room is becoming too negative because of the increased exhaust).

FIGURE 25.8.   DIFFERENTIAL SPACE PRESSURIZATION

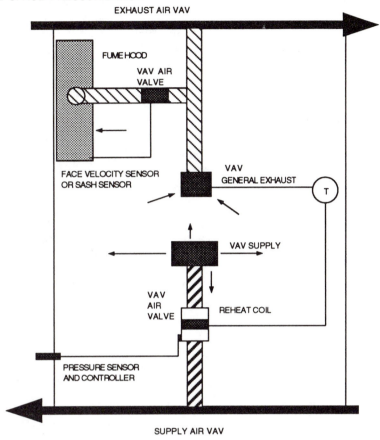

SUPPLY AIR VAV

- The room pressurization controller signals the supply air VAV controller to open allowing in more supply air (which reduces the negative pressure in the room and *cools* the room). If the room overcools, the general exhaust is closed and the supply air volume is reduced to maintain proper room differential pressure. If the room is still too cool, the reheat coil is energized.

OSHA STANDARD

In January 1990, the Occupational Safety and Health Administration (OSHA) published Standard 29 CFR Part 1910, *Occupational Exposures to Hazardous Chemicals in Laboratories*. This standard, which requires compliance within one year from the date of publication, defines a **laboratory** as a facility where relatively small quantities of hazardous chemicals are used on a nonproduction basis.

Laboratory use is the handling or use of such chemicals in which the chemical manipulations are carried out on a laboratory scale using multiple chemical procedures and/or chemicals. The procedures involved are not of a production process nor in any way simulate a production process. Adequate protective laboratory equipment is available and in common use to minimize the potential for employee exposure to hazardous chemicals. A *hazardous chemical* is a chemical for which there is significant evidence, based on at least one study conducted in accordance with established scientific principles, that acute or chronic effects may occur in exposed employees. Hazardous chemicals include carcinogens, toxic agents, irritants, corrosives and agents that damage the lungs, skin, eyes or mucous membrane.

This OSHA standard subscribes to the belief that laboratory workers are entitled to be kept informed as to the potential hazards of the chemicals with which they are working. It is the ultimate responsibility of management to minimize worker exposure to hazardous chemicals. With this in mind, the standard incorporates a plan called the *chemical hygiene plan*. The plan requires that employers establish, enforce and distribute to all laboratory workers a complete set of appropriate work practices and procedures to minimize the potential exposure to hazardous chemicals. The plan requires that fume hoods and other protective equipment be functioning properly and that specific measures shall be taken to ensure proper and adequate performance of such equipment. The plan also requires that workers be trained in the use of containment devices such as fume hoods.

The plan states that the best way to prevent exposure to airborne contaminants is to provide adequate ventilation. It also states the best way to prevent the escape of airborne substances into the work place is by the use of hoods and other ventilation devices. For instance:

- Quantity and quality of the ventilation and exhaust system should be evaluated on installation, and every three months thereafter.

- Quantity and quality of the ventilation and exhaust system should be reevaluated whenever there is a change in the ventilation or exhaust system.

- A laboratory hood with 2.5 linear feet of hood space should be provided for every two workers if they spend most of their time working with chemicals.

- Hood face velocity should be adequate.

- Confirm adequate hood performance before use.

- Keep hood closed at all times except when adjustments within the hood are being done.

- Keep materials stored in the hood to a minimum and do not allow them to block vents or air flow.

- Leave the hood on (when it is not actively used) if toxic substances are stored in it, or if it is uncertain whether adequate general laboratory ventilation will be maintained if the fan is off.

- Each hood should have a continuous monitoring device.

COMPANY LABORATORY POLICY

The following is an outline for a company policy for operation of a fume hood laboratory.

Standards are set for:

Laboratory design

Fume hood design

Fume hood operation and maintenance

Supply and exhaust system operation and maintenance

Additions or modifications to systems:

Ensure that there is adequate capacity

Ensure proper location of containment devices

Ensure that changes meet company design and safety standards

Training

Conduct ongoing training programs

Schedules are set for:

Inspection of ventilation and exhaust systems and laboratory fume hoods

Clutter in the hood

Corrosion in the hood or in the laboratory

Modifications to the hood or systems

Proper maintenance

Testing of systems and hoods (every 3 months)

Face velocities

Operation of visual and audible alarms

Smoke and dry ice tests

Static pressure tests

Policies are set to monitor laboratory operation and safety procedures:

The sash should be fully closed when hood is not attended.

The hood user should work at least 6 inches back in the hood (a stripe on the bench surface is a good reminder).

User should not put head into the hood when contaminants are being generated.

The hood should not be storing hazardous chemicals—store hazardous chemicals in a safety cabinet.

The hood should not be used as a waste disposal for hazardous chemicals.

Do not remove sash or panels except when necessary for apparatus set up or removal.

Replace sash or panels when changes are completed.

Do not place electrical receptacles or other spark producing sources inside the hood when flammable materials are present.

Do not allow permanent electrical receptacles in the hood.

Minimize foot traffic in the lab.

Keep lab doors closed unless designed to stay open.

Hoods for special research should be labeled as such.

Put up appropriate signs and barricades as needed.

To comply with OSHA's requirements to have a continuous monitoring device to allow convenient confirmation of adequate hood performance, every new and existing hood should have an air flow monitor installed. Use a low flow alarm with an audible and visual annunciation. Use either a direct reading, through-the-wall, intrinsically safe mass flow device that measures the face velocity, or an indirect reading cabling system that attaches to the sash and then to a potentiometer that measures the distance the sash moves.

APPENDIX A Terms

GENERAL TERMS

Absolute Pressure: Total or true pressure.

Absorb: To retain wholly, without reflection or transmission, that which is taken in. To take in, soak up.

Accelerated Motion: Changing velocities.

Acceleration: The time rate of change in velocity.

Adsorb: The taking up of one substance at the surface of another.

Adiabatic: Without loss or gain of heat. A process in which a gas is assumed to change its condition (pressure, volume and temperature) without the transfer of heat to or from the surroundings during the process.

Atmosphere: Normal barometric pressure at sea level, 14.7 psi, 29.92 inches of mercury.

Boiling: Occurs only at the saturation temperature and throughout the entire body of the liquid.

Btu: British thermal unit. The quantity of heat required to raise one pound of water 1°F.

Change of State: A change from solid to liquid, liquid to vapor, solid to vapor or vice versa.

Cold: Cold is a relative term to describe the temperature of an object or area compared to a known temperature. For instance, an outside air temperature of 50°F in the winter might be considered a warm temperature, while in the summer it would be a cool temperature.

Comfort Zone: The range of effective temperatures and humidities over which the majority of adults feel comfortable. Generally, between 68°F to 79°F and 20% to 60% relative humidity.

Condensation: Removing heat from a vapor so that it changes to the liquid state. Increasing the pressure on a vapor so that it changes to the liquid state. Removing heat from a vapor while increasing pressure on the vapor so that it changes to the liquid state.

Conduction: Heat energy is transmitted by direct contact.

Constant Speed: Unchanging rate of motion without regard for the direction of motion.

Constant Velocity: Unchanging rate of motion and unchanging direction of motion.

Convection: Heat that moves from one place to another by means of currents that are set up within some fluid medium, vapor or liquid.

Density: Mass per unit volume. Density = mass/volume.

Efficiency: Useful energy output divided by the power input.

Enthalpy: Total heat content. Thermodynamic property of a working substance. Enthalpy = Internal Energy + Pressure × Volume).

Evaporation: Occurs only at the surface of the liquid at any temperature below the saturation temperature. Heating of a liquid to convert it to a vapor.

Fluid: A liquid or a vapor. A vapor is a compressible fluid and a liquid is a noncompressible fluid.

Fluid Dynamics: The condition of a fluid in motion. The velocity of a fluid is based on the cross-sectional area of the conduit and the volume of fluid passing through the conduit.

Force: A force is a push or pull. A force is anything that has a tendency to set a body in motion, to bring a moving body to rest, or to change the direction of motion. Force = Mass × Acceleration.

Gauge Pressure: Pressure indicated by the gauge.

Heat of Compression: Mechanical energy of pressure transformed into heat energy.

Heating, Ventilating and Air Conditioning (HVAC): Heating, ventilating and conditioning a space using the fluids of air, water and refrigerants.

Hygroscopic: Water absorbing.

Isenthalpic: Of a process carried out at a constant enthalpy.

Isentropic: Of a process carried out at a constant entropy.

Isothermal: Occuring at a constant temperature.

Latent Heat: Hidden heat. Heat energy that causes or accompanies a change in the phase of a substance, but no change in temperature. Heat that causes a change in state of a substance. Expressed in Btu/lb.

Latent Heat of Fusion: Latent heat when a solid changes to a liquid or vice versa.

Log Mean Temperature Difference: The logarithmic average of the temperature differences when two fluids are used in a heat transfer process. The temperature difference after the process will be less than at the beginning. The exchange of heat will follow a logarithmic curve. The higher the LMTD number, the greater the heat transfer.

Mass: Weight of a body in pounds. The quantity of matter in a body measured by the ratio of the force required to produce given acceleration. Mass = Force/Acceleration. Mass = Volume/Specific Volume. Mass = Volume × Density.

Mass Flow Rate: Pounds per unit of time.

Mechanical Work: A force acting on a body moves the body some distance. Work = Force × Distance. Expressed in ft/lb.

Nonuniform Flow: Fluid flow varying in velocity across the plane perpendicular to flow.

Occupied Zone: The conditioned space from the floor to about six feet above the floor.

Optimize: To make the most efficient use of.

Power: The time rate of doing work. Power = Work/Time.

Pressure: Pressure is the force exerted per unit of area. Pressure = Force/Area.

Radiation: Heat energy is transmitted from one body to another without the need of intervening matter. Moves in waves.

Scalar: Magnitude only. Compare to vector.

Sensible Heat: Heat energy that causes or accompanies a change in temperature.

Specific Enthalpy: The enthalpy of one pound of a substance.

Specific Gravity: Also called relative density. The ratio of the density of the substance to some standard substance such as water. Water is 62.4 lb/cf. Specific Gravity = Density/Density of Water. The ratio of the mass of a substance to the mass of an equal volume of water at 4°C. Water at standard conditions has a specific gravity of 1.0. For temperatures between freezing (0°C) and boiling (100°C), a specific gravity of 1.0 is used.

Specific Heat: The amount of heat required to raise one pound of any substance one degree Fahrenheit (1.00 for water, 0.24 for air) Expressed in Btu.

Specific Volume: Volume per unit mass.

Standard Acceleration of Gravity: The speed of a body falling freely toward the earth through the action of gravity alone. Equal to 32.2 feet per second squared.

Throttling Characteristic: The relationship of the position of the valve disc or damper blade and its percent of flow. A valve or damper has a linear throttling characteristic when the disc or damper blade open percentage is the same as the flow percentage.

Total Enthalpy: The enthalpy of the entire mass of a substance.

Total Heat: Sensible heat plus latent heat.

Torque: The force which produces or tends to produce rotation. Measured in foot-pounds.

Turbulent Flow: Fluid flow in which the velocity varies in magnitude and direction in an irregular manner.

Uniform Flow: The smooth, straightline motion of a fluid across the area of flow.

Vapor (Dry Saturated): A saturated vapor completely free of liquid particles.

Vapor (Wet): A vapor with liquid particles.

Vaporization: Changing from a liquid to a vapor. Occurs by boiling or evaporation. Evaporation takes place only at the surface of a liquid and at any temperature below its saturation temperature. Boiling takes place throughout the liquid, but only occurs at the saturation temperature. Vaporization is accomplished by adding heat to a liquid, decreasing the pressure on the liquid or a combination of the two.

Vapor Pressure: The vapor pressure of a liquid at any given temperature is that pressure necessary to keep the liquid from boiling or flashing into a vapor.

Vector: A quantity that has magnitude and direction.

Velocity: Speed or rate of motion in feet per second or feet per minute. Velocity = Distance/Time.

Volume: Space taken up by a body. Expressed in cubic feet or cubic inches.

$$\text{Volume} = \frac{\text{Mass}}{\text{Density}}$$

$$\text{Volume} = \text{Mass} \times \text{Specific volume}$$

Volume Flow Rate: Cubic feet or cubic inches per unit of time.

Work: Work is force through a distance. Expressed in ft/lbs. Work = Force × Distance.

AIR TERMS

Air Changes Per Hour: A method of expressing the quantity of air exchanged per hour in terms of conditioned space volume.

Air Conditioning: Treating or conditioning the temperature, humidity and cleanliness of the air to meet the requirements of the conditioned space.

Air Entrainment: The induced flow of the secondary (or room) air by the primary (or supply outlet) air, creating a mixed air path.

Airflow Patterns: Airflow patterns are important for proper mixing of supply and room air. Cooled air should be distributed from ceilings or high sidewall outlets. The airflow pattern for cooled air distributed from low sidewall or floor outlets should be adjusted to direct the air up. Heated air should be distributed from low sidewall or floor outlets. The airflow pattern for heated air distributed from ceilings or high sidewall outlets should be directed down.

Since most HVAC systems handle both heated and cooled air, as the seasons change, a compromise is made. It usually favors the cooling season and ceiling or high sidewall outlets are specified. Generally, these outlets will be adjusted for a horizontal pattern. However, if the ceiling is very high or velocities are very low, such as when variable air volume boxes close off, it may be necessary to use a vertical pattern to force the air down to the occupied zone.

Ak Factor: The effective area of an air outlet or inlet. *See Effective Area.*

Ambient Air: The surrounding air.

Ambient Air Temperature: The temperature of the surrounding air.

Aspect Ratio: The ratio of duct width to duct height. Aspect ratio should not exceed 3:1.

Cold Deck: In a multizone or dual duct unit, the cold deck is the chamber after the air leaves the cooling coil.

Cubic Feet Per Minute: Air flow volume.

Diversity in Constant Air Volume Systems: The total output of the fan is greater than the maximum required volume through the cooling coil.

Diversity in Variable Air Volume Systems: The total volume (cubic feet per minute) of the VAV boxes is greater than the maximum output of the fan.

Draft: A localized feeling of coolness caused by high air velocity, low ambient temperature or direction of air flow.

Drop: The vertical distance that the lower edge of a horizontally projected air stream drops between the outlet and the end of its throw.

Dual Path: A system in which the air flows through heating and cooling coils essentially parallel to each other. The coils may be side-by-side or stacked. Multizone and dual duct systems are dual path. Some systems may not have a heating coil, but instead bypass return air or mixed air into the hot deck.

Dumping: The rapidly falling action of cold air caused by a variable air volume box (or other device), reducing airflow velocity.

Effective Area: The sum of the areas of all the vena contractas existing at the outlet. Is affected by the number of orifices and the exact location of the vena contractas, and the size and shape of the grille bars, diffuser rings, etc. Manufacturers have conducted air flow tests and, based on their findings, they've established flow factors or area correction factors for their products. Each flow factor, sometimes called "K-factor" or "Ak," applies to a specific type and size of grille, register or diffuser, a specific air measuring instrument and the correct positioning of that instrument.

Equivalent Duct Diameter: The equivalent duct diameter for a rectangular duct.

Face Velocity: The average velocity of the air leaving a coil, supply air outlet or entering a return or exhaust air inlet.

Fixed Duct System: A system in which there are no changes in the system resistance resulting from closing or opening of dampers, or changes in the condition of filters or coils. For a fixed system, an increase or decrease in system resistance results only from an increase or decrease in cubic feet per minute. This change in resistance will fall along the system curve.

Feet Per Minute: Air velocity.

High Pressure Systems: Static pressures above 6 in. wg with velocities above 2,000 feet per minute.

Hot Deck: In a multizone or dual duct unit, hot deck is the chamber after the air leaves the heating coil.

Laminar Air Flow: Air flow in parallel flow lines with uniform velocity and minimum eddies.

Low Pressure System: Static pressures to 2 in. wg with velocities to 2,000 feet per minute.

Make-up Air: Air introduced into the secondary air system for ventilation, pressurization, and replacement (or "make-up") of exhausted air.

Medium Pressure Systems: Static pressures between 2 and 6 in. wg with velocities between 2,000 and 4,000 feet per minute.

Mixed Air: Primary air plus secondary air.

Outlet Velocity: The average velocity of air emerging from a fan, outlet or opening.

Plenum: An air chamber or compartment.

Primary Air: The supply air.

Radius of Diffusion: The horizontal distance an air stream travels after leaving the outlet before it's reduced to its specified terminal velocity.

Residual Velocity: Room velocity.

Room Velocity: The air velocity in the room's occupied zone.

Secondary Air: The room air.

Single Path: A system in which the air flows through coils, essentially in series to each other. Single zone heating and cooling units and terminal reheat units are examples of a single path system.

Smudging: The black markings on ceilings and outlets usually made by suspended dirt particles in the room air, which is then entrained in the mixed air stream and deposited on the ceilings and outlets. Anti-smudge rings are available which lower the outlet away from the ceiling and cover the ceiling area a few inches beyond the diffuser.

Spread: The divergence of the air stream after it leaves the outlet.

Stratification: Layers of air at different temperatures or different velocities flowing through a duct or plenum. Also called stratified air.

Surface Effect: The effect caused by entrainment of secondary air when an outlet discharges air directly parallel and against a wall or ceiling. Surface effect is good for cooling applications, especially variable air volume systems, because it helps to reduce the dumping of cold air. Surface effect contributes to smudging.

Terminal Velocity: The maximum air velocity of the mixed air stream at the end of the throw.

Throw: The horizontal and vertical distance an air stream travels after leaving the supply air outlet before it's reduced to its specified terminal velocity.

Vena Contracta: The smallest area of an air stream leaving an orifice.

Ventilation: Supplying air to, or removing air from, a space by natural or mechanical means.

CLEAN ROOM TERMS

Class 10 Clean Room: A class 10 clean room will have an airflow at 90 fpmn or more based on gross ceiling area. The particle count in the clean room can not exceed 10 particles (particle size 0.5 μm to 5.0 μm) per cubic foot of air.

Class 100 Clean Room: A class 100 clean room will have an airflow at 90 fpm +/- 20 fpm based on gross ceiliiing area. The particle count in the clean room can not exceed 100 particles per cubic foot of air. Particle sizes 05. μm and larger.

Class 1,000 Clean Room: A class 1,000 clean room will have an airflow at 60 to 90 air changes per hour. The particle count in the clean room can not exceed 1000 particles per cubic foot of air. Particle sizes 0.5 μm and larger.

Class 10,000 Clean Room: A class 10,000 clean room will have an airflow at 50 to 60 air changes per hour. The particle count in the clean room can not exceed 10,000 particles per cubic foot of air. Particles sized 0.5 μm and larger. Also for class 10,000 rooms the particle count can not exceed 65 particles per cubic foot if the particle size is 5.0 μm and larger.

Class 100,000 Clean Room: A class 100,000 clean room will have an airflow at 20 to 30 air changes per hour. The particle count in the clean room can not exceed 100,000 particles per cubic foot of air. Particles sized 0.5 μm and larger. Also for class 100,000 rooms the particle count can not exceed 700 particles per cubic foot if the particle size is 5.0 μm and larger.

Clean Room: A specially constructed enclosed area that is environmentally controlled with respect to airborne particulates, temperature, humidity, air flow patterns, air pressure, air motion, vibration, noise and lighting.

Critical Surface: The surface to be protected from particulate contamination.

Design Conditions: The environmental conditions for which the clean room is designed.

DOP Test: A test using the oil dioctyl phthalate (DOP) to determine the effectiveness of air filters.

First Air: The air coming directly from the HEPA filter before it passes over any work location.

First Work Location: The location in the path of the first air stream.

Operational Conditions: The environment that exists within a clean room.

Particle Size: The maximum linear dimension of a particle.

Work Station: An open or closed work surface with direct air supply. A work station may be classified as a laminar (air) flow work station or a nonlaminar (air) flow work station.

CONTROL TERMS, GENERAL

Actuator: The device which receives the signal from the controller and positions the controlled device. Also called a motor or operator.

Alarm: A signal, either audible, visible or both, that warns of an abnormal and critical operating condition.

Alert: A form of alarm that warns of an abnormal, but not critical, operating condition.

Ambient Compensated: A control device designed to allow for ambient air temperatures.

Analog Signal: A type of signal whose level varies smoothly and continuously in amplitude or frequency. Traditionally, HVAC control has been performed by analog devices such as pneumatic controllers, transducers, relays and actuators.

Authority: The adjustment of a controller which determines the effect of the secondary signal as a percentage of the primary signal.

Averaging Element: A temperature sensing element that responds to the average temperature sensed at many different points in an air stream.

Closed Loop System: A control system in which information from the output is fed back and compared to the input to generate an error signal. This error signal is then used to generate the new output signal.

Control Loop: The basic control loop is comprised of a controlled variable, sensing element, controller, actuator and controlled device.

Control Point: The actual temperature, pressure or humidity which the controller is sensing.

Controlled Device: A flow control device, such as a damper for air control or a valve for water or steam control.

Controlled Medium: The fluid, air, water or steam whose controlled variable is being sensed and controlled.

Controlled Variable: The condition or quantity of the controlled medium—such as temperature, humidity, pressure or flow rate—being controlled.

Controller: A proportioning device designed to control

dampers or valves to maintain temperature (thermostat), humidity (humidistat), or pressure (pressurestat), or to control other controllers (master/submaster controller).

Cut-in: A control value to start a device, such as a compressor or boiler.

Cut-out: A control value to stop a device.

Deadband: The control set point when neither heating nor cooling occurs.

Deviation: The difference between the set point and the value of the controlled variable at any instant. Also called offset of error.

Differential: The difference between the cut-in value and the cut-out value. For example, if the cut-in temperature is 55°F and the cut-out temperature is 40°F the differential is 15°F.

Direct-Acting: A direct-acting controller increases its branch output as the condition it's sensing increases.

Economizer Control: A control system for the changeover between natural cooling, with outside air and refrigerated mechanical cooling.

Enthalpy Control: Control devices which compare the enthalpy of the return air with the enthalpy of the outside air, and then position the economizer dampers to admit the air with the lowest enthalpy.

Fail Safe: A design in which power supply or control is able to return to a safe condition in the event of failure or malfunction by automatic operation of protective devices. In HVAC, a fail safe system is a damper or valve which will go to its normal position to minimize damage to the HVAC equipment and components in case of a control failure. This could include placing heating valves open and humidifier valves closed upon loss of control power. In EMCS terminology, a fail safe system means returning all controlled devices to conventional control in case of load management panel failure.

Feedback: The transmission of information about the results of the control action.

Firestat: A device in the air handling unit or ductwork to shut down air conditioning or ventilating fans when air temperature goes above a preset limit.

Floating Control: With floating control, the controller moves the controlled device at a constant speed toward either its open or closed position. A neutral zone between the two positions allows the controlled device to stop at any position whenever the controlled variable is within the controller's differential. When the controlled variable is outside the differential, the controller moves the controlled device in the proper direction.

Freezestat: A device in the unit or ductwork to protect

against coil freeze up when the air temperature drops below a preset limit.

High Limit Control: A limit control that monitors the condition of the controlled medium and prevents the operation of equipment when it would cause dangerous or undesirably high temperatures, pressures or relative humidity.

Hunting: An undesirable condition where a controller is unable to stabilize the state of the controlled medium and cycles rapidly. The condition which happens when a controller changes or cycles continuously, resulting in fluctuation and loss of control. The desired set point condition can't be maintained.

Lag: A time delay between an initiating action and a desired effect. The delay in action of the sensing element of a control, due to the time required for the sensing element to reach equilibrium with the variable being controlled.

Limit Control: A temperature, pressure, humidity, dew point, or other control that is used as an override to prevent undesirable or unsafe conditions in a controlled system.

Limit Shutdown: A condition in which the system has been stopped because a preestablished limit value has been exceeded.

Low Limit Control: A limit control that monitors the condition of the controlled medium and interrupts system operation if the temperature, pressure or relative humidity drops below the desired minimum value.

Modulating Control: Tending to adjust by increments and decrements. A mode of automatic control in which the action of the controlled device is proportional to the deviation, from set point, of the controlled variable. *See Proportional Control.*

Morning Warm Up: A control system that keeps outside air dampers closed until a desired space temperature is achieved.

Night Setback: The control scheme which incorporates various control functions such as lowering space temperature, closing outside air dampers and intermittently operating blowers to reduce heating expense during unoccupied hours.

Normally Closed (NC): See *Normally Closed under Pneumatic and Electric/Electronic terms.*

Normally Open (NO): See *Normally Open under Pneumatic and Electric/Electronic terms.*

Offset: A sustained deviation between the actual control point and the set point under stable operating conditions.

On-Off Control: A simple control system, consisting basically of a switch, in which the device being con-

trolled is either on or off and no intermediate positions are available.

Open Loop System: A control system whose output is a function of only the input to the system.

Operator: See *Actuator.*

Override: A manual or automatic action taken to bypass the normal operation of a device or system.

Overshoot: The tendency to overcompensate for an error condition causing a new error in the opposite direction.

Process Plant: The controlled variable conditioning apparatus such as a coil, boiler, chiller, fan or humidifier being controlled.

Proportional Band: The range of values of a proportional controller through which the controlled variable must pass to move the final control element through its full operating range. The ratio of the controller throttling range to the sensor span. Also called modulating range. *See Throttling Range.*

Proportional Control: Proportional control is when a controlled device is positioned proportionally in response to changes in a controlled variable. The control signal is based on the difference between an actual condition and a desired condition. The difference in the conditions is called the "error" or the "offset." Some small amount of offset is always present. The controller creates an output signal related directly to the error's magnitude.

Range: The range on a control is the difference between the cut-in value and the cut-out value but it is not the same as the "differential." For example, if the cut-in temperature is 55°F and the cut-out temperature is 40°F, the differential is said to be 15°F. The range is said to be between 40°F and 55°F.

Relay: See *Relay under Pneumatic and Electric/Electronic terms.*

Reset: A process of automatically adjusting the set point of a given controller to compensate for changes in another variable. For example, in HVAC, the hot deck control point is normally reset upward as the outdoor temperature decreases, while the cold deck control point is normally reset downward as the outdoor temperature increases.

Reset Ratio: The ratio of change in outdoor temperature to the change in control point temperature. For example, a 2:1 reset ratio means that the control point will increase 1° for every 2° change in outdoor temperature.

Reverse-Acting: A reverse-acting controller decreases its branch output as the condition it's sensing increases.

Sensing Element: The sensing element is the component that measures the value of the controlled variable.

Sensitivity, Controller: The change in the output of a controller per unit change in the controlled variable. A pneumatic controller with a 12 psi range and a 4° throttling range has a sensitivity of 3 psi/°F.

Sensitivity, Sensor or Transmitter: The change in pressure in a remote sensor or transmitter per unit change in the sensed medium. All transmitters have a pressure range of 12 psi. A transmitter with a span of 100°F will have a sensitivity of 0.12 psi/°F.

Sensor: *See Sensing Element.*

Span: The difference between the lowest possible set point and the highest possible set point.

Setback: Reduction of heating or cooling at night or during hours when a building is unoccupied. *See Night Setback.*

Set Point: The point at which the controller is set to maintain the desired temperature, pressure or relative humidity.

Smoke Detector: A device in the air handling unit or ductwork to shut down air conditioning or ventilating fans when smoke is sensed.

Transducer: A device which converts signals from one physical form to another, such as mechanical to electronic.

Two Position Control: In two position control, the controlled device can only be positioned to either a minimum or maximum condition or an on or off condition.

Throttling Range: The change in the controlled condition necessary for the controller output to change over a certain range. For example, if a pneumatic thermostat has a 4° throttling range, the thermostat's branch line output will vary from 3 to 15 psi over a 4° change in temperature.

CONTROL TERMS, ELECTRIC AND ELECTRONIC

Adjustable Differential: The ability to change the difference between the cut-in and cut-out points.

Algorithm: A set of rules which specify a sequence of actions to be taken.

Amplifier: A device used to increase signal power or amplitude.

Analog: A type of signal whose level varies smoothly and continuously in amplitude or frequency. A control signal used to position actuators or actuate transducers or relays.

Analog Input: An electrical input of variable value provided by a sensing device such as temperature, humidity or pressure.

Analog Point: A point that has a variable value, such as temperature, humidity or pressure, which will be measured by a sensing device to provide input to the control.

Analog-to-Digital: The conversion of a sampled analog signal to a digital code that represents the voltage level of the analog signal at the instant of sampling.

Analog-to-Digital Converter: The part of a microprocessor based controller that changes analog input values to a digital number for use in a software program.

Anticipating Control: *See Anticipator.*

Anticipator: A method of reducing the operating differential of the system by adding a small resistive heater inside the thermostat to raise the internal temperature of the thermostat faster than the surrounding room temperature. This causes the thermostat to shut off the heating equipment and start the cooling equipment sooner than it would if affected only by the room temperature.

Automatic Control: A system that reacts to a change or unbalance in the controlled condition by adjusting the variables, such as temperature, pressure or humidity, to restore the system to the desired balance.

Auxiliary Authority: The amount of measured variable change at one sensing element compared to the change measured at the other sensing element, which causes the output signal to return to its original intermediate position, expressed as a percentage.

Auxiliary Contacts: A secondary set of electrical contacts mounted on a modulating motor or magnetic starter whose operation coincides with the operation of the motor or starter. Usually for pilot duty, low amp rating.

Auxiliary Potentiometer: A potentiometer mounted on a modulating motor. Used to control other modulating devices in response to the motor's operation. Also known as "follow-up pot."

Auxiliary Switch: A switch known as an "end switch."

Capacitor: An electrical component that stores electrical energy and prevents rapid changes of voltage across its terminals.

Central Processing Unit (CPU): The central control element of a computer.

Chip: *See Integrated Circuit.*

Comparator: A device whose output is a digital 0 or 1 depending on whether the input signal is above or below a given reference point.

Contactors: Electro-mechanical devices that "open" or "close" contacts to control motors.

Derivative Control: A control "braking" action to eliminate overshoot by anticipating the convergence of actual and desired conditions. This, in effect, counteracts the control signal produced by the proportional and integral control actions.

Digital: Representation of a numerical quantity by a number of discrete signals (not continuous) or by the presence or absence of signals in particular positions. Binary digital signals have one of two states (0 or 1) defined by voltage or current levels.

Digital-to-Analog: The conversion of a digital code into its equivalent analog signal level.

Digital-to-Analog Converter: The part of a microprocessor-based controller that changes digital values from a software program to analog output signals for use in the control system.

Digital Point: A point that has an "either/or" value, such as on/off, which will be sensed to provide direct input to the energy management system.

Diode: An electronic device that conducts current in one direction only.

Direct Digital Control: The sensing and control processes directed by digital control electronics. A control loop in which a digital controller periodically updates a process as a function of a set of measured control variables and a given set of control algorithms.

Electronic Controller Output Signal: The output signal from electronic controllers is direct current voltage (vdc). Typically the range is 3 volts from 6 to 9 vdc with a 7.5 vdc midpoint. In the following example, a controller that has a 6° throttling range from 72°F to 78°F will have a setpoint temperature of 75°F. The output signal will be as follows, depending on whether the controller is direct-acting or reverse-acting.

Temperature	D/A Output	R/A Output
72°F	6.0 vdc	9.0 vdc
75°F	7.5 vdc	7.5 vdc
78°F	9.0 vdc	6.0 vdc

Energy Management System: A system based on a microprocessor, microcomputer or minicomputer whose primary function is the controlling of energy using equipment so as to reduce the amount of energy used. Also called Energy Management Control System.

Gate: An electronic decision making circuit. An electronic circuit that performs a logical function (such as AND or OR) and passes signals when permitted by another independent source.

Hard Wiring: Any permanent wiring.

"IN" Contacts: Those relay contacts which complete circuits when the relay armature is energized. Also referred to as "Normally Open Contacts."

Integral Control: A control action designed to eliminate the offset in proportional control.

Integrated Circuit: A device in which all the components of an electrical circuit are fabricated on a single piece of semiconductor (silicon) material called a "chip."

Ladder Logic Diagram: The diagram that is used to describe the logical interconnection of the electrical wiring of control systems.

Light Emitting Diode: A low current and voltage light used as an indicator.

Limit: Control applied in the line or low voltage control circuit to break the circuit if conditions move outside a preset range. In a motor, a switch that cuts off power to the motor windings when the motor reaches its full open position.

Line Voltage: In the control industry, the normal electric supply voltages, which are usually 120 or 208 volts.

Load: An electric or electronic component or device that uses power.

Logic: The decision making circuitry on an integrated circuit.

Low Voltage: In the control industry, a power supply of 25 volts or less.

Magnetic Relay: Solenoid operated relay or contact; a switching relay that utilizes an electromagnet (solenoid) and an armature to provide the switching force.

Magnetic Starter: Contactors with overload protection relays. Sometimes called "mags" or "mag starters."

Manual Starter: A motor-rated switch that has provisions for overload protection. Generally, limited to motors of 10 horsepower or less.

Memory: A computer subsystem used to store instructions and data.

Microprocessor: A multifunctional integrated circuit that contains logic functions. A small computer used in load management to analyze energy demand and consumption so that loads are turned on and off according to a predetermined program.

Microprocessor-based Controller: A device consisting of a microprocessor unit, digital inputs and outputs, A/D and D/A converters, a power supply and software to perform direct digital, programmable logic control.

Modulating Motor: A reversible electric motor, used to drive a damper or valve, which can position the damper or valve anywhere between fully open or fully closed in proportion to a deviation of the controlled variable.

Motor Controller: Controllers are used for starting and stopping industrial motors. They can be grouped into three categories: manual starters, contactors and magnetic starters.

Multistage Thermostat: A temperature control that sequences two or more switches in response to the amount of heating or cooling demand.

Normally Closed: In relays, normally closed contacts are closed when the relay is deenergized.

Normally Open: In relays, normally open contacts are open when the relay is deenergized.

Optimum Start/Stop: A refined form of HVAC control that automatically adjusts the programmed start/stop schedule depending on inside and outside air temperature and humidity, resulting in the latest possible start and earliest possible stop of the HVAC equipment.

Pilot Duty Relay: A relay used for switching loads, such as another relay or solenoid valve coils. The pilot duty relay contacts are located in a second control circuit. Pilot duty relays are rated in volt-amperes (VA).

Potentiometer: An electro-mechanical variable resistance device having a terminal connected to each end of the resistive element, and third terminal connected to the wiper contact. The electrical input is divided as the contact moves over the element, thus making it possible to mechanically change the resistance.

Power Line Subcarrier: A device to allow the use of a building's existing electrical power system to carry the signals of the energy management system.

Processor: *See Central Processing Unit.*

Programmable Read Only Memory (PROM): A type of memory whose locations can be accessed directly and read. The operator can change parameters such as set points, limits and minimum off times within the control routines, but the program logic cannot be changed without replacement of the memory chips.

Proportional-Integral-Derivative: The type of action used to control modulating equipment such as valves, dampers and variable speed devices. *See Proportional Control, Integral Control and Derivative Control.*

Random Access Memory (RAM): A type of memory that can be read from or written into, and in which any location can be accessed directly and as fast as any other location.

Read Only Memory (ROM): A type of memory whose locations can be accessed directly and read, but cannot be written into.

Relay: An electro-mechanical device with a coil and an isolated set of contacts that opens or closes contacts in response to some controlled action. Relay contacts are normally open (NO) and normally closed (NC).

Resistance Temperature Detector (RTD): A sensing element whose resistance varies significantly and predictably with temperature.

Resistor: An electronic component that slows current. The current is proportional to the voltage applied and to the electrical conductivity of the base material.

Sequencer: An electronic device that may be programmed or set to initiate a series of events and make the events follow in sequence.

Sequencing Control: A control that energizes successive stages of heating or cooling equipment as its sensor detects the need for increased heating or cooling capacity. May be electronic or electro-mechanical.

Shielded Cable: Special cable used with equipment that generates a low voltage output. Used to minimize the effects of frequency "noise" on the output signal.

Silicon Controlled Rectifier (SCR): A three terminal electronic semiconductor switching device.

Step Controller: An electro-mechanical device used with electric or pneumatic systems which may be set to initiate a series of events and to make the events follow in sequence.

Switching Relay: General purpose switching relays are used to increase switching capability and isolate electric circuits.

System-level Controller: A microprocessor-based controller that controls centrally located HVAC equipment such as air handlers, chillers, etc. These controllers typically have an expandable input/output device capability, a library of control programs and may control more than one mechanical system from a single controller.

Timed Two-position Control: Timed two-position control is a variation of straight electrical two-position action. It is typically used in room electric/electronic thermostats to reduce operating differential. An anticipator control is used in the electric thermostat. The electronic thermostat uses a programmable clock to set the minimum on and off times.

Thermistor: A sensing element whose resistance varies significantly and predictably with temperature.

Thermocouple: A junction between two dissimilar materials that produces a voltage that is a function of the temperature of the junction.

Time Based Scheduling: The process of scheduling electrical loads on and off based on the time of day, the day of the week, month and the date.

Time Clock: A mechanical, electrical or electronic timekeeping device connected to electrical equipment for the purpose of turning the equipment on and off at selected times.

Voltage Range: The change in the voltage necessary for the controller output to change over the throttling range. For example, a 6 to 9 volt system will have a 3 VR.

Watt Transducer: A device that converts a current signal into a proportional millivolt signal. Used to interface between current transformers and a load management panel.

Zone Level Controller: A microprocessor-based controller that controls various HVAC equipment such as VAV terminals, fan coil units and heat pumps. These controllers typically have relatively few input/output device capability, standard control sequences and are dedicated to specific applications.

CONTROL TERMS, PNEUMATIC

Air Compressor: A reciprocating, positive displacement pump which compresses air to relatively high pressures, commonly 60 to 100 psig.

Air Dryer: A refrigerated device for removing moisture from compressed air.

Air Filter: A device installed in the compressor's air intake to keep dirt and other contaminants from entering the compressor.

Air Lines: Air lines are generally either copper or polyethylene (plastic) tubing. The air lines going to controlling devices such as thermostats, humidistats, etc., are called mains. The air lines leading from controlling devices to the actuator of controlled devices, such as dampers or valves, are called branches.

Automatic Drain: A drain installed on the refrigerated air dryer which automatically removes moisture from the dryer.

Deadband Pressure: The deadband pressure is the output pressure at which no heating or cooling occurs.

Dual Pressure System: A system which requires two different main air pressures.

High Pressure Gauge: A pressure gauge installed in the main supply line before the pressure reducing

valve which indicates the pressure of the air stored in the receiver tank.

Low Pressure Gauge: A pressure gauge installed in the main supply line after the pressure reducing valve which indicates the pressure of the main air.

Manual Drain: A manual device on the receiver tank to remove moisture and contaminants.

Normally Closed: The position of a controlled device when the power source is removed. A controlled device that moves toward the closed position as the branch line pressure decreases is normally closed. The position of the damper or valve when the actuator is deenergized.

Normally Open: The position of a controlled device when the power source is removed. A controlled device that moves toward the open position as the branch line pressure decreases is normally open. The position of the damper or valve when the actuator is deenergized.

Pneumatic Controller Output Pressure: The output pressure from pneumatic controllers is in pounds per square inch (psi). Typically, the range is 12 psi from 3 to 15 psi with a 9 psi midpoint. In the following example, a controller that has a 6°F throttling range from 72°F to 78°F will have a setpoint temperature of 75°F. The output pressure will be as follows depending on whether the controller is direct-acting or reverse-acting.

Temperature	D/A Output	R/A Output
72°F	3 psi	12 psi
75°F	9 psi	9 psi
78°F	12 psi	3 psi

Positive Positioner: A positive positioner is used where accurate positioning of the controlled device is required. Positive positioners provide up to full main control air to the actuator for any change in position required by the controller.

Pressure Range: The change in the pressure necessary for the controller output to change over the throttling range. For example, in a 3 to 15 psi system, the pressure range is 12 psi.

Pressure Reducing Valve: The pressure reducing valve (PRV) reduces the pressure of the air in the receiver tank to the main air pressure to be used in the controlling devices. Main air pressure is normally set for 18 to 20 psig.

Pressure Switch: A switch which starts and stops the compressor at predetermined set points. For example, the switch may be set to start the compressor when the pressure in the receiver tank falls to 65 psig and stop the compressor when the pressure in the tank reaches 85 psig.

Receiver Tank: A tank which receives and stores the compressed air from the compressor for use throughout the system.

Relay: A pneumatic switching device.

Relief Valve: Generally, there are two safety relief valves. One high pressure relief valve is installed on the receiver tank to protect the tank from excessive pressure. A second relief valve is installed in the supply line to protect the control devices from excessive line pressure.

Single Pressure System: A system which requires only one main air pressure.

Spring Range: The spring range of an actuator restricts movement of the controlled device within set limits. Typical spring ranges for pneumatic actuators are 3 to 7 psig, 3 to 8 psig, 8 to 13 psig and 9 to 13 psig.

ELECTRICAL TERMS

Alternating Current: The type of electrical circuit in which the current constantly reverses flow.

Amperage: Electron flow of one coulomb per second past a given point.

Ampere: A measure of electrical flow rate. One ampere is equal to one coulomb per second or 6.3×10^{18} electrons per second.

Conductor: A material which readily passes electrons. Good conductors are silver, copper and aluminum wire. Copper, the second best conductor, is the most often used in electrical wiring because of price and availability.

Current: The transfer of electrical energy through a conductor.

Direct Current: The type of electrical circuit in which the current always flows in one direction.

Electromotive Force: A measure of electric force or potential, also called voltage.

Frequency: The number of complete cycles per second of alternating current.

Ground: The lowest potential or voltage of an electrical system.

Hertz: The number of complete cycles per second of alternating current.

Hot Lead: See Hot Wire.

Hot Leg: See Hot Wire.

Hot Wire: Any wire which is at a higher voltage than neutral or ground.

Impedance: The total opposition to the flow of alternating current by any combination of resistance, inductance and capacitance. Impedance is measured in Ohms.

Induction: The process of producing electron flow by the relative motion of a magnetic field across a conductor. In a transformer, current flowing through the primary coil sets up a magnetic field. This magnetic field produces current flow in the secondary coil.

Insulator: A material which doesn't readily pass electrons. Some of the best insulators are rubber, glass and plastic.

Intrinsically Safe Circuit: A circuit in which any spark or thermal effect, produced either normally or in specified fault conditions, is incapable, under the test conditions of causing ignition of a mixture of flammable or combustible material in the air in the mixture's most easily ignitable concentration. Source: *Underwriters Laboratory Standard UL 913.*

Intrinsically Safe Device: An apparatus in which all the circuits are intrinsically safe. Source: *Underwriters Laboratory Standard UL 913.*

Kilo-volt-Ampere: 1,000 volt-amperes.

Kilowatt: 1,000 watts.

Kilowatt-hour: 1,000 watt-hours.

Neutral: That part of an electrical system which is at zero voltage difference with respect to ground.

Ohm: A measure of resistance in an electrical circuit.

Open Circuit: The condition which exists when (either deliberately or accidentally) an electrical conductor or connection is broken or opened with a switch, safety or interlock. An open circuit stops the flow of electricity.

Phase: The number of separate highest voltages alternating at different intervals in the circuit.

Short Circuit: The condition which occurs when a hot wire comes in contact with neutral or ground.

Voltage: A measure of electric force or potential. Also called electromotive force.

Volt-Ampere: A unit of apparent power.

Watt: A unit of actual power.

Watt-hour: A measure of energy.

ENERGY TERMS

Btu: A unit of energy.

Energy: A measure of power consumed. The ability to do work. Stored work. The units of energy are foot-pound, Btu, and kilowatt-hour.

Entropy: The extent to which the energy of a system is available for conversion to work.

Foot-pound: A unit of energy.

Heat: Energy in transit from one body to another as the result of a temperature difference between the two bodies. A form of energy transferred by a difference in temperature. Heat (energy) always flows from a higher temperature to a lower temperature. In HVAC systems, fluids such as air, water and refrigerants are used to carry or transfer heat from one place to another.

Kilowatt-hour: 1,000 watt-hours. A unit of energy.

Kinetic Energy: The energy in a body due to its motion.

Law of Conservation of Energy: Energy is neither created nor destroyed. Energy can be converted from one form to another.

Mechanical Energy Equivalent: 778 ft-lb = 1 Btu.

Potential Energy: Stored energy.

Total Energy: Total energy = kinetic energy + potential energy.

Watt-hours: A unit of energy.

FAN TERMS

Drive Side: On single inlet, single wide fans, the drive side is the side opposite the inlet. On double inlet, double wide fans, the drive side is the side that has the drive.

Fan Air Volume: The rate of flow expressed at the fan inlet in cubic feet per minute of air produced, independent of air density.

Fan Blast Area: The fan outlet area less the area of the cutoff.

Fan Efficiency: The output of useful energy divided by the power input; air horsepower divided by brake horsepower.

Fan Outlet Area: The gross inside area of the fan outlet expressed in square feet.

Fan Outlet Velocity: The theoretical velocity of the air as it leaves the fan outlet. The velocity that would exist at the fan outlet if the velocity was uniform. Fan outlet velocity is calculated by dividing the air volume by the fan outlet area.

Fan Performance Curve: A fan performance curve is a graphic representation of the performance of a fan from free delivery to no delivery.

Fan Static Efficiency: Static air horsepower divided by brake horsepower.

Fan Static Pressure: The fan total pressure less the fan velocity pressure.

Fan Total Efficiency: Total air horsepower divided by brake horsepower.

Fan Total Pressure: The rise in total pressure from the fan inlet to the fan outlet. The measure of total mechanical energy added to the air by the fan.

Fan Velocity Pressure: The pressure corresponding to the average air velocity at the fan outlet.

Nonoverloading Fan: A fan that is selected where the horsepower curve increases with an increase in air quantity, but only to a point to the right of maximum efficiency, and then gradually decreases. If a motor is selected to handle the maximum brake horsepower shown on the fan performance curve, the motor will not overload in any condition of fan operation. Backward curved and backward inclined fans are "nonoverloading" fans.

Static Pressure: The pressure or force within the unit or duct that exerts pressure against all the walls and moves the air through the system.

Tip Speed: The velocity in feet per minute at the tip of the fan blade.

Total Pressure: The sum of the static pressure and the velocity pressure taken at a given point of measurement.

Total Static Pressure: The static pressure rise across the fan calculated from static pressure measurements at the fan inlet and outlet.

Velocity Pressure: Velocity pressure is the pressure caused by the air being in motion and has a direct mathematical relation to the velocity of the air.

HVAC COMPONENT AND EQUIPMENT TERMS

Adjustable Sheave: The belt grooves in the sheave are adjustable. Adjustable groove sheaves are also known as variable speed sheaves or variable pitch sheaves.

Air Conditioning Unit: An assembly of components for the treatment of air. Also called: Air handing unit (AHU) for larger systems or fan-coil unit (FCU) for smaller systems.

Air Separator: Air separators free the air entrained in the water in a hydronic system. There are several types of air separators. The centrifugal type of air separator works on the action of centrifugal force and low velocity separation. The centrifugal motion of the water circulating through the air separator creates a vortex or whirlpool in the center of the tank and sends heavier, air-free water to the outer part of the tank and allows the lighter air-water mixture to move to a low velocity air separation and collecting screen located in the vortex. The entrained air, being lighter, collects and rises into the compression tank. The boiler dip tube type of air separator is a tube in the top or top side of the boiler. When the water is heated, air is released and collects at a high point in the boiler. The dip tube allows this collected air to rise into the compression tank. The in-line, low velocity air separating tank with dip tube type of air separator is used when a boiler isn't available or is not usable as the point of air separation.

Annular Flow Meter: Annular flow meters have a multiported flow sensor installed in the water pipe. The holes in the sensor are spaced to represent equal annular areas of the pipe in the same manner as the Pitot tube traverse is for round duct. The flow meter is designed to sense the velocity of the water as it passes the sensor. The upstream ports sense high pressure and the downstream ports sense low pressure. The resulting difference, or differential pressure, is then measured with an appropriate differential pressure gage. Calibration data which shows flow rate in gallons per minute (gpm) versus measured pressure drop is furnished with the flow meter.

Automatic Air Vent: There are two types of automatic air vents. The hydroscopic type contains material that expands when wet and holds the air vent valve closed. Air in the system dries out the material, causing it to shrink and open the vent valve. The float type contains a float valve that keeps the air vent closed while there's water in the system. If air is in the system, it rises to the air vent. When it reaches the float, the float drops and opens the air vent valve.

Automatic Temperature Control Damper: Dampers controlled by temperature requirements of the system. Automatic temperature control dampers are usually opposed or parallel bladed dampers and can be either two-position or modulating. Two-position control means the damper is either open or closed. Modulating control provides for the gradual opening or closing of a damper. Automatic temperature control dampers should have a tight shutoff when closed.

Automatic Temperature Control Valve: Automatic control valves are used to control flow rate or to mix or divert water streams. They're classified as two-way or three-way construction and either modulating or two-position.

Backdraft Damper: A damper which opens when there's a drop in pressure across the damper in the direction of airflow and closes under the action of gravitational force when there's no airflow.

Balancing Station: An assembly to measure and control air or water flow. It has a measuring device and a volume control device with the recommended lengths of straight duct or pipe entering and leaving the station.

Ball Valve: The ball valve is a manual valve used for regulating water flow. It is similar to the plug valve and is often used for water balancing. It has a low pressure drop and good flow characteristics.

Butterfly Valve: The butterfly valve is a manual valve used for regulating water flow. It has a heavy ring enclosing a disc which rotates on an axis and, in principle, is similar to a round single blade damper. It has a low pressure drop and is used as a balancing valve. It doesn't have the good throttling characteristics of a ball or plug valve.

Calibrated Balancing Valve: With the other types of flow meters, such as the orifice plate, Venturi and annular type, a balancing valve is also needed to set the flow. Calibrated balancing valves are designed to do both duties of a flow meter and a balancing valve. These valves are similar to ordinary balancing valves, except the manufacturer has provided pressure taps in the inlet and outlet and has calibrated the device by measuring the resistance at various valve positions against known flow quantities. The valve has a graduated scale or dial to show the degree open. Calibration data which shows flow rate in gallons per minute (gpm) versus measured pressure drop is provided with the valve. Pressure drop is measured with any appropriate differential gage.

Ceiling Diffuser: A diffuser which typically provides a horizontal flow pattern that tends to flow along the ceiling, producing a high degree of surface effect. Typical square or rectangular ceiling diffusers deliver air in a one-, two-, three- or four-way pattern. Round ceiling diffusers deliver air in all directions.

Check Valve: The check valve is a manual valve which limits the direction of flow. The check valve allows water to flow in one direction and stops its flow in the other direction. The swing check valve has a gate which opens when the system is turned on and pressure from the water flowing in the proper direction is applied. The gate closes due to its own weight and gravity when the system is off. The spring loaded check valve has a spring that keeps the valve closed. Water pressure from the proper direction against the spring opens the valve.

Closed Impeller: An impeller having shrouds or side walls. Designed primarily for handling clear liquids, such as water.

Coil: Coils are heat transfer devices (heat exchangers). They come in a variety of types and sizes and are designed for various fluid combinations. In hydronic applications, coils are used for heating, cooling or dehumidifying air. Hydronic coils are most often made of copper headers and tubes with aluminum or copper fins and galvanized steel frames.

Combination Valve: The combination valve regulates flow and limits direction. It's a combination of a check valve, calibrated balancing valve and shutoff valve. It's made in a straight or angle pattern. The valve acts as a check valve, preventing backflow when the pump is off and can be closed for tight shutoff or partly closed for balancing. The valve has pressure taps for connecting flow gages and reading pressure drop across it. A calibration chart is supplied with the valve for conversion of pressure drop to gpm. The valve also has a memory stop. Combination valves are sometimes called multi-purpose or triple-duty valves.

Compression Tank: A tank which compensates for the normal expansion and contraction of water in a hydronic system. When the system is filled with water, the air in the compression tank (about $2/3$ of the tank is water, $1/3$ is air) will act as a spring or cushion to keep the proper pressure on the system to accommodate the fluctuations in water volume and control pressure change in the system. Compression tanks should be installed on the suction side of the pump. The point where the compression tank connects to the system is called "the point of no pressure change."

Constant Volume Box: A terminal box which delivers a constant quantity of air. The boxes may be single duct or dual duct.

Constant Volume Dual Duct Box: A terminal box supplied by separate hot and cold ducts through two inlets. The boxes mix warm or cool air as needed to properly condition the space and maintain a constant volume of discharge air. Dual duct boxes may use a mechanical constant volume regulator with a single damper motor to control the supply air. The mixing damper is positioned by the motor in response to the room thermostat. As the box inlet pressure increases, the regulator closes down to maintain a constant flow rate through the box. Another type of constant flow regulation uses two motors, two mixing dampers and a pressure sensor to control flow and temperature of the supply air. The motor connected to the hot duct inlet responds to the room thermostat and opens or closes to maintain room temperature. The motor on the cold duct inlet is also connected to the room thermostat, but through a relay which senses the pressure difference across the sensor. This motor opens or closes the damper on cold duct inlet to maintain room temperature and maintain a constant pressure across the sensor (a constant volume through the box).

Constant Volume Single Duct Box: A single inlet terminal box supplied with air at a constant volume and temperature (typically cool air). Air flowing through the box is controlled by a manually operated damper or a mechanical constant volume regulator. The mechanical volume regulator uses springs and perforated plates or damper blades which decrease or increase the available flow area as the pressure at the inlet to the box increases or decreases. A reheat coil or cooling coil may be installed in the box or immediately downstream from it. A room thermostat controls the coil.

Cooling Coil: A chilled water or refrigerant coil.

Cooling Coil Tubes: Cooling coil tubes are usually made of copper, but other materials used include carbon steel, stainless steel and brass. For special applications, cupro-nickel is used. For applications where the air stream may contain corrosives, there are various protective coatings available. The number of tubes varies in both depth and height, but is usually 1 to 12 rows in the direction of airflow (depth) and 4 to 36 tubes per row (height). The more tubes, the more heat transfer, but also the more resistance to airflow and initial cost of the coil. Tube diameters are usually five-eighths of an inch.

Counter Flow Coil: Counter flow means that the flow of air and water are in opposite directions to each other. The water enters on the same side that the air leaves. Coils are piped counter flow for the greatest heat transfer for a given set of conditions.

Cutoff Plate: The cutoff plate is located at the outlet of a centrifugal fan. Incorrect positioning of the cutoff plate will allow air to be drawn back into the fan wheel and will reduce fan performance.

Damper: A device used to regulate air flow.

Diffuser: A supply air outlet generally found in the ceiling. It has various deflectors arranged to promote mixing of primary air with secondary air. Types of diffusers are: round, square, rectangular, linear and troffers. Some diffusers have a fixed airflow pattern, while others have field adjusted patterns.

Diverting Tee: A device to create the proper resistance in a one-pipe system to direct water to the terminal.

Double Inlet, Double Wide Fan: Fans that have two single wide fan wheels mounted back to back on a common shaft in a single housing. Air enters both sides of the fan. On double inlet, double wide fans, the bearings are in the air stream and they are large in relation to the inlet. They will reduce fan performance if the fan is small. Therefore, double inlet, double wide fans are less common in smaller sizes. Generally, because of the double inlet, DIDW fans are more suited to open inlet plenums. They're used most often in high volume applications.

Double Suction Pump: A pump in which the liquid enters the impeller inlet from both sides. The impeller is similar to two single suction impellers, back to back. Double suction pumps have fixed suction and discharge openings. The suction connections are normally one or two pipe sizes larger than the discharge connection.

Dual Current Motor: A motor that will operate safely at either of two nameplate currents. The operating current depends on the voltage supplied.

Dual Duct Box: A dual inlet terminal box supplied with any combination of heated, cooled or dehumidified air. The hot duct supplies warm air which may be either heated air or return air from the conditioned space. The cold duct supplies cool air which may be either cooled and dehumidified when the refrigeration unit is operating, or cool outside air brought in by the economizer cycle. A room thermostat controls a mixing damper arrangement in the boxes which determines whether the discharge air will be cool air, warm air or a mixture of both. Dual duct boxes are pressure independent and may be constant or variable volume.

Dual Voltage Motor: A motor that will operate safely from either of two nameplate voltages. Typical single-phase dual voltage motors are: $110/220$ volts or $115/230$ volts. Typical three-phase motor dual voltage motors are: $220/440$ volts, $230/460$ volts or $240/480$ volts.

Duct: A passageway made of sheet metal or other suitable material used for conveying air.

Equalized Spring Coupling: The equalized spring-type coupling is used where quiet, smooth operation is required. The motor drives the pump shaft through four springs. The tension on the springs is balanced by an equalizing bar. This coupling needs no maintenance when the alignment is correct. Check alignment if the coupling breaks or noisy operation is observed.

Evaporator Coil: A coil containing a refrigerant other than water. Used for cooling the air.

Expansion Tank: A tank which compensates for the normal expansion and contraction of water in a hydronic system. Water expands when heated in direct proportion to its change in temperature. In hydronic systems, an allowance must be made for this expansion, otherwise when the system is completely filled, the water has nowhere to go and there's the possibility of a pipe or piece of equipment breaking.

In an open system, an expansion tank which is open to the atmosphere, is provided about 3 feet above the highest point in the system. Then, as the water temperature increases (and therefore, the total volume of water in the system increases), the water level rises in the tank. Open expansion tanks are limited to installations having operating tem-

peratures of 180°F or less because of water boiling and evaporation problems. Being open to the atmosphere also has the drawback of the continual exposure to the air and its possible corrosive effects.

Exhaust Air Inlet: An exhaust air grille, register or other opening to allow air from the conditioned space into the exhaust air duct.

Extractor: A device used in low pressure systems to divert air into branch ducts.

Fan: A power driven, constant volume machine which moves a continuous flow of air by converting rotational mechanical energy to increase the total pressure of the moving air.

Fan Sheave: The driven pulley on the fan shaft.

Fins: Fins on a coil increase the area of heat transfer surface and improve the efficiency and rate of transfer. Air conditioning coils are generally spaced from 11 to 14 fins per inch (fpi).

Fixed Sheave: The belt grooves are fixed.

Flexible Connectors: Flexible connectors are used between piping and pumps and other pieces of equipment to reduce noise and vibration. Flexible connectors can be made of rubber, plastic or braided metal.

Flexible Disc Coupling: The flexible disc is used in heavy duty applications where extremely quiet conditions aren't needed. The motor drives the pump shaft through the flexible disc. If rough operation or noise is observed, check the coupling for wear. The flexible disc should never be tightly bound between the two coupler halves. Check with the manufacturer for clearance specifications.

Flow Meter: Flow meters such as orifice plates, venturis, annular type and calibrated balancing valves are permanently installed devices used for flow measurement of pumps, primary heat exchange equipment, distribution pipes and terminals. For flow meters to give accurate, reliable readings they should be installed far enough away from any source of flow disturbance to allow the turbulence to subside and the water flow to regain uniformity. The manufacturers of flow meters usually specify the lengths of straight pipe upstream and downstream of the meter needed to get good readings. Straight pipe lengths vary with the type and size of flow meter, but typical specifications might be between 5 to 25 pipe diameters upstream and 2 to 5 pipe diameters downstream of the flow meter.

Fractional Horsepower V-belts: A light-duty, flexible belt in sizes 2L through 5L. Fractional horsepower belts are generally used on smaller diameter sheaves because they're more flexible than the industrial belt for the same equivalent cross-sectional size.

Gate Valve: The gate valve is a manual valve used for

tight shutoff to service or remove equipment. It has a straight flow through passage which results in a low pressure drop. It regulates flow only to the extent that it's either fully open or fully closed.

Globe Valve: The globe valve is a manual valve used in water make-up lines. It can be used in partially open positions and therefore, can be used for throttling flow. However, the globe valve has a high pressure drop, even when fully open, which unnecessarily increases the pump head. It should not be used for balancing.

Grille: A wall, ceiling or floor mounted louvered or perforated covering for an air opening. To control air flow pattern, some grilles have a removable louver. Reversing or rotating the louver changes the air direction. Grilles are also available with adjustable horizontal or vertical bars so the direction, throw and spread of the supply air stream can be controlled.

Heat Exchanger: A device specifically designed to transfer heat between two physically separated fluids. The term heat exchanger can describe any heat transfer device such as a coil or a particular category of devices. Heat exchangers are made in various sizes and types (shell and tube, U-tube, helical and plate) and are designed for several fluid combinations such as: steam to water (converter), water to steam (generator) refrigerant to water (condenser), water to refrigerant (chiller), water to water (heat exchanger), air to refrigerant (coil) and air to water or water to air (coil).

Heating Coil: A hot water coil, steam coil or electric coil.

High Efficiency Particulate Air (HEPA) Filters: A filter with an efficiency in excess of 99.97% for 0.3 micrometer particles as determined by Dioctyl Phthalate (DOP) test.

Induction Box: Constant temperature primary supply air is forced through a discharge device, such as a nozzle or venturi, to induce secondary room or return air into the box where it mixes with the supply air. The high velocity of the primary air creates a low pressure region which draws in (or induces) the higher pressure secondary air. The mixed air (primary + secondary) is then supplied to the conditioned space. Some induction boxes have heating or cooling coils through which the secondary air is induced. Induction boxes may be constant or variable volume, pressure dependent or pressure independent.

Industrial V-belts: Belt sizes "A" through "E."

Light Troffer: A type of ceiling diffuser which fits over a fluorescent lamp fixture and delivers air through a slot along the edge of the fixture. They're available in several types. One type delivers air on both sides of the lamp fixture and another type provides air on only one side of the fixture.

Linear Slot Diffuser: This type of diffuser is manufactured in various lengths and numbers of slots and may be set for different throw patterns.

Manual Air Vent: A manual valve which is opened to allow entrained air to escape.

Manual Valve: Manual valves are used to regulate flow rate or limit the direction of flow. Gate, globe, plug, ball and butterfly valves regulate flow rate, while a check valve limits the direction of flow. A valve called a combination valve does both.

Memory Stop: Some plug valves, calibrated balancing valves and combination valves have adjustable memory stops. The memory stop is set during the final balance so that, if the valve is closed for any reason, it can later be reopened to the original setting.

Mechanical Seal: Mechanical seals have a stationary ring, usually made of hard ceramic material and a rotating graphite ring. The stationary ring fits into a recess in the pump body and has a rubber gasket behind it which forms a watertight seal. Behind the molded graphite ring is a rubber bellows and a seal spring. It's this spring that keeps the rotating graphite ring tight against the face of the ceramic ring, making the water seal. No maintenance or adjustments are needed because the spring continually pushes the graphite ring forward to make up for wear. It's important that a water film is between the two surfaces to provide lubrication and cooling. Running or bumping a pump without water in the system will damage mechanical seals. When the seal leaks, replace it.

Motor Sheave: The driver pulley on the motor shaft. The motor sheave may be either a fixed or adjustable groove sheave.

Multistage Pump: A pump having two or more impellers on a common shaft, acting in series in a single casing. The liquid is conducted from the discharge of the preceding impeller through fixed guide vanes to the suction of the following impeller, causing a pressure increase at a given flow rate as it passes through each stage.

Opposed Blade Damper: A multibladed damper with a linkage that rotates the adjacent blades in opposite directions, resulting in a series of openings that become increasingly narrow as the damper closes. This type of blade action results in a straight, uniform flow pattern sometimes called "nondiverting." Generally, opposed blade dampers are used in a volume control application.

Orifice Plate: An orifice plate is essentially a fixed circular opening in a duct. A measurable "permanent" pressure loss is created as the air passes from the larger diameter duct through the smaller opening. This abrupt change in velocity creates turbulence and a measurable amount of friction, resulting in a pressure drop across the orifice. Calibration data which shows flow rate in cubic feet per minute (cfm) versus measured pressure drop is furnished with the orifice plate. A differential pressure gage (such as a manometer) is connected to the pressure taps as the flow is read.

Parallel Blade Damper: Generally, parallel blade dampers are used in a mixing application. Because the blades rotate parallel to each other, a parallel blade damper produces a "diverting" type of air pattern. When in a partially closed position, the damper blades throw the air to the side, top or bottom of the duct. This flow pattern may adversely affect coil or fan performance or the airflow into branch ducts if the damper is located too close upstream.

Parallel Flow Coil: Parallel flow means that the flow of air and water are in the same direction to each other. The water and air enter on the same side.

Perforated Face Diffuser: Perforated face diffusers are used with lay-in ceilings and are similar in construction to the standard square ceiling diffuser with an added perforated face plate. They're generally equipped with adjustable vanes to change the flow pattern to a one-, two-, three-, or four-way throw.

Pipe: A passageway made of steel, iron or other suitable material that is used for conveying water.

Plug Valve: Plug valves are manual valves used for balancing water flow. Plug valves have a low pressure drop and good throttling characteristics and therefore, add little to the pumping head. Plug valves can also be used for tight shutoff.

Pressure Dependent Box: The quantity of air passing through this terminal box is dependent on the inlet static pressure.

Pressure Independent Box: The quantity of air passing through this terminal box is independent (within design limits) of the inlet static pressure.

Pressure Reducing Valve: The pressure reducing valve reduces the pressure from the city water main to the proper pressure needed to completely fill the system and maintain this pressure. Pressure reducing valves are installed in the piping that supplies make-up water to the system.

Pressure Relief Valve: Pressure relief valves are safety devices used to protect the system and human life. The pressure relief valve opens on a preset value so the system pressure can't exceed this amount.

Pump: A machine for imparting energy to a fluid. The addition of energy to a fluid makes it do work such as rising to a higher level or causing it to flow. In a hydronic system, the pump is the component which provides the energy to overcome system resistance and produce the required flow.

Register: A grille with a built-in or attached damper assembly.

Return Air Inlet: A return air grille, register or other inlet, such as a perforated face opening, linear slot, light troffer or other opening, to allow air from the conditioned space or return air plenum into a return air duct or mixed air plenum. Return inlets are generally chosen and located to suit architectural design requirements for appearance and compatibility with supply outlets.

Shaft Couplings: Shaft couplings compensate for small deviations in alignment between the pump and motor shafts within the tolerances established by the manufacturer. Couplings are made in "halves" so the pump and motor may be disconnected from each other. It's important that shafts be aligned as close as possible for quiet operation and the least coupling and bearing wear. The coupling can accommodate small variations in alignment, but its function is coupling, not compensating, for misalignment. Severe misalignment between the shafts will lead to noisy operation, early coupling failure and possible pump or motor bearing failures.

Sheave: The pulley on a shaft.

Single Duct Box: A terminal box which is usually supplied with cool air through a single inlet duct. The box may be constant or variable volume, pressure dependent or pressure independent. It can also have a water coil (heating or cooling), steam coil or electric reheat.

Single Inlet, Single Wide Fan: Single inlet, single wide fans have one fan wheel and a single entry. They're more suited to having inlet duct attached to them than double inlet, double wide fans.

Single Phase Motor: A motor supplied with single phase current.

Single Stage Pump: A pump with one impeller.

Single Suction Pump: A pump in which the liquid enters the impeller inlet from one side. Single suction pumps are usually built with the inlet at the end of the impeller shaft. The casing is made so the discharge may be rotated to various positions. The suction connection is normally one or two pipe sizes larger than the discharge connection.

Splitter Damper: A device used in low pressure systems to divert airflow.

Strainer: Strainers catch sediment or other foreign material in the water.

Strainer Screen: A strainer contains a fine mesh screen

formed into a sleeve or basket that fits inside the strainer body at the water pump. This sleeve must be removed and cleaned. A strainer with a dirty sleeve, or a sleeve with a screen that's too fine, means there will be excessive pressure drop across the strainer and lower water flow. Automatic control valves or spray nozzles which operate with small clearances and need protection from materials which might pass through the pump strainer may also have fine mesh sleeve strainers installed before the valve or nozzle.

Stuffing Box: The stuffing box seal has a "packing" which has rings made of graphite impregnated cord, molded lead foil or some other resilient material formed into fitted split rings. These packing rings are compressed into the stuffing box by a packing gland. The tension on the packing gland is critical to the proper operation of the pump. If there's too much tension, the proper water leakage won't occur. This will cause scoring of the shaft and overheating of the packing. Another problem is that as the seal gets older and the packing gland has been tightened over time, the packing becomes compressed, loses its resiliency and overheats the stuffing box. When the packing gland is backed off to allow cooler operation, there's excessive leakage. When this happens, replace the packing.

Supply Air Outlet: A supply air diffuser, grille, register or other opening to allow supply air into the conditioned space to mix with the room air to maintain a uniform temperature throughout the occupied zone. Supply air diffusers, grilles and registers are chosen and located to control airflow patterns to avoid drafts and air stagnation, and to complement the architectural design of the building.

Temperature Well: Temperature wells are installed at specific points in the piping so a test thermometer can be inserted to measure the temperature of the water in the pipe. Generally, temperature wells are installed on the entering and leaving sides of chillers, condensers, boilers and coils. Thermometer wells must be long enough to extend into the pipe so good contact is made with the water. The well forms a cup to hold a heat conducting liquid (usually an oil), so good heat transfer is made from the water in the pipe to the liquid in the well to the thermometer. Therefore, wells should be installed vertically, or not more than 45° from vertical, so they'll hold the liquid. An accurate reading cannot be obtained from inserting a thermometer into a dry well because air will act as an insulator.

Terminal Box: A device or unit which regulates supply airflow, temperature and humidity to the conditioned space. Terminal boxes are classified as single duct, dual duct, constant volume, variable volume,

medium pressure, high pressure, pressure dependent, pressure independent, system powered, fan powered, induction, terminal reheat and bypass. They may also contain a combination of heating or cooling coils, dampers and sound attenuation. The airflow through the box is normally set at the factory, but can also be adjusted in the field. Terminal boxes also reduce the inlet pressures to a level consistent with the low pressure, low velocity duct connected to the discharge of the box. Any noise that's generated within the box in the reduction of the pressure is attenuated. Baffles or other devices are installed which reflect the sound back into the box where it can be absorbed by the box lining. Commonly, the boxes are lined with fiberglass, which also provides thermal insulation, so the conditioned air within the box won't be heated or cooled by the air in the spaces surrounding the box. Terminal boxes work off static pressure in the duct system. Each box has a minimum inlet static pressure requirement to overcome the pressure losses through the box plus any losses through the discharge duct, volume dampers and outlets.

Thermal Overload Protection: Thermal overload protection devices, sometimes called "heaters" or "thermals," prevent motors from overheating. If a motor becomes overloaded, or one phase of a three-phase circuit fails (single phasing), there will be an increase in current through the motor. If this increased current draw through the motor lasts for any appreciable time and it's greatly above the full load current rating, the windings will overheat and damage may occur to the insulation, resulting in a burned out motor. Because most motors experience various load conditions, from no load to partial load to full load to short periods of being overloaded, their overload protection devices must be flexible enough to handle the various conditions under which the motor and its driven equipment operate. Single-phase motors often have internal thermal overload protection. This device senses the increased heat load and breaks the circuit, stopping the motor. After the thermal overload relays have cooled down, a manual or automatic reset is used to restart the motor. Other single-phase and three-phase motors require external overload protection.

Thermals: *See Thermal Overload Protection.*

Three-phase Motor: A motor supplied with three-phase current. For the same size, three-phase motors have a capacity of about 150% greater, are lower in first cost, require less maintenance and generally do better than single-phase motors.

Three-way Automatic Control Valve: Three-way ACV's are used to mix or divert water flow and are generally classified as mixing or diverting valves. They may be either single-seated (mixing valve) or double-seated (diverting valve). The single-seated mixing valve is the most common. The terms "mixing" and "diverting" refer to the *internal construction of the valve and not the application.* The internal difference is necessary so the valve will seat against flow (*See Two-Way Valves*). A mixing valve has two inlets and one outlet. The diverting valve has one inlet and two outlets. Either valve may be installed for a flow control action (bypassing application) or a temperature control action (mixing application), depending on its location in the system. Diverting valves, however, shouldn't be substituted for mixing valves and vice versa. Using either design for the wrong application would tend to cause chatter.

Two-way Automatic Control Valve: ACV's control flow rate. They may be either single-seated or double-seated (balanced valve). The single-seated valve is the most common. The valve must be installed with the direction of flow opposing the closing action of the valve plug. The double-seated (or balanced valve) is generally recommended when high differential pressures are encountered and tight shutoff isn't required. The flow through this valve tends to close one port while opening the other port. This design creates a balanced thrust condition which enables the valve to close off smoothly without water hammer despite the high differential pressure.

V-belts: V-belts are rated by horsepower per belt, by length and minimum recommended pitch diameter. Two types of V-belts are generally used on HVAC equipment. They are the light-duty, fractional horsepower belts and the standard industrial belt. The general practice in HVAC design is to use belts of smaller cross-sectional size with smaller sheaves, instead of large belts and large sheaves for the drive components. Multiple belts are used to avoid excessive belt stress.

Variable Air Volume Box: VAV boxes are available in many combinations that include pressure dependent, pressure independent, single duct, dual duct, cooling only, cooling with reheat, induction, bypass and fan powered. VAV boxes can be classified by volume control (throttling, bypass or fan powered) intake controls and sensors (pneumatic, electric, electronic or system powered), thermostat action (direct-acting or reverse-acting) and the condition of the box at rest (normally open or normally closed). The basic VAV box has a single inlet duct. The quantity of air through the box is controlled by throttling an internal damper or air valve. If the box is pressure dependent, the volume control device will be controlled just by a room thermostat. The pressure independent version will also have a regulator to limit the air volume between a preset maximum and minimum.

Inside the pressure independent box is a sensor. Mounted on the outside is a controller with connections to the sensor, volume damper and room thermostat. The quantity of air will vary from a design maximum cfm down to a minimum cfm. The main feature of the VAV box is its ability to vary the air delivered to the conditioned space as the heat load varies. Then, as the total required volume of air is reduced throughout the system, the supply fan will reduce its cfm output. This means savings of energy and cost to operate the fan. The exception to this is the VAV bypass box.

The types of controls used to regulate the flow of air through VAV boxes are as varied as the types of boxes. Many boxes are designed to use external sources of power, such as pneumatic, electric or electronic. These boxes are sometimes called non-system powered. Other boxes are system powered, which means that the operating controls are powered by the static pressure from the main duct system. System powered boxes don't need a separate pneumatic or electric control system. This reduces costs at first—however, they usually have a higher required minimum inlet static pressure, which means that the supply fan will be required to produce higher static pressures, resulting in increased operating costs. All controllers, except for the bypass box type, reduce airflow from the fan.

Valve: Valves are used in hydronic systems for regulating water flow and isolating part or all of the system.

VAV Bypass Terminal Box: A bypass box uses a constant volume supply fan, but provides variable air volume to the conditioned space. The supply air comes into the box and can exit either into the conditioned space through the discharge ductwork or back to the return system through a bypass damper. The conditioned space receives either all the supply air or only a part of it, depending on what the room thermostat is calling for. Since there's no reduction in the main supply air volume feeding the box, this type of system has no savings of fan energy.

VAV Ceiling Induction Box: A VAV ceiling induction box has a primary damper at the box inlet and an induction damper in the box which allows air in from the ceiling plenum. On a call for cooling, the primary damper is full open and the induction damper is closed. As the conditioned space cools down, the primary damper throttles back and the induction damper opens to maintain a constant mixed air flow to the conditioned space. At some point, the induction damper is wide open and the primary damper is throttled (about 75%) to allow for the maximum induction ratio. Another type of induction box has a constant pressure nozzle inducing either primary air from the main supply system or return air from the ceiling plenum. The room thermostat opens or closes a primary air bypass damper to allow for the induction of primary or return air. This box uses volume regulators to reduce the airflow to the conditioned space. Some boxes may contain a reheat coil.

VAV Dual Duct Box: A VAV dual duct terminal box is supplied by separate hot and cold ducts through two inlets. A variety of control schemes vary the air volume and discharge air temperature. One type uses a temperature deadband, which supplies a varying quantity of either warm or cool air, but not mixed air, to the conditioned space.

VAV Fan Powered Box: A VAV fan powered box has the advantage of the energy savings of a conventional, single duct VAV system with the addition of several methods of heating and a constant airflow to the conditioned space. The box contains a fan and a return air opening from the ceiling space. When the room thermostat is calling for cooling, the box operates as would the standard VAV box. However, on a call for heat, the fan draws warm (secondary) air from the ceiling plenum and recirculates it into the rooms. Varying amounts of cool (primary) air from the main system is introduced into the box on either the inlet or discharge side of the fan and mixes with the secondary air. A system of dampers, backdraft or motorized, control the airflow and mixing of the air streams. As the room thermostat continues to call for heat, the primary air damper closes off and more secondary air is drawn into the box, and it alone is recirculated. Therefore, the airflow to the conditioned space stays constant. If more heat is needed, reheat coils may be installed in the boxes. The fan may operate continuously, or it may shut off. A common application of fan powered boxes is around the perimeter or other areas of a building where air stagnation is a problem when the primary air throttles back, zones have seasonal heating and cooling requirements, heat is needed during unoccupied hours when the primary fan is off, or when heating loads can be offset mainly with recirculated return air.

VAV Fan Powered Bypass Box: This box acts the same as the conventional bypass box with the addition of a secondary fan in the box. The bypass box uses a constant volume supply primary fan, but provides variable air volume to the conditioned space. The supply air comes into the box and can exit either into the conditioned space through the secondary fan and the discharge ductwork or back to the return system through a bypass damper. The fan in the box circulates the primary air, or return air, into the room. The conditioned space receives either all primary air, all return air or a mixture of the two, depending on what the room thermostat is calling for.

Since there's no reduction in the main supply air volume feeding the box, this type of system has no savings of primary fan energy.

VAV Pressure Dependent Box: A VAV pressure dependent box is essentially a pressure reducing and sound attenuation box with a motorized damper that's controlled by a room thermostat. These boxes simply position the damper in response to the signal from the thermostat. Because the airflow to these boxes is in direct relation to the box inlet static pressure, it's possible for the boxes closest to the supply fan (where the static pressure is the greatest), to get more air than is needed. The boxes farther down the line will be getting little or no air. Therefore, pressure dependent boxes should only be installed in systems where there's no need for limit control and the system static pressure is stable enough not to require pressure independence.

Pressure dependent maximum regulated volume boxes may be used where pressure independence is required only at maximum volume and the system static pressure variations are only minor. These boxes regulate the maximum volume, but the flow rate at any point below maximum varies with the inlet static pressure. This may cause "hunting."

VAV Pressure Independent Box: VAV pressure independent boxes can maintain airflow at any point between maximum and minimum, despite the box inlet static pressure, as long as the pressure is within the design operating range. Flow sensing devices regulate the flow rate through the box in response to the room thermostat's call for cooling or heating.

VAV Single Duct Pressure Independent Box: To maintain the correct airflow in a pressure independent box over the entire potential range of varying inlet static pressure, a sensor reads the differential pressure at the inlet of the box and transmits it to the controller. The room thermostat, responding to the load conditions in the space, also sends a signal to the controller. The controller responds by actuating the volume damper and regulating the airflow within the preset maximum and minimum range. For example, as the temperature rises in the space, the damper opens for more cooling. As the temperature in the space drops, the damper closes. If the box also has a reheat coil, the volume damper, on a call for heating, would close to its minimum position (usually not less than 50% of maximum) and the reheat coil would be activated. Because of its pressure independence, the airflow through the boxes is unaffected. Other VAV boxes in the system modulate and change the inlet pressures throughout the system.

VAV System Powered Box: System powered VAV boxes use the static pressure from the supply duct to power the VAV controls. The minimum inlet static pressure with this type of box is usually higher than other VAV systems in order to operate the controls and provide the proper airflow quantity.

Venturi: A venturi operates on the same principle as the orifice plate, but its shape allows gradual changes in velocity and the "permanent" pressure loss is less than is created by an orifice plate. Calibration data which shows flow rate in cubic feet per minute (cfm) versus measured pressure drop is furnished with the venturi. The pressure drop is measured with a differential gage.

Volume Dampers: Manual dampers used to control the quantity of airflow in the system by introducing a resistance to flow. If not properly selected, located, installed and adjusted they don't control the air as intended, they add unnecessary resistance to the system and they can create noise problems. The resistance a volume damper creates in a duct system is determined by how complicated the system is. For instance, if the system is very simple and the damper is a large part of that resistance, then any movement of the damper will change the resistance of the entire system and good control of the airflow will result. If, however, the damper resistance is very small in relation to the entire system, poor control will be the case. For instance, partial closing of a damper will increase its resistance to airflow, but depending on the resistance of the damper to the overall system resistance, the reduction in airflow may or may not be in proportion to closure. In other words, closing a damper 50% doesn't necessarily mean that the airflow will be reduced to 50%. For example, a damper (when open) might be 10% of the total system resistance. When this damper is half closed, the airflow is reduced to 80% of maximum flow. However, a similarly built damper in another duct system is 30% of the total system resistance when open. When this damper is half closed, the airflow is reduced to 55% of maximum.

The relationship between the position of a damper and its percent of airflow is termed its "flow characteristic." Opposed blade dampers are generally recommended for large duct systems because they introduce more resistance to airflow for most closed positions. Therefore, they have a better flow characteristic than parallel blade dampers. However, flow characteristics of dampers aren't consistent and may vary from one system to another. The actual effect of closing a damper can only be determined in the field by measurement.

Proper location of balancing dampers not only permits maximum air distribution, but also equalizes the pressure drops in the different airflow paths within the system. Manual dampers should be provided in each takeoff to the runout to control the air to grilles and diffusers. They should also be in the main, each submain, each branch and each sub-branch duct. Manually operated opposed blade or single blade quadrant type volume dampers should also be installed in every zone duct of a multizone system.

Single blade or opposed blade volume dampers placed immediately behind diffusers and grilles shouldn't be used for balancing. When throttled, they create noise at the outlet and change the effective area of the outlet, so the flow (Ak) factor is no longer valid. Proper installation and location of balancing dampers in the takeoffs eliminate the need for volume controls at grilles and diffusers.

Manual volume dampers may need to be installed in the outside, relief and return air connections to the mixed air plenum in addition to any automatic dampers. These volume control dampers balance the pressure drops in the various flow paths so the pressure drop in the entire system stays constant as the proportions of return air and outside air vary to satisfy the temperature requirements.

Volume dampers and handles should have enough strength and rigidity for the operating pressures of the duct system in which they will be installed. For a small duct, a single blade damper is satisfactory. For a large duct, dampers should be opposed blade. Each damper should have a locking quadrant handle or regulator.

INSTRUMENTATION TERMS

Compound Pressure Gage: A gauge measuring pressures both above and below atmospheric (pounds per square inch and inches of mercury).

Pressure Gage: A gauge measuring pressure above atmospheric (pounds per square inch, psi).

Vacuum Pressure Gage: A gauge measuring pressure below atmospheric (inches of mercury, in. Hg).

Laboratory Terms

Access Opening: That part of the fume hood through which work is performed.

AI: As installed (after construction but before occupancy). The purpose of an "as installed" fume hood test is to determine the hood performance rating when many of the variables affecting fume hood performance can be controlled. A well designed and located fume hood in a properly air balanced laboratory should be able to achieve a performance rating of 4.0 AI 0.10 (ASHRAE).

Airfoil: Curved or angular member(s) at the fume hood

entrance. Used to counteract the effect of the vena contracta generated at the opening of plain entrance hoods.

Air Volume: Quantity of airflow expressed in cubic feet per minute (cfm).

AM: As manufactured. The purpose of an "as manufactured" fume hood test is to evaluate a fume hood in the relatively ideal conditions of the manufacturer's test lab where many of the factors that adversely affect hood performance can be eliminated. New fume hoods should be able to achieve a performance rating of 4.0 AM 0.05 (ASHRAE).

Air Lock: An anteroom with airtight doors between a controlled and uncontrolled space.

AU: As used. The "as used" test is intended to be conducted within the hood as it is typically used. The exposure to the user at the hood will be dependent upon various factors, including apparatus used in the hood, whether the hood is used as a storage cabinet, whether the apparatus forces the user to conduct his tests too far forward, etc.

Auxiliary Air: Supply or supplemental air delivered to a laboratory fume hood to reduce room air consumption. Exhaust air requirements for laboratory fume hoods and other containment devices often exceed the supply air needed for the normal room air conditioning. To meet these needs, auxiliary air can be ducted directly into the fume hood or supplied to the room. The auxiliary air should be conditioned to meet the temperature and humidity requirements of the lab. Energy savings may be obtained when the heating, cooling and humidifying requirements are kept to a minimum. In some laboratory buildings, a corridor separates the offices from the labs. The offices are under positive pressure. Air leaves the offices and enters into the corridor. The corridor acts as a plenum (NFPA 90A). The air then flows into the negatively pressurized lab. The result is that it may be possible to reduce the auxiliary air or even eliminate it.

Baffle: A panel located across the fume hood interior back which controls the pattern of air moving into and through the fume hood. The baffle should be constructed so that it is impossible, by adjustment, to restrict air flow through the fume hood by more than 20%.

Bench Mounted Hood: A fume hood that rests on a counter top.

Biological Safety Cabinet: A special safety enclosure which uses air currents to protect the user. Used to handle pathogenic microorganisms. Biological safety cabinets are also called safety cabinets, laminar flow cabinets and glove boxes.

Bypass: A compensating opening that maintains a rel-atively constant volume of exhaust through a fume hood, regardless of sash position. The bypass functions to limit the maximum face velocity as the sash is closed.

California Hood: A rectangular enclosure used to house distillation and other large research apparatus. It can provide visibility from all sides with horizontal sliding access doors along the length of the assembly. The California hood is not considered a laboratory fume hood. Sizes may range from about 6 feet wide × 8 feet high × 3 feet deep.

Canopy Hood: A suspended ventilating device used to exhaust only heat, water vapor and odors. The canopy hood is not considered a laboratory fume hood.

Capture Velocity: The air velocity at the hood face necessary to overcome opposing air currents and contain contaminated air within the laboratory fume hood.

Cross Draft: A flow of air that blows across or into the face of the hood. Cross drafts, whether created by people moving about, the room ventilation system, or an open door (if located adjacent to the fume hood), can drastically disturb the flow of air into the hood face and even cause reverse flow of air out the front of the hood.

Dead Air Space: Lack of air movement in the hood.

Deflector Vane: An airfoil-shaped vane along the bottom of the hood face which deflects incoming air across the work surface to the lower baffle opening. The opening between the work surface and the deflector vane is open, even with the sash fully closed.

Diversity Factor: A diversity permits the exhaust system to have less capacity than that required for the full operation of all units.

Effluent: Ouflow.

Exhaust Collar: The connection between the duct and the fume hood through which all exhaust air passes.

Face Velocity: The average velocity of air moving into the fume hood opening. Expressed in fpm.

Fume Hood Exhaust System: An arrangement consisting of a laboratory fume hood, its adjacent room environment and the equipment (such as the ductwork and the blower) required to make the hood and system operable.

Glove Box: An enclosure used to confine and contain hazardous materials with user access by means of gloved portals or other limited openings. Glove boxes provide greater protection but are more restrictive than laboratory fume hoods or other biological safety cabinets. Glove boxes require far less exhaust air than laboratory fume hoods or other biological safety cabinets.

Hazardous Chemical: A chemical for which there is

significant evidence, based on at least one study conducted in accordance with established scientific principles, that acute or chronic effects may occur in exposed employees. Hazardous chemicals include carcinogens, toxic agents, irritants, corrosives and agents that damage the lungs, skin, eyes or mucous membrane.

Hood Face: The plane of minimum area at the front portion of the hood through which air enters when the sash is fully open.

HOPEC Laboratory Fume Hood: Hand Operated Positive Energy Control laboratory fume hood. A hood with a combination vertical-horizontal sash. *See Sash.*

Infiltrated Air: Auxiliary air induced from the corridor or other spaces into the lab.

Laboratory: A facility where relatively small quantities of hazardous chemicals are used on a nonproduction basis.

Laboratory Fume Hood: A laboratory fume hood is a ventilated box-like structure enclosing a work space. It is intended to capture, contain, and exhaust fumes, vapors, particulate matter, and other contaminants generated inside the enclosure. The fume hood consists of side, back and top enclosure panels and one side—called the face—which is open or partially open. The fume hood also has a sash and an exhaust plenum equipped with a baffle system for regulation of airflow. The fume hood is generally mounted on a bench or table. Air is brought into the fume hood to contain and exhaust the contaminants.

The conventional constant air volume fume hood has a sash which moves vertically up and down. At the sash's full open position the free area at the hood face is generally about 10 to 13 square feet with the minimum face velocity being about 100 feet per minute. The volume of air through the hood is therefore about 1000 to 1300 cfm. As the sash is lowered the volume of air is reduced but the velocity of the air through the opening is increased (sometimes as high as 400 feet per minute). These high face velocities can cause unwanted turbulence which cna induce contaminants out of the hood into the laboratory space. In order to try to maintain constant face velocities around 100 fpm, a bypass hood is used. The bypass hood also has a vertically moveable sash. The construction of the hood is similar to the conventional hood described above with the addition of the bypass to allow for a constant volume of airflow through the fume hood as the sash is closed. In other words, as the sash is pulled down the air volume through the hood face is reduced. At the same time the sash is being closed the bypass is being opened, and more air is then drawn through

the bypass. This keeps the volume of air through the hood and the hood face velocities relatively constant. Another variation of the conventional hood uses a combination horizontal/vertical sash.

The auxiliary air type fume hood also has a vertically movable sash. With the sash open, auxiliary air is ducted to the hood and distributed across the face area prior to its passage into the hood. With the sash closed, the auxiliary air is introduced directly into the fume hood interior. Auxiliary air fume hoods are designed to reduce the amount of conditioned laboratory air required for exhaust air. For instance, a hood might use 70% auxiliary air and only 30% room air as the total fume hood exhuast. In theory, this makes this type of hood more energy efficient because the auxiliary air is only nomimally heated or cooled as compared to the room air. However, tests have shown that the best safety performance for this type of hood occurs when the auxiliary air is slightly warmer than the laboratory room air. Therefore, a concern about this type of fume hood is that the auxiliary air may not enter the hood properly if it is only nomimally treated. Another concern is that the auxiliary air enters the hood in such a way that it may create turbulence, which can cause the air to reverse flow back out the hood face.

Variable air volume fume hoods may use either a vertical sash or a combination sash. This type of hood is equipped with special controls to allow the volume of exhuast air to vary while still maintaining a constant face velocity.

Laboratory Fume Hood System: Laboratory fume hood systems may be constant air volume, variable air volume or a combination of the two. They may be auxiliary air systems, partial return air systems or 100% room air systems. With the 100% room air system the exhausted air is replaced entirely by the conditioned air from the laboratory space. Since there is no return air there is an added energy burden placed on the heating and cooling components to condition the outside air. To try to reduce energy costs variable air volume systems are used.

Laboratory Use: The handling or use of such chemicals in which the chemical manipulations are carried out on a laboratory scale using multiple chemical procedures and/or chemicals. The procedures involved are not of a production process. They do not, in any way, simulate a production process. Adequate protective laboratory equipment is available and in common use to minimize the potential for employee exposure to hazardous chemicals.

Laminar Flow Cabinet: A clean bench or biological safety cabinet that uses smooth directional airflow to capture and carry away airborne particles. The laminar flow cabinet is not considered a laboratory fume hood.

Liner: Interior lining of a laboratory fume hood used for the side, back and top enclosure panels, exhaust plenum and baffles.

Lintel: The portion of a laboratory fume hood located directly above the access opening.

Lpm: Liters per minute. For the ASHRAE tracer test:

1 liter per minute approximates pouring volatile solvent back and forth.

4 liters per minute is an intermediate between 1 and 8 liters per minute.

8 liters per minute approximates violently boiling water on a 500 watt hotplate.

Particulate: Of, pertaining to, or formed of separate particles.

Perchloric Acid Hood: A special purpose hood designed primarily to be used with perchloric acid. Perchloric acid fume hoods are mandatory for research in which perchloric acid is used because of the explosion hazard associated with this chemical.

Perforated Ceiling: An air distribution device. Perforated ceiling panels or filter pads are used to distribute the air uniformly throughout the ceiling or a portion of the ceiling.

Performance or Test Rating: The performance or test rating of a fume hood is a series of numbers consisting of a two digit number, which indicates the release rate of the tracer gas in liters per minute. The letters AM indicate "as manufactured" and a two or three digit number indicates the level of control of the tracer gas in parts per million of air by volume as established by the test. For example, a rating of 1.0 AM 10 would indicate that the hood controls to 10 ppm at a release rate of 1 lpm.

Plenum: An air chamber or compartment.

PPM: Parts Per Million. Parts of tracer gas per million parts of air by volume.

Radioisotope Hood: A special purpose hood design primarily used with radiochemicals or radioactive isotopes. Special filtering and shielding are required. Radioisotope hoods have exhaust ducts with flanged, neoprene gaskets with quick disconnect fasteners that can be quickly dismantled for decontamination.

Release Rate: The rate of release in liters per minute of tracer gas during a fume hood test.

Reverse Air Flow: Air movement toward the front of the hood.

Sash: The moveable, normally transparent panel set in the fume hood entrance. Sashes may be vertical or a combination of horizontal and vertical. The combination horizontal/vertical sash has horizontally sliding sashes set in a vertical rising sash. With the com-

bination sash, the vertical sash allows for easier setup or removal of hood equipment or apparatus, while the horizontal sash facilitates user operations and also reduces total exhaust air volume.

Slot Velocity: The speed of the air moving through the fume hood baffle openings.

Smoke Candle: A smoke producing device used to allow visual observation of airflow.

Specified Rating: The hood performance rating as specified, proposed or guaranteed.

Spot Collector Hood: A small, localized ventilation hood usually connected by a flexible duct to an exhaust fan. The spot collector hood is not considered a laboratory fume hood.

Stretched String Distance: The shortest distance from exhaust opening to intake opening over and along the building surface.

Supply Air Devices: Devices or openings through which air flows into the laboratory space.

Threshold Limit Values (TLV): The values for airborne toxic materials which are to be used as guides in the control of health hazards. They represent time weighted concentrations to which workers may be exposed 8 hours a day over extended periods of time without adverse effects.

Titanium Tetrachloride (TiCL$_4$): Titanium tetrachloride is a chemical that generates white smoke. It is used to test the air pattern in laboratory fume hoods. Titanium tetrachloride is corrosive and irritating. It can stain the hood and will produce a residue that must be cleaned up. Care must be taken to minimize the effects on the hood. Skin contact or inhalation should be avoided.

Zone Pressurization: Zone pressurization is a means of isolating spaces that generate harmful contaminants. Zone pressurization means that the air distribution system is designed so the hazardous areas have negative pressure and any airborne contaminants are contained in the negative pressurized areas.

MOTOR TERMS

Full Load Amperage: The full load operating current at rated voltage and horsepower.

Locked Rotor Amperage: Locked rotor amperage occurs between zero and full motor speed when the starting current is drawn from the line with the rotor locked and with rated voltage supplied to the motor. During this short time (a fraction of a second for small motors to a second or longer for large motors), the locked rotor amperage far exceeds the full load operating current. Locked rotor amperage will generally be 5 to 6 times the full load amperage. This inrush of current will continue to decrease until the motor reaches full operating speed.

Motor rpm: The rated motor speed. Motors operate at different speeds according to their type, construction and the number of magnetic poles in the motor. Some single phase motors are designed for multiple rpm by switching the winding connections, so two to four different speeds are available. Wiring diagrams are usually provided on the motor.

Motor Voltage: The rated operating voltage.

Nonoverloading Fan Motor: A motor selected for a fan where the horsepower curve increases with an increase in air quantity, but only to a point to the right of maximum efficiency, and then gradually decreases. If a motor is selected to handle the maximum brake horsepower shown on the fan performance curve, the motor will not overload in any condition of fan operation.

Service Factor: The number by which the horsepower or amperage rating is multiplied to determine the maximum safe load that a motor may be expected to carry continuously at its rated voltage and frequency. Typical service factors are: 1.0, 1.10 and 1.15 for large motors and 1.20, 1.25, 1.30, and 1.40 for small motors.

Single Phasing: The condition which results when one phase of a three-phase motor circuit is broken or opened. Motors won't start under this condition, but if already running when it goes into the single phase condition, the motor will continue to run with a lower power output and possible overheating.

POWER TERMS

Air Horsepower: The theoretical horsepower required to drive a fan if the fan were 100% efficient.

Brake Horsepower: The total horsepower applied to the drive shaft of any piece of rotating equipment. The actual power required to drive a fan or pump. Air horsepower divided by fan efficiency. Water horsepower divided by pump efficiency.

Btuh or Btu/hr: A unit of power. Btu per hour.

Horsepower: A unit of power. One horsepower equals 746 watts.

Kilowatt: A unit of power.

Power: The rate of doing work. Electrical power is measured in watts or kilowatts. Other units of power are horsepower and Btuh.

Power Factor: The ratio of actual power to apparent power.

Volt-Ampere: A unit of apparent power.

Water Horsepower: The theoretical horsepower required to drive a pump if the pump were 100% efficient.

Watt: A unit of actual power.

PSYCHROMETRIC TERMS

Btu: A unit of heat. Btu stands for British thermal unit. The amount of heat required to raise one pound of water 1°F. The term Btu/hr or Btuh is used to quantify heat losses and heat gains in the conditioned space and to identify the heating and cooling capacities of various types of equipment.

Coil Bypass Factor: The amount of air passing through the coil without being affected by the coil temperature. It is a ratio of the leaving air dry bulb temperature minus the coil temperature, divided by the entering air dry bulb temperature minus the coil temperature. A more efficient coil (a coil with more fins per inch and/or more rows) has a lower coil bypass factor.

Dew Point: The temperature at which moisture will start to condense from the air.

Dry Bulb Temperature: The temperature of the air read on an ordinary thermometer.

Enthalpy: The measurement of the heat content of the air in Btu per pound of dry air.

Evaporative Cooling: The adiabatic exchange of heat between air and a water spray or wetted surface. The wet bulb temperature of the air remains constant, but the dry bulb temperature is decreased.

Latent Heat: The heat, which when supplied to or removed from a substance, causes a change of state without any change in temperature. The units of latent heat are Btu per pound of dry air.

Relative Humidity: The ratio of the amount of moisture present in the air to the total amount of moisture that the air can hold at a given temperature. Relative humidity is expressed as a percentage.

Sensible Heat: The heat which causes a temperature change in a substance. The units of sensible heat are Btu per pound of dry air.

Sensible Heat Ratio: Sensible heat ratio is the ratio of the sensible heat to the total heat in the air.

Specific Heat: The ratio of the amount of heat required to raise the temperature of one pound of substance 1°F as compared to the amount required to raise one pound of water 1°F. The specific heat of air is 0.24 Btu/lb °F. The specific heat of water is 1 Btu/lb °F.

Specific Humidity: The weight of water vapor associated with one pound of dry air. Specific humidity is measured in grains of moisture per pound of dry air or pounds of moisture per pound of dry air. 7,000 grains equals one pound. Also called humidity ratio.

Specific Volume: The volume of a substance per unit weight. For air, specific volume is in cubic feet per pound. Specific volume is the reciprocal of density.

Standard Air Conditions: Dry air having the following properties: 70°F, 29.92 inches of mercury, 13.33 cubic feet per pound and 0.075 pounds per cubic foot.

Total Heat: The sum of latent heat and sensible heat. The units of total heat are Btu per pound of dry air.

Wet Bulb Depression: The temperature difference between the dry bulb temperature and the wet bulb temperature.

Wet Bulb Temperature: Wet bulb temperature is obtained by an ordinary thermometer with the sensing bulb covered with a wet wick and exposed to air moving at a velocity above 700 feet per minute. For wet bulb temperatures below 32°F, the wick is frozen.

PUMP TERMS

Cavitation: The phenomena occurring in a flowing liquid when the pressure falls below the vapor pressure of the liquid, causing the liquid to vaporize and form bubbles. The bubbles are entrained in the flowing liquid and are carried through the pump impeller inlet to a zone of higher pressure where they suddenly collapse or implode with terrific force. The following are symptoms of a cavitating pump: snapping and crackling noises at the pump inlet, severe vibration, a drop in pressure and brake horsepower, a reduction in flow or the flow stops completely.

Dynamic Discharge Head: Static discharge head plus friction head plus velocity head.

Dynamic Suction Head: Static suction head minus friction head loss and velocity head.

Dynamic Suction Lift: Static suction lift plus friction head loss plus velocity head.

Friction Head: The pressure required to overcome the resistance to flow, expressed in psi or feet of head.

Net Positive Suction Head: The minimum suction pressure at the pump to overcome all the factors limiting the suction side of the pump; internal losses, elevation of the suction supply, friction losses, vapor pressure and altitude of the installation.

Net Positive Suction Head Available: Net positive suction head available (NPSHA) is a characteristic of the system in which the pump operates. The factors influencing NPSHA are elevation of the suction supply in relation to the pump, the friction loss in the suction pipe, the altitude of the installation or the pressure on the suction supply and vapor pressure. In determining NPSHA these considerations must be evaluated and a pump selected for the worst conditions likely to be encountered in the installation. As a safety factor, the NPSHA should always exceed the NPSHR by two feet or more.

Net Positive Suction Head Required: Net Positive Suction Head Required (NPSHR) is the actual absolute pressure needed to overcome the pump's internal losses and allow the pump to operate satisfactorily. NPSHR is determined by the pump manufacturer through laboratory tests. NPSHR is a fixed value for a given capacity and does not vary with altitude or temperature. It does, however, vary with each pump capacity and speed change. NPSHR for a specific pump is available from the manufacturer either on submittal data, from a pump curve or from a catalog. A pump curve will give the full range of NPSHR values for each impeller size and capacity.

Nonoverloading Pump: A pump that is selected where the horsepower curve increases with an increase in water quantity—but only to a point near maximum capacity—and then gradually decreases. If a motor is selected to handle the maximum brake horsepower shown on the pump performance curve, the motor will not overload in any condition of pump operation.

Pump Performance Curve: A pump performance curve is a graphic representation of the performance of a pump. Pumps are generally selected so their design operating point falls about midway, plus or minus 1/3, of the published curve. This allows for changes in installation conditions.

Primary-Secondary Pump: The function of the primary pump in a primary-secondary circuit is to circulate water around the primary circuit. The function of the secondary pump is to supply the terminals.

Pump Efficiency: The output of useful energy divided by the power input. Water horsepower divided by brake horsepower.

Shut-off Head: The pressure developed by the pump when its discharge valve is shut. On the pump curve, it's the intersection of the head-capacity curve with the zero capacity line.

Static Head: The static pressure of a fluid expressed in terms of the height of a column of the fluid.

Static Discharge Head: The vertical distance from the centerline of the pump to the free discharge liquid level.

Static Suction Head: The vertical distance from the centerline of the pump to the suction liquid free level.

Static Suction Lift: The vertical distance from the centerline of the pump down to the suction liquid free level.

Suction Head: When the source of supply is above the pump centerline.

Suction Lift: When the source of supply is below the pump centerline.

Total Discharge Head: The static discharge head plus friction losses plus velocity head.

Total Dynamic Head: The total discharge head minus the total suction head or the total discharge head plus suction lift. Suction head is when the water source is above the pump centerline. Suction lift is when the water source is below the pump centerline. For test and balance purposes, TDH is the difference between the gage pressure at the pump discharge and the gage pressure at the pump suction.

Total Head: In a flowing fluid, the sum of the static and velocity heads at the point of measurement.

Total Static Head: The vertical distance in feet from the suction liquid level to the discharge liquid level. The sum of static suction lift and static discharge head. The difference between static suction head and static discharge head.

Velocity Head: The head required to create flow. The height of the fluid equivalent to its velocity pressure.

Volute Pump: A pump having a casing made as a spiral or volute curve. The volute casing starts with a small cross sectional area near the impeller and increases gradually to the pump discharge.

REFRIGERATION TERMS

Absorbent: Used in the refrigerant system as a dehydrator. A substance which has the ability of taking in or absorbing another substance. The absorbent removes moisture by chemical action which converts the water to some other substance or compound. In some absorption systems, lithium bromide is the absorbent. In other systems, water is the absorbent.

Absorber: A vessel containing liquid for absorbing refrigerant vapor.

Adsorbent: Used in the refrigerant system as a dehydrator. The adsorbent removes moisture as water and holds it as water and no chemical action takes place.

Back Pressure: See Evaporator Pressure.

Bottom Dead Center: When the piston in a reciprocating compressor is at the bottom of its stroke.

Clearance: The space between the piston head and the end of the cylinder when the piston is at the top of its stroke.

Coefficient Of Performance (COP): Ratio of work performed to energy used.

Cooling Tower: A cooling tower cools heated water from a water-cooled condenser. As outdoor air passes through the cooling tower, it removes heat from the condenser water. The water is cooled to the wet bulb temperature of the air.

Compressor: The pump in the refrigeration system. It takes the low temperature, low pressure refrigerant gas from the evaporator and compresses it to a high temperature, high pressure gas.

Compression Ratio: Compression ratio is the ratio of the discharge pressure to the suction pressure. By equation, it is the absolute discharge pressure (psia) divided by the absolute suction pressure (psia). Compression ratio is of primary importance when it approaches a high limit. A high compression ratio (a high head pressure and low suction pressure) results in a loss of efficiency and excessive superheating of the discharge vapor which can damage the compressor.

Concentrated Solution: A solution with a large concentration of absorbent as compared to the amount of dissolved refrigerant.

Concentrator: A vessel containing a solution of absorbent and refrigerant to which heat is applied for the purpose of boiling away some of the refrigerant.

Condenser: The condenser receives the high temperature, high pressure refrigerant gas from the compressor and cools the gas to a high temperature, high pressure liquid. Condensers may be air-cooled or water-cooled.

Condenser Rise: The difference between the temperature of the water leaving the condenser and the temperature of the water entering the condenser.

Condensing Pressure: The saturation pressure corresponding to the temperature of the liquid-vapor mixture in the condenser.

Condensing Temperature: The temperature at which the refrigerant vapor condenses in the condenser and is the saturation temperature of the vapor corresponding to the pressure in the condenser.

Cylinder Bypass: A method of compressor capacity control.

Cylinder Unloaders: See Unloaders.

Cylinder Unloading: A method of compressor capacity control.

Dilute Solution: A solution with a small concentration of absorbent as compared to a large amount of dissolved refrigerant.

Discharge Temperature: The temperature at which the refrigeration vapor is discharged from the compressor.

Enthalpy: The measurement of the heat content of the refrigerant in Btu per pound.

Evaporation: The heating of a liquid refrigerant to convert it to a vapor.

Evaporator: The part of the refrigeration system in which the refrigerant is vaporized and absorbs heat. The evaporator receives the liquid from the condenser by way of the metering device. In the evaporator, the low temperature, low pressure liquid is heated and changes state to a low temperature, low pressure gas. If the evaporator is in an air conditioning unit, it's also called an evaporator coil, an evap coil, or a DX (dry-expansion/direct expansion) coil.

Evaporator Coil: A coil containing a refrigerant other than water which is used for cooling the air. Also called a refrigerant coil, an evaporator or evap coil.

Evaporator Pressure: The pressure of the refrigerant vapor in the suction line. Also called suction pressure, back pressure or low-side pressure.

Filter-Drier: See Strainer-Drier.

Flash Gas: Instantaneous evaporation of some liquid refrigerant in the metering device which cools the remaining liquid refrigerant to a desired evaporator temperature. The liquid refrigerant, which is boiled off or flashed, changes state to a vapor. Unwanted flashing may occur in the liquid line or in the metering device (excess flash gas).

Generator: See Concentrator.

Head Pressure: Pressure of the refrigerant vapor on the discharge of the compressor. Also called discharge pressure or high-side pressure.

Heat of Absorption: The heat released when two liquids are mixed.

Heat of Compression: The energy equivalent of the work done on the refrigerant vapor to compress it. Heat of compression occurs in the compressor. It is added to the heat absorbed in the evaporator plus any superheat to calculate the total amount of heat that needs to removed from the refrigeration system.

Heat of Dilution: See Heat of Absorption.

Hot Gas Bypass: A method of compressor capacity control.

Latent Heat of Condensation: The amount of heat released by a pound of refrigerant to change its state from a vapor to a liquid.

Latent Heat of Vaporization: The amount of heat required by a pound of refrigerant to change its state from a liquid to a vapor. The heat that a liquid will absorb when going from the liquid state to the vapor state. The heat that a vapor will give up when going from the vapor state to the liquid state.

Liquid Slugging: When liquid refrigerant enters the compressor. Liquid slugging causes noisy operation, loss of capacity, an increase in power required and possible damage to the compressor.

Metering Device: A capillary tube, thermostatic expansion valve, automatic expansion valve or other device that reduces the refrigerant pressure and corresponding temperature and controls the flow of refrigerant into the evaporator coil. The metering device reduces the temperature and pressure of the liquid.

The metering device controls the flow of refrigerant and changes the high temperature, high pressure liquid refrigerant from the condenser to a low temperature, low pressure liquid.

Moisture Content: Percent by weight of the liquid in any mixture of liquid and vapor.

Quality of Vapor: Percent by weight of the vapor in any mixture of liquid and vapor.

Receiver: The apparatus that receives and holds the liquid refrigerant from the condenser.

Refrigerant: Any substance that acts as a cooling agent by absorbing heat from another body or substance. Refrigerant fluids that are used in the refrigeration system to absorb heat through vaporization and release heat through condensation. The refrigerant for absorption chillers will normally be water with lithium bromide as the absorbent, or ammonia as the refrigerant with water as the absorbent.

Refrigerating Effect: The quantity of heat that each unit mass of refrigerant absorbs from the conditioned space. Refrigerating effect per unit mass of liquid refrigerant is potentially equal to its latent heat of vaporization. Due to the flash gas process, the refrigerating effect per unit mass of liquid refrigerant circulated is always less than the total latent heat of vaporization.

Refrigeration: The branch of science that deals with the process of reducing and maintaining the temperature of a space or material below the temperature of the surroundings. Refrigeration is the transfer of heat from one place where it is not wanted to another place where it is unobjectionable. This transfer of heat is through a change in state of a refrigerant.

Refrigeration Load: The rate at which heat must be removed from the refrigerated space or material in order to maintain the desired temperature conditions.

Saturation Temperature: The temperature at which a fluid will change from a liquid to a vapor or vice versa. Increasing the pressure on the fluid raises the saturation temperature. The saturation temperature depends on the pressure on the fluid. Saturation temperature increases with an increase in pressure and decreases with a decrease in pressure.

Saturated Liquid: A liquid at the saturation temperature. A liquid cannot exist as a liquid at any temperature above its saturation temperature corresponding to the pressure. A liquid cannot exist as a liquid at any pressure below its saturation pressure.

Saturated Vapor: A vapor at the saturation temperature. A vapor cannot exist as a vapor at any temperature below its saturation temperature corresponding to the pressure. A vapor cannot exist as a vapor at any pressure above its saturation pressure.

Slugging: See Liquid Slugging.

Strainer-Drier: A combination device used as a strainer and moisture remover. Removes moisture and solid particles from the refrigerant before entering the metering device. Usually found in the liquid line. Also called filter-drier.

Suction Pressure: Low-side pressure, evaporator pressure or back pressure.

Subcooling: Cooling the liquid refrigerant out of the condenser below the condensing temperature.

Subcooled Liquid: A liquid at any temperature below the saturation temperature. When the temperature of a liquid has been decreased below its saturation temperature. Subcooling is sensible heat.

Superheat: The sensible heat added to a vapor after vaporization. The sensible temperature of the vapor above its boiling temperature.

Superheated Vapor: A vapor at any temperature above the saturation temperature. When the temperature of a vapor has been increased above its saturation temperature. Superheating is sensible heat.

Sublimination: When a substance goes directly from the solid state to a vapor state without passing through the liquid state.

Top Dead Center: When the piston is at the top of its stroke.

Total Heat Rejected: The total heat rejected at the condenser includes both the heat absorbed in the evaporator and the heat of compression, plus any superheat.

Tower Approach: The difference between the temperature of the water leaving the cooling tower and the wet bulb temperature of the air entering the tower.

Tower Range: The difference between the temperature of the water leaving the cooling tower and the temperature of the water entering the cooling tower.

Unloaders: Cylinder unloaders reused for capacity control in reciprocating compressors.

Vapor: A gas.

Vapor-Compression Cycle: A four-step process which includes expansion, vaporization, compression, and condensation.

Vaporization: The conversion of a solid or a liquid into a vapor.

WATER TERMS

Closed System: A closed system is when there's no break in the piping circuit and the water is closed to the atmosphere.

Coil Face Area: The area (width × height) across which the air flows.

Direct Return: In a direct return system, the return is routed to bring the water back to the pump by the shortest possible path. The terminals are piped "first in, first back; last in, last back." The direct return arrangement is popular because generally, less main pipe is needed. However, since water will follow the path of least resistance, the terminals closest to the pump will tend to receive too much water, while the terminals farthest from the pump will starve. To compensate for this, balancing valves are required in the branches.

Feet Per Second (fps): Water velocity.

Fixed Pipe System: A system in which there are no changes in resistance resulting from changes in the condition of the system, such as opening or closing of valves or changes in the condition of strainers or coils. For a fixed system, an increase or decrease in system resistance results only from an increase or decrease in gallons per minute. This change in resistance will fall along the system curve.

Four-Pipe System: The four-pipe system is two separate two-pipe arrangements. One two-pipe arrangement is for chilled water and one is for heating water. No mixing occurs. The return connections from the terminals can be made either direct or reverse return.

High Temperature Water: Temperature range of 350 to 450°F.

Hydronics: The science of heating and cooling with liquids.

Low Temperature Water: Temperature range to 250°F.

Make-up Water: The water that replaces the water lost through leakage and evaporation. To prevent air problems, the make-up water to a closed system should be introduced into the system at some point either in the air line to the compression tank or at the bottom of the compression tank.

Medium Temperature Water: Temperature range 250 to 350°F.

One-pipe System: The one-pipe system is used for individual space control in residential and small commercial and industrial heating applications. This piping arrangement uses a single loop main, but differs from the series loop arrangement since each terminal is connected by a supply and return branch pipe to the main. Because the terminal has a higher pressure drop than the main, the water circulating in the main will tend to flow through the straight run of the tee fittings. This starves the terminal. To overcome this problem, a diverting tee is installed in either the supply branch, return branch or sometimes, both branches. The advantage of the one-pipe main arrangement over the series loop is that each terminal can be separately controlled and serviced by installing the proper valves in the branches. However, as with the series loop arrangement, if there are too many terminals, the water temperature at the terminals farthest from the boiler may not be adequate. To have the water temperature to each terminal equal to the temperature at which it's generated, two-pipe arrangements are used.

Open System: An open system is when there's a break in the piping circuit and the water is open to the atmosphere.

Operating Point: The intersection of the system curve with the pump curve.

Primary-Secondary Circuit: Primary-secondary circuits reduce pumping horsepower requirements while increasing system control. The primary pump and the secondary pump have no effect on each other when the two circuits are properly interconnected. The secondary flow may be less than, equal to or greater than the primary flow.

Reverse Return: In a reverse return system, the return is routed so the length of the circuit to each terminal and back to the pump is essentially equal. The terminals are piped "first in, last back; last in, first back." Because all the circuits are essentially the same length, reverse return systems generally need more piping than direct return systems, but are considered more easily balanced. However, balancing valves in the branches are still required.

Series Loop: The series loop piping arrangement is generally limited to residential and small commercial heating applications. In a series loop, supply water is pumped through each terminal in series and then returned back to the boiler. The advantage to this type of piping arrangement is it's simple and inexpensive. The disadvantages are that if repairs are needed on any terminal, the whole system must be shut down. Also, it's not possible to provide a separate capacity control to any individual terminal, since valving down one terminal reduces flow to the terminals down the line. However, space heating can be controlled through dampering airflow. These disadvantages can be partly remedied by designing the piping with two or more circuits and installing balancing valves in each circuit. This type of arrangement is called a split series loop.

The series loop circuit length and pipe size are also important because they directly influence the water flow rate, temperature and pressure drop. For instance, as the heating supply water flows through the terminals, its temperature drops continuously as it releases heat in each terminal. If there are too many terminals in series, the water temperature in the last terminal may be too cool.

Three-pipe System: A three-pipe arrangement has two supply mains and one return main. One supply circulates chilled water from the chiller and the other supply circulates heating water from the boiler.

Two-Pipe System: Two-pipe arrangements have two mains—one for supply and one for return. Each terminal is connected by a supply and return branch to its main. This design not only allows separate control and servicing of each terminal, but because the supply water temperature is the same at each terminal, two-pipe systems can be used for any size application.

Wiredrawing: Occurs when a high velocity water stream through a valve causes the erosion of the valve seat. Eventually, this erosion will cause leakage when the valve is fully closed.

a	acceleration
A/D	analog to digital
ADP	appartus dew point
AEV, AXV	automatic expansion valve
AHU	air handling unit
ATM	atmosphere
B	boiler
bhp	brake horsepower
Btu	British thermal unit
Btuh, But/hr	British thermal unit per hour
BTUHl	British thermal unit per hour latent
BTUHs	British thermal unit per hour sensible
BTUHt	British thermal unit per hour total
Btum	Brithsh thermal unit per minute
c	coil
C	condenser
CD	ceiling diffuser
cf	cubic feet
CFC	chloroflurocarbons
cfm	cubic feet per minute
CH	chiller
CHWR, CHR	chilled water return
CHWS, CHS	chilled water supply
C-HWS	chilled-heating water supply

C-HWR	chilled-heating water return
ci	cubic inches
comp	compressor
cond	condenser
CWR, CR	condenser water return
CWS, CS	condenser water supply
D	density
DA	direct acting
D/A	digital to analog
DB	dry bulb temperature, decibels
DIDW	double inlet, double wide
DDC	Direct digital control
DP	delta (Δ) P, pressure difference, differential
DP, dp	dew point temperature
DTW	dual temperature water
DX	direct exchange or direct expansion
EA	exhaust air, exhaust air duct or inlet
EAT	entering air temperature
econ	economizer
EF	exhaust fan
eff	efficiency
EMCS	Energy Mangement Control System
EMS	Energy Management System
EP, E-P	electric to pneumatic
evap	evaporator
EWT	entering water temperature
FCU, F-C	fan coil unit
ft. hd	feet of head
ft. H_2O	feet of water
ft-lb	foot-pounds
fpm	feet per minute
fps	feet per second
FSP	fan static pressure
FSE	fan static efficiency
FTE	fan total efficiency
FTP	fan total pressure

ft. wc	feet of water column
ft. wg	feet of water gauge
G	Standard Acceleration of Gravity
G	grille, grains
Gr, gr	grains
h	enthalpy
hp	horsepower
HTR	high temperature return
HTS	high temperature supply
HTW	high temperature water
HWR, HR	heated water return
HWS, HS	heated water supply
HX	heat exchanger
IAQ	indoor air quality
in. Hg	inches of mercury
in. wc	inches of water column
in. wg	inches of water gauge
K	Kilo, 1,000
kw	kilowatt, 1,000 watts
kwh	kilowatt-hour, 1,000 watt-hours
LAD	linear air diffuser
LAT	leaving air temperature
lbs	pounds
lbs/cf	pounds per cubic feet
lbs/hr	pounds per hour
lbs/min	pounds per minute
lbs/sec	pounds per second
LT	light troffer
LTR	low temperature return water
LTS	low temperature supply water
LTW	low temperature water
LWT	leaving water temperature
M	Roman numeral, 1,000
m	mass
MBH	1,000 British thermal unit per hour
MD	metering device

MTR	medium temperature return water
MTS	medium temperature supply water
MTW	medium temperature water
NC	normally closed
NO	normally open
OA	outside air
OSA	outside air
Pa	pressure absolute
PD	pressure drop or pressure difference
PE	professional mechanical engineer
PE, P-E	pneumatic to electric
PID	proportional-integral-derivative
PPM	parts per million
PROM	programmable read only memory
RAM	random access memory
ROM	read only memory
RTD	resistance temperature detector
psf	pounds per square foot
psia	pounds per square inch absolute
psi	pounds per square inch
psig	pounds per square inch gauge
OV	outlet velocity
RA	return air, return air duct or inlet
RA	reverse acting
RAF, RF	return air fan, relief air fan
RH	relative humidity
SA	supply air, supply air duct or outlet
SAF, SF	supply air fan
sf	square feet
SHF	sensible heat factor
SHR	sensible heat ratio
SISW	single inlet, single wide
SP	static pressure
SW	side wall grille
TD	temperature difference
TEV	thermostatic expansion valve

TLV	threshold limit value
TP	total pressure
TS	tip speed
TSP	total static pressure
TXV	thermostatic expansion valve
VP	velocity pressure
V	volts
vdc	direct current voltage
W	watt
wb	wet bulb temperature
Wh	watt-hour

APPENDIX C *Tables*

AIR CHANGES PER HOUR TABLE

Air Changes Per Hour	Ceiling height	CFM/SF
7.5	8 ft.	1
6	9 ft.	0.9

AIR VELOCITY TABLE—RECOMMENDED AIR VELOCITIES

Apparatus	FPM
Coil, Chilled water	500–600
Coil, Hot water	400-700
Duct, Branch, Office	800–1,600
Duct, Fume Exhaust	1,500–2,000
Duct, Main, Office	1,200–2,400
Filter, Electronic	500
Filter, Fiber, Dry	750
Filter, Fiber, HEPA	250
Filter, Fiber, Viscous	700
Filter, Renewable, Dry	200
Filter, Renewable, Viscous	500
Louvers, Exhaust	500
Louvers, Intake	400
Outlets	400–800
Stack, Fume Exhaust	2,500–3,000

AIR VOLUME TABLE (APPROXIMATE)

Description	CFM
Cooling, General	400 cfm per ton
Cooling, General	30 Btuh per cfm
Cooling, General	1 cfm per square foot
Cooling, Theater, Church, Auditorium	20–30 cfm per seat
Heating, General	12 cfm per 1,000 Btu input
Outside Air Requirements, General	15–25 cfm per person

ALTITUDE—PRESSURE TABLE

Altitude in feet	Inches of mercury	Feet of water
−1,000	31.02	35.1
0	29.92	33.9
1,000	28.86	32.8
2,000	27.72	31.6
3,000	26.81	30.5
4,000	25.84	29.4
5,000	24.89	28.2
6,000	23.98	27.3
7,000	23.09	26.2
8,000	22.22	25.2
9,000	21.38	24.3
10,000	20.58	23.4

CONVERSION TABLE

Unit	Equals
One atmosphere	33.9 ft. wc 407 in. wc 14.7 psi 29.92 in. Hg
One boiler horsepower	33,475 Btuh 34.5 pounds of water
One Btuh	0.000393 horsepower 0.000293 kilowatts
One cubic foot	1,728 cubic inches
One cubic foot of water	7.5 gallons 62.4 pounds
One foot of water	0.833 in. Hg 12 in. wg 0.433 psi
778 foot-pounds	0.000393 horsepower-hours 0.000293 kilowatt-hours
One gallon of water	231 cubic inches 8.33 pounds
One horsepower	2,545 Btuh 42.42 Btu per minute 550 foot-pounds per second 33,000 foot-pounds per minute 0.746 kilowatts 746 watts
One inch of mercury	1.13 ft. wg 13.6 in. wg 0.4391 psi 0.03945 mmHg
One square foot of mercury	70.73 pounds

CONVERSION TABLE (*Continued*)

Unit	Equals
One square foot of mercury	70.73 pounds
One inch of water	0.036 psi 5.2 psf
One kilowatt	3,413 Btuh 1.34 horsepower 56.9 Btum
One mile per hour	88 feet per minute
One pound	7,000 grains
One pound per square inch	2.04 in. Hg 2.31 ft. wg 27.7 in. wg
Standard air	0.075 lb/cf 29.92 in. Hg 70°F 14.7 psi
One ton of refrigeration	12,000 Btuh 200 Btum
One watt	3.41 Btuh 0.00134 horsepower 44.26 foot-pounds per minute
One year	8,760 hours 4,620 hours of light 4,140 hours of darkness

DUCT FRICTION LOSS CORRECTION FOR VARIOUS MATERIALS

Duct Material	Correction Factor
Galvanized Duct	1.00
Fiberglass Duct	1.35
Fibewrglass Lined Duct	1.08–1.42
Flex Duct, Fully Extended	1.85
Flex Duct, Compressed 10%	3.65

Multiply correction factor times calculated friction loss per 100 feet.

DUCT PRESSURE CLASS TABLE

Pressure Class	Static Pressure Inches of Water	Velocity
Low	To 2 inches	To 2,500 fpm
Medium	Between 2 and 6 inches	Between 2,000 and 4,000 fpm
High	Above 6 inches	Above 2,000 fpm

INSULATION TABLE

Material	R-11	R-13	R-19	R-22	R-30	R-33	R-38
				Inches			
Mineral Wool	3.5	3.875	6	6.5	9	11	12
Fiberglass Batt	3.5	3.875	6	6.5	9	11	12
Mineral Blowing Wool	3.75		6.5	7.5	10.25	11.25	13
Fiberglass Blowing Wool	5		8.875	10	13.75	15	18
Cellulose	3		5	6	8	9	10

MOTOR AMPERAGE RATING TABLE

Motor	Three-phase			Single-phase	
Hp	230V	460V	575V	115V	230V
	FLA	FLA	FLA	FLA	FLA
1/2	2	1	0.8	9.8	4.9
3/4	2.8	1.4	1.1	13.8	6.9
1	3.6	1.8	1.4	16	8
1.5	5.2	2.6	2.1	20	10
2	6.8	3.4	2.7	24	12
3	9.6	4.8	3.9	34	17
5	15.2	7.6	6.1	56	28
7.5	22	11	9		40
10	28	14	11		50
15	42	21	17		
20	54	27	22		
25	68	34	27		
30	80	40	32		
40	104	52	42		
50	130	75	52		
60	154	77	62		
75	192	96	77		
100	248	124	99		
125	312	156	125		
150	360	180	144		
200	480	240	192		

Note: Locked rotor amps (LRA) are approximately 6 times full load amps (FLA).

MOTOR EFFICIENCY AND POWER FACTOR TABLE (APPROXIMATE)

Motor	Three-Phase	
Hp	Eff	PF
1/2	70	69.2
3/4	72	72.0
1	79	76.5
1.5	80	80.5
2	80	85.3
3	81	82.6
5	83	84.2

MOTOR EFFICIENCY AND POWER FACTOR TABLE (APPROXIMATE)
(*Continued*)

Motor		*Three-Phase*
Hp	*Eff*	*PF*
7.5	85	85.5
10	85	88.8
15	86	87.0
20	87	87.2
25	88	86.8
30	89	87.2
40	89	88.2
50	89	89.2
60	89	89.5
75	90	89.5
100	90	90.3
125	90	90.5
150	91	90.5
200	91	90.5

Note: Motor efficiency and power factor curves remain fairly flat until the motor load falls below 50%.

MOTOR MAGNETIC STARTER SIZE TABLE

Hp	*230 V*	*460 V*	*575 V*	*115 V*	*230 V*
1/2	00	00	00	0	00
3/4	00	00	00	0	00
1	00	00	00	0	00
1.5	00	00	00	1	0
2	0	00	00	1	0
3	0	0	0	1P/2	1
5	1	0	0	2.5	1P
7.5	1	1	1		2
10	1.75	1	1		2.5
15	2	1.75	1.75		
20	2.5	2	2		
25	3	2	2		
30	3	2.5	2.5		
40	3.5	3	3		
50	4	3	3		
60	4.5	3.5	3.5		
75	4.5	3.5	3.5		
100	5	4	4		
125	6	4.5	4.5		
150	6	4.5	4.5		
200	6	5	5		

MOTOR WIRE SIZE TABLE

Hp	230 V	460 V	575 V	115 V	230 V
1/2	14	14	14	14	14
3/4	14	14	14	12	14
1	14	14	14	12	14
1.5	14	14	14	10	14
2	14	14	14	10	14
3	14	14	14	8	10
5	12	14	14	4	8
7.5	10	14	14		6
10	8	12	14		6
15	6	10	10		
20	4	8	10		
25	4	8	8		
30	3	6	8		
40	1	6	6		
50	2/0	4	6		
60	3/0	3	4		
75	250	1	3		
100	350	2/0	1		
125	2-3/0	3/0	2/0		
150	2-4/0	4/0	3/0		
200	2-350	350	250		

POWER OF 10

	Power	Symbol	Prefix	Value
10	−18	a	atto	0.000,000,000,000,000,001
10	−15	f	femto	0.000,000,000,000,001
10	−12	p	pico	0.000,000,000,001
10	−9	n	nano	0.000,000,001
10	−6	u	micro	0.000,001
10	−3	m	milli	0.001
10	−2	c	centi	0.01
10	−1	d	deci	0.1
10	1	da	deka	10
10	2	h	hecto	100
10	3	k	kilo	1,000
10	6	M	mega	1,000,000
10	9	G (b)	giga	1,000,000,000
10	12	T (t)	tera	1,000,000,000,000
10	15	quadrillion (quad)		1,000,000,000,000,000
10	18	quintillion		1,000,000,000,000,000,000

(b) billion
(t) trillion

PRESSURE TABLE: ABSOLUTE AND GAGE (GAUGE)

Pressure Above Atmospheric (PSI)

Gage Pressure	Absolute Pressure		
50 psig	64.7 psia		
40 psig	54.7 psia		
30 psig	44.7 psia		
20 psig	34.7 psia		
10 psig	24.7 psia		

Atmospheric Pressure

0 psig	14.7 psia		
	29.92 in. Hg	33.9 ft. wg	407 in. wg

Pressure Below Atmospheric (Inches of Mercury)

Gage Pressure	Absolute Pressure			
10 in. Hg	−4.9 psig/9.8 psia	19.92 in. Hg	22.6 ft. wg	271 in. wg
20 in. Hg	−9.8 psig/4.9 psia	9.92 in. Hg	11.3 ft. wg.	136 in. wg
29.92 in. Hg	−14.7 psig/0 psia	0 in. Hg	0 ft. wg.	0 in. wg

REFRIGERATION TABLE

Chilled Water TD	GPM/TON
8	3
10	2.4
12	2
20	1.2

V-BELT TABLE

Standard Industrial V-Belt	Width (inches)
A	1/2
B	21/32
C	7/8
D	1 1/4
E	1 1/2

Fractional Horsepower Belt	Width (inches)
2L	9/32
3L	3/8
4L	1/2
5L	21/32

WATER PROPERTY TABLE

Temp.	Den.	Wt.	VP	SG
50	62.38	8.34	0.41	1.002
60	62.35	8.33	0.59	1.001
70	62.27	8.32	0.84	1.000
80	62.19	8.31	1.17	0.998
90	61.11	8.30	1.62	0.997
100	62.00	8.29	2.20	0.995
110	61.84	8.27	2.96	0.993
120	61.73	8.25	3.95	0.990
130	61.54	8.23	5.20	0.988
140	61.40	8.21	6.78	0.985
150	61.20	8.18	8.74	0.982
160	61.01	8.16	11.20	0.979
170	60.00	8.12	14.20	0.975
180	60.57	8.10	17.85	0.972
190	60.35	8.07	22.30	0.968
200	60.13	8.04	27.60	0.965
210	59.88	8.00	34.00	0.961

Temp. = temperature, degree Fahrenheit

Den. = density, pounds per cubic foot

Wt = weight, pounds per gallon

VP = vapor pressure, feet of water

SG = specific gravity

APPENDIX D *Equations*

GENERAL EQUATIONS

$$\text{Den (lbs/cu. ft.)} = \frac{\text{Mass (lbs)}}{\text{Vol (cu. ft.)}}$$

$$\text{SpV (cu. ft./lb)} = \frac{\text{Vol (cu. ft.)}}{\text{Mass (lbs)}}$$

$$\text{SpV (cu. ft./lb)} = \frac{1}{\text{Den (lbs/cu. ft.)}}$$

$$\text{Den (lb/cu. ft.)} = \frac{1}{\text{SpV (cu. ft./lb)}}$$

$$\text{Mass (lb)} = \text{Vol (cu. ft.)} \times \text{Den (lb/cu. ft.)}$$

$$\text{Mass (lb)} = \frac{\text{vol (cf)}}{\text{SpV}\left(\dfrac{\text{cf}}{\text{lb}}\right)}$$

Mass flow rate = lb/min

$$\text{Vol (cu. ft.)} = \text{Mass (lb)} \times \text{SpV (cu. ft./lb)}$$

$$\text{Vol (cf)} = \frac{\text{mass (lb)}}{\text{Den}\left(\dfrac{\text{lb}}{\text{cf}}\right)}$$

Volume flow rate = cf/min

$$\text{SpG} = \frac{\text{Den}}{\text{Dw}} \quad \text{Dw} = 62.4 \text{ lbs/cu. ft.}$$

$$\text{Work (ft/lb)} = \text{Dist (ft.)} \times \text{F (lb)}$$

Den = density
Vol = volume
Mass = mass
SpV = specific volume
SpG = specific gravity
Dw = density of water
Dist = distance
F = force

AFFINITY LAWS

Subscript 1 is the original field measurement.
Subscript 2 is the desired final condition.

FAN LAWS

$$\text{rpm}_2 = \text{rpm}_1 \times \frac{\text{cfm}_s}{\text{cfm}_1}$$

$$\text{cfm}_2 = \text{cfm}_1 \times \frac{\text{rpm}_2}{\text{rpm}_1}$$

$$\text{cfm}_2 = \text{cfm}_1 \times \frac{Pd_2}{Pd_1}$$

$$Pd_2 = Pd_1 \times \frac{\text{cfm}_2}{\text{cfm}_1}$$

$$\text{rpm}_2 = \text{rpm}_1 \times \frac{Pd_2}{Pd_1}$$

$$Pd_2 = Pd_1 \times \frac{\text{rpm}_2}{\text{rpm}_1}$$

$$\text{cfm}_2 = \text{cfm}_1 \times \sqrt[3]{\frac{\text{amp}_2}{\text{amp}_1}}$$

$$\text{SP}_2 = \text{Sp}_1 \times \left(\frac{\text{rpm}_2}{\text{rpm}_1}\right)^2$$

$$\text{amp}_2 = \text{amp}_1 \times \left(\frac{Pd_2}{Pd_1}\right)^3$$

$$\text{rpm}_2 = \text{rpm}_1 \times \sqrt{\frac{\text{SP}_2}{\text{SP}_1}}$$

$$Pd_2 = Pd_1 \times \sqrt[3]{\frac{\text{amp}_2}{\text{amp}_1}}$$

$$\text{SP}_2 = \text{SP}_1 \times \left(\frac{\text{cfm}_2}{\text{cfm}_1}\right)^2$$

$$\text{bhp}_2 = \text{bhp}_1 \times \left(\frac{\text{SP}_2}{\text{SP}_1}\right)^3$$

$$\text{cfm}_2 = \text{cfm}_1 \times \sqrt{\frac{\text{SP}_2}{\text{SP}_1}}$$

$$\text{bhp}_2 = \text{bhp}_1 \times \left(\frac{\text{SP}_2}{\text{SP}_1}\right)^{1.5}$$

$$\text{SP}_2 = \text{SP}_1 \times \left(\frac{Pd_2}{Pd_1}\right)^2$$

$$\text{SP}_2 = \text{SP}_1 \times \frac{d_2}{d_1}$$

$$Pd_2 = Pd_1 \times \sqrt{\frac{\text{SP}_2}{\text{SP}_1}}$$

$$\text{bhp}_2 = \text{bhp}_1 \times \frac{d_2}{d_1}$$

$$\text{bhp}_2 = \text{bhp}_1 \times \left(\frac{\text{rpm}_2}{\text{rpm}_1}\right)^3$$

cfm = air volume
SP = static pressure (in. wg)
rpm = fan speed
amp = amperage
Pd = pitch diameter of the motor sheave
bhp = brake horsepower
d = air density (lb/cf)

$$\text{rpm}_2 = \text{rpm}_1 \times \sqrt[3]{\frac{\text{bhp}_2}{\text{bhp}_1}}$$

$$\text{bhp}_2 = \text{bhp}_1 \times \left(\frac{\text{cfm}_2}{\text{cfm}_1}\right)^3$$

PUMP LAWS

$$\text{cfm}_2 = \text{cfm}_1 \times \sqrt[3]{\frac{\text{bhp}_2}{\text{bhp}_1}}$$

$$\text{gpm}_2 = \text{gpm}_1 \times \frac{D_2}{D_1}$$

$$\text{bhp}_2 = \text{bhp}_1 \times \left(\frac{Pd_2}{Pd_1}\right)^3$$

$$D_2 = D_1 \times \frac{\text{gpm}_2}{\text{gpm}_1}$$

$$Pd_2 = Pd_1 \times \sqrt[3]{\frac{\text{bhp}_2}{\text{bhp}_1}}$$

$$\text{TDH}_2 = \text{TDH}_1 \times \left(\frac{D_2}{D_1}\right)$$

$$\text{amp}_2 = \text{amp}_1 \times \left(\frac{\text{rpm}_2}{\text{rpm}_1}\right)^3$$

$$D_2 = D_1 \times \sqrt{\frac{\text{TDH}_2}{\text{TDH}_1}}$$

$$\text{rpm}_2 = \text{rpm}_1 \times \sqrt[3]{\frac{\text{amp}_2}{\text{amp}_1}}$$

$$\text{TDH}_2 = \text{TDH}_1 \times \left(\frac{\text{gpm}_2}{\text{gpm}_1}\right)^2$$

$$\text{amp}_2 = \text{amp}_1 \times \left(\frac{\text{cfm}_2}{\text{cfm}_1}\right)^3$$

$$\text{gpm}_2 = \text{gpm}_1 \times \sqrt{\frac{\text{TDH}_2}{\text{TDH}_1}}$$

$$bhp_2 = bhp_1 \times \left(\frac{D_2}{D_1}\right)^3$$

$$D_2 = D_1 \times \sqrt[3]{\frac{bhp_2}{bhp_1}}$$

$$bhp_2 = bhp_1 \times \left(\frac{gpm_2}{gpm_1}\right)^3$$

$$gpm_2 = gpm_1 \times \sqrt[3]{\frac{bhp_2}{bhp_1}}$$

gpm = volume of water flow
D = diameter of the impeller (inches)
TDH = total dynamic head (feet of water)
bhp = brake horsepower

AIRFLOW EQUATIONS

Air Velocity

$$v = 4{,}005\sqrt{VP}$$

v = velocity (fpm)
4,005 = constant
VP = velocity pressure (in. wg)

$$V = \frac{Q}{A}$$

V = velocity (fpm)
Q = volume of air (cfm)
A = cross-sectional area of the duct (sf)

Air Velocity Pressure

$$VP = \left(\frac{v}{4{,}005}\right)^2$$

v = velocity (fpm)
4,005 = constant
VP = velocity pressure (in. wg)

Air Volume—Cubic Feet per Minute

$$Q = AV$$

Q = volume of air (cfm)
A = cross sectional area of the duct (sf)
V = velocity (fpm)

Air Density

$$d_c = 1.325\frac{Pb}{T}$$

d_c = calculated air density (lb/cf)
1.325 = constant, 0.075 lb/cf divided by (29.92 in. Hg/530°F)
Pb = barometric pressure (in. Hg)
T = absolute temperature (indicated temperature in °F plus 460)

Instrument Correction Factor for Air Density

For instruments calibrated to standard air conditions:

$$F_c = \sqrt{\frac{0.075}{d}}$$

F_c = correction factor
0.075 = density of standard air (lb/cf)
d = calculated air density (lb/cf)

Air Velocity with Instrument Corrected for Density

$$Vc = Vm \times Fc$$

Vc = corrected velocity (fpm)
Vm = measured velocity (fpm)
F_c = correction factor

Air Volume with Instrument Correction for Density

$$Q = A \times Vc$$

Q = volume of air (cfm)
A = cross sectional area of the duct (sf)
Vc = corrected velocity (fpm)

Air Volume—Pounds per Hour

$$lbs/hr = cfm \times 4.5$$

$$cfm = \frac{lb/hr}{4.5}$$

lbs/hr = pounds per hour of air
cfm = quantity of airflow
4.5 = constant (60 minutes per hour × 0.075 lb/cf)

Air Volume Through Orifice Plate

$$cfm = 3{,}144\ CD_2\sqrt{PD}$$

cfm   =   quantity of airflow
3,144 =   constant
  C   =   airflow constant for orifice
  D   =   diameter of orifice in feet
 PD   =   measured pressure drop across the orifice, inches of water

$$\text{cfm} = 21.8 \, CD_2 \sqrt{PD}$$

cfm   =   quantity of airflow
21.8 =   constant
  C   =   airflow constant for orifice
  D   =   diameter of orifice in inches
 PD   =   measured pressure drop across the orifice, inches of water

$$\text{cfm} = 861 \, CD^2 \sqrt{\frac{PD}{d}}$$

cfm   =   quantity of airflow
861   =   constant
  C   =   airflow constant for orifice
  D   =   diameter of orifice in feet
 PD   =   measured pressure drop across the orifice, inches of water
  d   =   0.075 density of standard air

Air Volume for Furnaces

Gas Furnace

$$\text{cfm} = \frac{HV \times \text{cfh} \times \text{eff}}{1.08 \times TR}$$

Oil Furnace

$$\text{cfm} = \frac{HV \times \text{gph} \times \text{eff}}{1.08 \times TR}$$

Electric Furnace

$$\text{cfm} = \frac{V \times A \times 3.41}{1.08 \times TR}$$

cfm =   volume of air flow
HV   =   heating value (gas = Btu/cf; oil = Btu/gal)
cfh =   cubic feet per hour
gph =   gallons per hour
 V   =   volts
 A   =   amps
eff =   efficiency (70% to 80%)
3.41 =   constant, Btuh/watt
1.08 =   constant for heating
TR   =   temperature rise across the heat exchanger

Air Changes per Hour

$$\text{AC/hr} = \frac{\text{cfm} \times 60}{\text{Vol}}$$

$$\text{cfm} = \frac{\text{AC/hr} \times \text{Vol}}{60}$$

$$\text{Vol} = \frac{\text{cfm} \times 60}{\text{AC/hr}}$$

Ac/hr =   air changes per hour
 cfm =   quantity of airflow
  60 =   constant, minutes per hour
 Vol =   room volume, length × width × height (cf)

AREA EQUATIONS

Area of Circle

$$A = \pi R^2$$

$$A = \frac{\pi D^2}{4}$$

$$A = 0.7854 \, D^2$$

Duct Area

$$A = \frac{Q}{V}$$

A   =   cross-sectional area of the duct (sf)
Q   =   volume of air (cfm)
V   =   velocity (fpm)

Rectangular Duct Area

$$A = \frac{ab}{144}$$

 A   =   cross sectional area of the duct (sf)
 a   =   length of one side of rectangular duct (inches)
 b   =   length of adjacent side of rectangular duct (inches)
144 =   constant, si/sf

Round Duct Area

$$A = \frac{\pi R^2}{144}$$

A   =   cross sectional area of the duct (sf)
π   =   3.14

R^2 = radius, in inches, squared

144 = constant, square inches per square foot

V-BELT EQUATIONS

Belt Length

$$L = 2C + 1.57\,(D + d) + \frac{(D - d)^2}{4C}$$

L = belt pitch length
C = center to center distance of the shafts
D = pitch diameter of the large sheave
d = pitch diameter of the small sheave
1.57 = constant (0.5 pi)

Approximate Belt Length with Sheave Change

$$L = 1.57(\Delta\,Pd)$$

L = approximate new pitch length
1.57 = constant (0.5 pi)
ΔPD = difference between old sheave pitch diameter and new sheave pitch diameter

CIRCULAR EQUIVALENTS OF RECTANGULAR DUCT

Approximate Equivalent

$$d = \sqrt{\frac{4ab}{\pi}}$$

Circular Equivalent of Rectangular Duct for Equal Friction

$$d = 1.30\,\frac{ab^{0.625}}{(a + b)^{0.25}}$$

d = equivalent duct diameter
a = length of one side of rectangular duct, inches
b = length of adjacent side of rectangular duct, inches
π = 3.14

CONTROL EQUATIONS

Pneumatic Systems

1. Proportional band is equal to the throttling range divided by the span, expressed as a percentage.

$$\text{Proportional band} = \frac{\text{Throttling range}}{\text{Span}} \times 100$$

2. Sensitivity is equal to the pressure range divided by the throttling range.

$$\text{Sensitivity} = \frac{\text{Pressure range}}{\text{Throttling range}}$$

3. The equation to calculate output pressure for the measured value of a controller or sensor is the difference between the measured value and the set point value divided by the throttling range times the pressure range. This value is added to or subtracted from the setpoint pressure.

$$P_o = P_{sp} + \text{or} - \left(\frac{M - SP}{TR} \times PR\right)$$

4. The equation to determine the measured value of a direct acting controller is the pressure output minus the midpoint pressure corresponding to set point temperature, pressure or humidity divided by the pressure range times the throttling range plus the set point value.

$$M = \left(\frac{P_o - P_{sp}}{PR} \times TR\right) + SP$$

5. The equation to determine the measured value of a reverse acting controller is the midpoint pressure corresponding to set point temperature, pressure or humidity minus pressure output divided by the pressure range times the throttling range plus the set point value.

$$M = \left(\frac{P_{sp} - P_o}{PR} \times TR\right) + SP$$

6. The equation to calculate pressure output for controllers with sensors with fixed ranges is the difference between the measured value and the set point value divided by the proportional band times the sensitivity, times the pressure range. This value is added to or subtracted from the set point pressure.

$$P_o = P_{sp} + \text{or} - \left(\frac{M - SP}{PB \times S} \times PR\right)$$

P_o = Calculated pressure output from the controller in pounds per square inch (psi)

P_{sp} = midpoint pressure corresponding to set point temperature, pressure, or humidity (e.g., a 3 to 15 psi controller will have a midpoint or set point value of 9 psi)

+/− = sign for pressure change due to control action. (+) for direct acting, (−) for reverse acting

M = measured value either temperature, pressure, or humidity

SP = set point

TR = throttling range
PR = pressure range (12 psi on a 3 to 15 psi controller)
PB = proportional band, % of sensor span
S = span

Electronic Systems

1. The calculated output voltage for the measured value is the difference between the measured value and the set point divided by the throttling range times the voltage range. This value is added to or subtracted from the set point voltage.

$$V_o = V_{sp} + \text{or} - \left(\frac{M - SP}{TR} \times VR \right)$$

V_o = Calculated voltage output from the controller in volts direct current (vdc)

V_{sp} = midpont voltage corresponding to set point temperature, pressure, or humidity (e.g., a 6 to 9 vdc controller will have a midpoint or set point value of 7.5 vdc)

+/– = sign for voltage change due to control action. (+) for direct acting, (–) for reverse acting

M = measured value either temperature, pressure, or humidity

SP = setpoint

TR = throttling range

VR = voltage range (3 volts in a 6 to 9 vdc controller)

COOLING COIL EQUATIONS

Chilled Water Coil Bypass Factor

$$CBF = \frac{ADP - LAT}{ADP - EAT} \times 100$$

Leaving Air Temperature

$$LAT = EAT - [(EAT - ADP) \times (1 - CBF)]$$

CBF = coil bypass factor
ADP = apparatus dew point (°F) from psychrometric chart
LAT = leaving air temperature (dry bulb °F)
EAT = entering air temperature (dry bulb °F)
ET = evaporator temperature (refrigerant saturated vapor temperature, °F). Substitute ET for ADP when using refrigerant coil.

Cooling Tower and Condenser Equations

$$\frac{\text{Cooling tower range}}{\times \text{Cooling tower gpm}} = \frac{\text{Condenser rise}}{\times \text{Condenser gpm}}$$

$$CTR = \frac{CR \times Cgpm}{CTgpm}$$

$$CTgpm = \frac{CR \times Cgpm}{CTR}$$

$$CR = \frac{CTR \times CTgpm}{Cgpm}$$

$$Cgpm = \frac{CTR \times CTgpm}{CR}$$

Cooling tower range = entering water temperature – leaving water temperature
CTR = Cooling tower range
Cooling tower gpm = volume of water flow through the tower
CTgpm = Cooling tower gpm
Condenser rise = entering water temperature – leaving water temperature
CR = Condenser rise
Condenser gpm = volume of water flow through the condenser
Cgpm = Condenser gpm

Drive Equations

$$rpm_m \times D_m = rpm_f \times D_f$$

$$rpm_m = \frac{rpm_f \times D_f}{D_m}$$

$$D_m = \frac{rpm_f \times D_f}{rpm_m}$$

$$rpm_f = \frac{rpm_m \times D_m}{D_f}$$

$$D_f = \frac{rpm_m \times D_m}{rpm_f}$$

rpm_m = speed of the motor shaft
D_m = pitch diameter of the motor sheave
rpm_f = speed of the fan shaft
D_f = pitch diameter of the fan sheave

$$rpm = \frac{fpm \times \frac{12}{C}}{Pd \times 0.262}$$

rpm = shaft speed
fpm = belt speed
12 = constant (in/ft)

C = circumference of instrument measuring wheel (inches)
Pd = pitch diameter of driven sheave (inches)
0.262 = constant (3.14/12)

ELECTRICAL EQUATIONS

Amps

$$\text{Amps} = \frac{\text{Volts}}{\text{Ohms}}$$

$$\text{Amps} = \frac{\text{Watts}}{\text{Volts}}$$

$$\text{Amps} = \sqrt{\frac{\text{Watts}}{\text{Ohms}}}$$

Brake Horsepower—Single-Phase Circuit

$$\text{bhp} = \frac{V \times A \times \text{Eff} \times \text{PF}}{746}$$

Brake Horsepower—Three-Phase Circuit

$$\text{bhp} = \frac{V \times A \times \text{Eff} \times \text{PF} \times 1.73}{746}$$

bhp = brake horsepower
V = volts (For three-phase circuits, this is average volts)
A = amps (For three-phase circuits, this is average amps)
Eff = motor efficiency
PF = power factor
746 = constant (watts/horsepower)
1.73 = constant for three-phase circuits ($\sqrt{3}$)

Brake Horsepower—No Load

$$\text{FLA}_c = \frac{V_n \times \text{FLA}_n}{V_m}$$

$$\text{bhp} = \frac{\text{RLA} - 0.5\,\text{NLA}}{\text{FLA}_c - 0.5\,\text{NLA}} \times \text{hp}_n$$

RLA = running load amps, field measured
NLA = no load amps (motor sheave in place, belts removed)
FLA_c = full load amps, field corrected
hp_n = nameplate horsepower
V_n = nameplate volts
FLA_n = nameplate full load amps
V_m = volts, field measured

Capacitors—Connected in Parallel

$$C_T = C_1 + C_2 + C_3$$

Capacitors—Connected in Series

$$C_T = \frac{1}{\dfrac{1}{C_1} + \dfrac{1}{C_2} + \dfrac{1}{C_3}}$$

C_T = total capacitance, ohms
C_1 = ohms capacitor 1
C_2 = ohms capacitor 2
C_3 = ohms capacitor 3

Current Unbalance

$$\%C = \frac{\Delta D_{max}}{C_{avg}} \times 100$$

$\%C$ = % current unbalance (should not exceed 10%)
ΔD_{max} = maximum deviation from average amps
C_{avg} = average amps

Ohms

$$\text{Ohms} = \frac{\text{Amps}}{\text{Volts}}$$

$$\text{Ohms} = \frac{\text{Volts}^2}{\text{Watts}}$$

$$\text{Ohms} = \frac{\text{Watts}}{\text{Amps}^2}$$

Power Factor—Single-Phase Circuit

$$\text{PF} = \frac{W}{VA}$$

$$\text{PF} = \frac{KW}{KVA}$$

Power Factor—Three-Phase Circuit

$$\text{PF} = \frac{W}{VA \times 1.73}$$

$$\text{PF} = \frac{KW}{KVA \times 1.73}$$

PF = power factor
W = watts

$$\begin{aligned}
V &= \text{volts} \\
A &= \text{amps} \\
KW &= \text{kilowatts} \\
KVA &= \text{kilovolt amps} \\
1.73 &= \text{constant for three-phase circuits } (\sqrt{3})
\end{aligned}$$

Volts

$$\text{Volts} = \sqrt{\text{Watts} \times \text{Ohms}}$$

$$\text{Volts} = \text{Amps} \times \text{Ohms}$$

$$\text{Volts} = \frac{\text{Watts}}{\text{Amps}}$$

Voltage Unbalance

$$\%V = \frac{\Delta D_{max}}{V_{avg}} \times 100$$

$$\begin{aligned}
\%V &= \text{\% voltage unbalance (should not exceed 2\%)} \\
\Delta D_{max} &= \text{maximum deviation from average voltage} \\
V_{avg} &= \text{average voltage}
\end{aligned}$$

Watts

$$\text{Watts} = \text{Volts} \times \text{Amps}$$

$$\text{Watts} = \text{Amps}^2 \times \text{Ohms}$$

$$\text{Watts} = \frac{\text{Volts}^2}{\text{Ohms}}$$

ENERGY COST/SAVINGS EQUATIONS

Cost or Savings per Year for Change in Brake Horsepower

$$\$/yr = (bhp1 - bhp2) \times \frac{0.746 \text{ kw/bhp}}{\text{Motor}_{eff}} \times Hr/Yr \times \$/kwh$$

$$\begin{aligned}
\$/yr &= \text{dollar cost or savings per year} \\
bhp1 &= \text{original brake horsepower} \\
bhp2 &= \text{changed brake horsepower} \\
0.746 &= \text{constant, kilowatt per brake horsepower} \\
eff &= \text{motor efficiency} \\
Hr/Yr &= \text{hours per year of operation} \\
\$/kwh &= \text{dollar cost of electrical energy per kilowatt-hour}
\end{aligned}$$

Cost or Savings per Year for Change in a System Component

Fan System

$$\$/YR = \frac{\$/kwh \times cfm \times P \times 0.746 \text{ kw/bhp} \times Hr/Yr}{6{,}356 \times \text{Motor}_{eff} \times \text{Fan}_{eff}}$$

Pump System

$$\$/YR = \frac{\$/kwh \times gpm \times H \times 0.746 \text{ kw/bhp} \times Hr/Yr}{3{,}960 \times \text{Motor}_{eff} \times \text{Pump}_{eff}}$$

$$\begin{aligned}
\$/YR &= \text{dollar cost or savings per year} \\
\$/kwh &= \text{dollar cost of electrical energy per kilowatt-hour} \\
cfm &= \text{volume of airflow} \\
gpm &= \text{volume of water flow} \\
P &= \text{air pressure (inches of water gauge) loss across} \\
&\quad \text{component} \\
H &= \text{water pressure (feet of head) loss across component} \\
0.746 &= \text{constant, kilowatt per brake horsepower} \\
Hr/Yr &= \text{hours per year of operation} \\
6{,}356 &= \text{constant for air systems} \\
3{,}960 &= \text{constant for water systems} \\
eff &= \text{motor efficiency} \\
eff &= \text{pump efficiency} \\
eff &= \text{fan efficiency}
\end{aligned}$$

FAN EQUATIONS

Air Horsepower

$$ahp = \frac{cfm \times P}{6{,}356}$$

Brake Horsepower

$$bhp = \frac{cfm \times FSP}{6{,}356 \times FSE}$$

$$bhp = \frac{cfm \times FTP}{6{,}356 \times FTE}$$

$$bhp = \frac{cfm \times TSP}{6{,}356 \times 0.70}$$

Fan Efficiency

$$FSE = \frac{cfm \times FSP}{6{,}356 \times bhp}$$

$$FTE = \frac{cfm \times FTP}{6{,}356 \times bhp}$$

$$\text{Eff} = \frac{\text{cfm} \times \text{TSP}}{6,356 \times \text{bhp}}$$

ahp = air horsepower
cfm = airflow volume
P = fan pressure (in. wg)
6,356 = constant (33,000 ft-lb/min ÷ 5.19)
Eff = efficiency (percent)
bhp = brake horsepower
FSP = fan static pressure (in. wg)
FSE = fan static efficiency (percent)
FTP = fan total pressure (in. wg)
FTE = fan total efficiency (percent)
TSP = total static pressure
0.70 = use if fam efficiency is unknown

Temperature Rise Across the Fan

Motor Out of the Air Stream

$$\text{Temperature rise} = \frac{\text{TSP} \times 2,545}{6,356 \times 1.08 \times \text{Eff}_f}$$

Motor in the Air Stream

$$\text{Temperature rise} = \frac{\text{TSP} \times 2,545}{6,356 \times 1.08 \times \text{Eff}_f \times \text{Eff}_m}$$

Temperature rise = °F
TSP = total static pressure rise across the fan (in. wg)
2,545 = constant
6,356 = constant
1.08 = constant
Eff_m = motor efficiency (percent)
Eff_f = fan efficiency (percent)

Tip Speed

$$\text{Tip speed} = \frac{\text{D} \times \text{rpm} \times 3.14}{12}$$

Tip speed = fpm
D = fan wheel diameter (inches)
rpm = revolutions per minute of the fan
3.14 = constant
12 = constant (in/ft)

FLUID FLOW EQUATIONS

$$v = \sqrt{2gh}$$

$$h = \frac{v^2}{2g}$$

v = velocity of fluid (fps)
g = gravitational acceleration (32.2 fps squared)

h = head (feet)
$$v = \sqrt{2gh}$$
$$v = 60\sqrt{2 \times 32.2 \times 5.19 \times 13.33}$$
$$v = 4,005$$

v = velocity (fpm)
60 = sec/min
5.19 = density of water (62.3 lb/cf) divided by 12 in/ft
13.33 = 1 inch of water column divided by 0.075 lb/cf

HEAT TRANSFER EQUATIONS

Air

$$\text{Btuh} = \text{cfm} \times 1.08 \times \text{TD}$$

$$\text{cfm} = \frac{\text{Btuh}}{1.08 \times \text{TD}}$$

$$\text{TD} = \frac{\text{Btuh}}{1.08 \times \text{cfm}}$$

Btuh = Btu per hour sensible heat (heating coil or dry cooling coil load or room load)
cfm = volume of airflow
1.08 = constant, 60 min/hr × 0.075 lb/cu ft × 0.24 Btu/lb F
TD = dry bulb temperature difference of the air entering and leaving the coil. In applications where cfm to the conditioned space needs to be calculated, the TD is the difference between the supply air temperature dry bulb and the room temperature dry bulb.

$$\text{Btuhl} = \text{cfm} \times 4.5 \times \Delta\text{hl}$$

$$\Delta\text{hl} = \frac{\text{Btuhl}}{4.5 \times \text{cfm}}$$

$$\text{cfm} = \frac{\text{Btuhl}}{4.5 \times \Delta\text{hl}}$$

Btuhl = Btu per hour latent heat
cfm = volume of airflow
4.5 = constant, 60 min/hour × 0.075 lb/cu ft
Δhl = change in latent heat content of the supply air, btu/lb (from dew point and table of properties of mixtures of air and saturated water vapor) (See Chap. 10)

$$\text{Btuht} = \text{cfm} \times 4.5 \times \Delta\text{ht}$$

$$\Delta\text{ht} = \frac{\text{Btuht}}{4.5 \times \text{cfm}}$$

$$\text{cfm} = \frac{\text{Btuht}}{4.5 \times \Delta\text{ht}}$$

Btuht = Btu per hour total heat (a wet cooling coil)

cfm = volume of airflow
 4.5 = constant, 60 min/hour × 0.075 lb/cu ft
 Δht = change in total heat content of the supply air, btu/lb
          (from wet bulb temperatures and psychrometric chart
          or table of properties of mixtures of air and saturated
          water vapor) (Chap. 10)

Water

$$\text{Btuh} = \text{gpm} \times 500 \times \text{TD}$$

$$\text{gpm} = \frac{\text{Btuh}}{500 \times \text{TD}}$$

$$\text{TD} = \frac{\text{Btuh}}{500 \times \text{gpm}}$$

Btuh = Btu per hour
 gpm = water volume
  500 = constant, 60 min/hour × 8.33 lbs/gallon × 1 Btu/lb °F
   TD = temperature difference between the entering and
          leaving water

Power Equations

Fan Power

$$\text{KW} = \frac{\text{cfm} \times \text{TSP}}{8{,}520 \times \text{Eff}_m \times \text{Eff}_f}$$

Pump Power

$$\text{KW} = \frac{\text{gpm} \times \text{TDH}}{5{,}308 \times \text{Eff}_p \times \text{Eff}_m}$$

  KW = kilowatts
 cfm = airflow volume
 TSP = total static pressure (in. wg)
8,520 = constant, 6356/.746
 gpm = water volume
 TDH = total dynamic head (feet of water)
5,308 = constant, 3960/.746
Eff_m = motor efficiency (percent)
Eff_f = fan efficiency (percent)
Eff_p = pump efficiency (percent)

Pump Equations

Water Horsepower

$$\text{WHP} = \frac{\text{gpm} \times \text{H} \times \text{SpG}}{3{,}960}$$

Pump Brake Horsepower

$$\text{bhp} = \frac{\text{gpm} \times \text{TDH}}{3{,}960 \times \text{Eff}}$$

Pump Efficiency

$$\text{Eff} = \frac{\text{gpm} \times \text{TDH}}{3{,}960 \times \text{bhp}}$$

WHP = water horsepower
 gpm = water flow rate
   H = pressure against which the pump operates (feet of
          water)
 SpG = specific gravity
3,960 = constant (33,000 ft-lb/min divided by 8.33 lb/gal)
 bhp = brake horsepower
 TDH = total dynamic head, against which the pump operates
          (feet of water)
 Eff = pump efficiency (percent)
Note: For water temperatures between freezing and boiling, the
specific gravity is taken to be 1.0, and is therefore dropped from
the equations for bhp and efficiency.

Refrigeration Equations

Tonnage

$$\text{Approximate condenser tones} = \frac{\text{Chiller gpm} \times \text{Chiller TD} \times 1.25}{24}$$

$$\text{Condenser tons} = \frac{\text{Condenser gpm} \times \text{Condenser TD}}{24}$$

$$\text{Chiller tons} = \frac{\text{Chiller gpm} \times \text{Chiller TD}}{24}$$

$$\text{Approximate chiller tons} = \frac{\text{Condenser gpm} \times \text{Condenser TD}}{30}$$

  TD = temperature difference, entering and leaving water
 1.25 = approximate condenser load based on evaporator
          load plus 25% for heat of compression
   24 = 12,000 Btuh/ton ÷ 500 constant for water flow
   30 = 24 × 1.25

Refrigerant Flow and Capacity

$$\frac{\text{Btu}}{\text{min}} = \frac{\text{lb}}{\text{min}} \times \frac{\text{Btu}}{\text{lb}}$$

$$\frac{\text{lb}}{\text{min}} = \frac{\dfrac{\text{Btu}}{\text{min}}}{\dfrac{\text{Btu}}{\text{lb}}}$$

$$\frac{\text{Btu}}{\text{lb}} = \frac{\dfrac{\text{Btu}}{\text{min}}}{\dfrac{\text{lb}}{\text{min}}}$$

$$Btuh = \frac{lb}{min} \times \frac{Btu}{lb} \times \frac{60\ min}{hr}$$

$$Tons = \frac{Btuh/min}{200\ Btu/min/ton}$$

$$Tons = \frac{Btuh}{12,000\ Btuh/ton}$$

$$lbs/min/ton = \frac{200\ Btu/min/ton}{Btu/lb}$$

$$lb/min = cfm \times Density\ (suction\ vapor)$$

$$lbs/min = lbs/min/ton \times tons$$

$$cfm/ton = lbs/min/ton \times Specific\ Volume\ (suction\ vapor)$$

$$cfm = cfm/ton \times tons$$

Btu = British thermal unit
min = minutes
lb = pound(s)
hr = hour
ton = ton of refrigeration, 12,000 Btu
cfm = volume of refrigerant vapor, cubic feet per minute

Liquid-Vapor Refrigerant Mixture

$$Specific\ Volume_{mixture} = (\%\ liquid \times Specific\ Volume_{liquid}) + (\%\ vapor \times Specific\ Volume_{vapor})$$

$$Enthalpy_{mixture} = (\%\ liquid \times Enthalpy_{liquid}) + (\%\ vapor \times Enthalpy_{vapor})$$

Net Refrigeration Effect (Btu/lb)

$$NRE = h_g - h_f$$

NRE = net refrigeration effect

hg = enthalpy of vapor leaving the evaporator (point C on the pressure-enthalpy diagram)

hf = enthalpy of liquid entering the evaporator (point A′ on the pressure-enthalpy diagram)

Heat of Compression (Btu/lb)

$$HC = h_{gl} - h_{ge}$$

HC = heat of compression
hgl = enthalpy of vapor leaving the compressor (point D′ on the pressure-enthalpy diagram)
hge = enthalpy of vapor entering the compressor (point C′ on the pressure-enthalpy diagram)

Heat Rejected (Btu/lb)

$$HR = NRE + HC$$

$$THR = NRE + SH + HC$$

HR = heat rejected
NRE = net refrigeration effect
HC = heat of compression
THR = total heat rejected
SH = superheat

Heat Rejected (Btuh)

Open Compressor

$$Btuh = Evaporator\ load + (2545\ Btuh/BHP \times BHP)$$

Hermetic Compressor

$$Btuh = Evaporator\ load + (3413\ Btuh/kw \times kw)$$

Btuh = British thermal units per hour
Evaporator load = refrigeration load in Btuh
BHP = brake horsepower
kw = kilowatts

If brake horsepower or kilowatts is unknown, multiply 1.25 times the evaporator load for an approximation.

Coefficient of Performance (no units)

$$COP = \frac{RE}{HC}$$

COP = coefficient of performance
NRE = net refrigeration effect
HC = heat of compression

Horsepower per Ton

$$HP/ton = \frac{4.71}{COP}$$

HP/ton = horsepower per ton
4.71 = Constant (200 Btu/min/ton divided by 42.42 Btu/min/hp)
COP = coefficient of performance

RECIPROCATING COMPRESSOR EQUATIONS

Piston Velocity

$$Piston\ velocity = \frac{rpm \times Stroke \times 2}{12}$$

Piston Displacement

$$\text{cfm} = \frac{(0.7854 D^2)(L)(N)(\text{rpm})}{1,728}$$

Bore Diameter

$$D = \sqrt{\frac{(1,728)(\text{cfm})}{0.7854\,(L)(N)(\text{rpm})}}$$

cfm = refrigerant flow volume, cubic feet per minute
0.7854 = constant (3.14/4)
D = piston bore diameter (inches)
L = length of stroke (inches)
N = number of pistons
rpm = compressor speed
1,728 = constant (cubic inches per cubic foot)
12 = inches/foot

Compressor Displacement

Compressor Displacement cfm = piston displacement cfm × volumetric efficiency

cfm = refrigerant flow volume, cubic feet per minute

Compression Ratio

$$\text{Compression ratio} = \frac{\text{Compressor discharge (psia)}}{\text{Compressor suction (psia)}}$$

psia = pounds per square absolute

TEMPERATURE EQUATIONS

Temperature Conversions

$$°\text{Fahrenheit} = 1.8\ \text{Celsius} + 32$$
$$°\text{Celsius} = \frac{(\text{Fahrenheit} - 32)}{1.8}$$
$$°\text{Rankin} = \text{Fahrenheit} + 460$$
$$°\text{Kelvin} = \text{Celsius} + 273$$

Log Mean Temperature Difference

$$\text{LMTD} = \frac{\Delta T_L - \Delta T_S}{\text{Ln}\left(\dfrac{\Delta T_L}{\Delta T_S}\right)}$$

Temperature Difference—Refrigerant Coil

$$\Delta T = \text{EAT} - \text{ET}$$
$$\Delta T = \text{LAT} - \text{ET}$$

Temperature Difference—Parallel Flow Water Coil

$$\Delta T = \text{EAT} - \text{EWT}$$
$$\Delta T = \text{LAT} - \text{LWT}$$

Temperature Difference —Counter Flow Water Coil

$$\Delta T = \text{EAT} - \text{LWT}$$
$$\Delta T = \text{LAT} - \text{EWT}$$

LMTD = log mean temperature difference
ΔT_L = the larger temperature difference
ΔT_S = the smaller temperature difference
Ln = natural log
EAT = entering air temperature
LAT = leaving air temperature
EWT = entering water temperature
LWT = leaving water temperature
ET = evaporator temperature

Mixed Air Temperature

$$\text{MAT} = (\%\text{OA} \times \text{OAT}) + (\%\text{RA} \times \text{RAT})$$

Percent of Outside Air

$$\%\text{OA} = \frac{\text{MAT} - \text{RAT}}{\text{RAT} - \text{OAT}} \times 100\%$$

MAT = mixed air temperature
%OA = percent of outside air
OAT = outside air temperature
%RA = percent of return air
RAT = return air temperature

Fan Discharge Air Temperature

$$\text{FDAT} = (\%\text{OA} \times \text{OAT}) + (\%\text{RA} \times \text{RAT}) + 0.5\,(\text{TSP})$$

Percent of Outside Air

$$\%\text{OA} = \frac{\text{RAT} - [\text{FDAT} - 0.5(\text{TSP})]}{\text{RAT} - \text{OAT}} \times 100$$

FDAT = fan discharge air temperature
%OA = percent of outside air
OAT = outside air temperature
%RA = percent of return air
RAT = return air temperature
TSP = total static pressure rise across the fan (in. wg)
0.5 = 1/2° per inch of static pressure for fan heat of compression
0.6 = 0.6° per inch of static pressure for fan heat of compression and motor heat

Use 0.5 if motor is out of the air stream.

Use 0.6 if motor is in the air stream.

Supply Air Temperature

$$SAT = (\%BA \times BAT) + (\%CA \times CAT)$$

Percent of Bypassed Air

$$\%BA = \frac{(SAT - CAT) \times 100}{(CAT - BAT)}$$

SAT = supply air temperature
%BA = percent of bypassed air (cfm)
BAT = bypassed air temperature
%CA = percent of air through the coil (cfm)
CAT = temperature of the air leaving the coil

Mixed Water Temperature

$$MWT = (\%BW \times CWRT) + (\%CWS \times CWST)$$

Percent of Bypassed Water

$$\%BW = \frac{(MWT - CWST) \times 100}{(CWRT - CWST)}$$

MWT = mixed water temperature
%BW = percent of bypassed water
CWST = condenser water supply temperature
%CWS = percent of condenser water supply
CWRT = condenser water return temperature

WATER FLOW EQUATIONS

Flow Through Control Valves

$$gpm = C_v\sqrt{\Delta P}$$

$$C_V = \frac{gpm}{\sqrt{\Delta P}}$$

$$\Delta P = \left(\frac{gpm}{C_V}\right)^2$$

gpm = volume of water flow
ΔP = pressure drop (psi)
C_V = valve coefficient

Flow Through Coils

$$gpm_C = gpm_R \sqrt{\frac{\Delta P_M}{\Delta P_R}}$$

$$\Delta P_C = \Delta P_R \times \left(\frac{GPM_M}{GPM_R}\right)^2$$

GPM_R = rated water flow
GPM_C = calculated water flow
GPM_M = measured water flow
ΔP_R = rated pressure drop
ΔP_C = calculated pressure drop
ΔP_M = measured pressure drop

The following system and component information should be gathered and recorded, as applicable, on the appropriate reporting form. (A) is Actual, (D) is design, (M) is measured.

Air Handling Unit and System

☐ Cooling only
☐ Heating only
☐ Heating and cooling
☐ Reheat
☐ Constant Air Volume
☐ Variable Air Volume
☐ Single duct
☐ Double duct
☐ Single zone
☐ Multizone

Fan Housing Condition:
☐ Extensive air leakage
☐ Nominal air leakage
☐ Negligible air leakage

Fan Wheel and Blade:
☐ Rotation correct
☐ Clearance

Plenum Condition:
☐ Extensive air leakage
☐ Nominal air leakage
☐ Negligible air leakage

Flexible Connection Condition:
☐ Extensive air leakage
☐ Nominal air leakage
☐ Negligible air leakage

Fans

Supply Fan Type:
☐ Forward curved
☐ Backward curved
☐ Backward inclined
☐ Airfoil
☐ Tubeaxial
☐ Vaneaxial
☐ Inline Centrifugal
☐ Other _____

Supply Fan Manufacturer/Model Number:
_____

Fan Speed:
RPM   (D) _____   (M) _____

Fan Pressures:
Static Pressure:
Suction   (D) _____   (M) _____
Discharge   (D) _____   (M) _____

Total Pressure:
Suction   (D) _____   (M) _____
Discharge   (D) _____   (M) _____

Total Static Pressure   (D) _____   (M) _____
External Static Pressure   (D) _____   (M) _____
Fan Static Pressure   (D) _____   (M) _____
Fan Total Pressure   (D) _____   (M) _____

Discharge Air Temperature:
Dry Bulb   (D) _____   (M) _____
Wet Bulb   (D) _____   (M) _____

Return/Relief Fan Type:
☐ Forward curved
☐ Backward curved
☐ Backward inclined
☐ Airfoil
☐ Propeller
☐ Tubeaxial
☐ Vaneaxial
☐ Inline Centrifugal
☐ Other _____

Return/Relief Fan Manufacturer/Model Number:
_____

Fan Speed:
RPM   (D) _____   (M) _____

Fan Pressures:
Static Pressure:

Suction   (D) _____   (M) _____
Discharge   (D) _____   (M) _____

Total Pressure:
Suction   (D) _____   (M) _____
Discharge   (D) _____   (M) _____

Total Static Pressure   (D) _____   (M) _____
External Static Pressure   (D) _____   (M) _____
Fan Static Pressure   (D) _____   (M) _____
Fan Total Pressure   (D) _____   (M) _____

Return Air Temperature:
Dry Bulb   (D) _____   (M) _____
Wet Bulb   (D) _____   (M) _____

Exhaust Fan Type:
☐ Forward curved
☐ Backward curved
☐ Backward inclined
☐ Airfoil
☐ Radial
☐ Propeller
☐ Tubeaxial
☐ Vaneaxial
☐ Inline Centrifugal
☐ Other _____

Exhaust Fan Manufacturer/Model Number:

_____

Fan Speed:
RPM   (D) _____   (M) _____

Fan Pressures:
Static Pressure:
Suction   (D) _____   (M) _____
Discharge   (D) _____   (M) _____

Total Pressure:
Suction   (D) _____   (M) _____
Discharge   (D) _____   (M) _____

Total Static Pressure   (D) _____   (M) _____
External Static Pressure   (D) _____   (M) _____
Fan Static Pressure   (D) _____   (M) _____
Fan Total Pressure   (D) _____   (M) _____

Exhaust Air Temperature:
Dry Bulb   (D) _____   (M) _____
Wet Bulb   (D) _____   (M) _____
Air Volume:
Outside Air Cubic Feet Per Minute   (D) _____   (M) _____
Exhaust Air Cubic Feet Per Minute   (D) _____   (M) _____
Return Air Cubic Feet Per Minute   (D) _____   (M) _____
Supply Air Cubic Feet Per Minute   (D) _____   (M) _____

Outside Air

Outside Air Temperature:
Dry Bulb (D) _____ (M) _____
Wet Bulb (D) _____ (M) _____

Louver and Screen Condition:
☐ Clean
☐ Clogged

Outside Air Damper:
Position:
☐ Minimum
☐ Full open
☐ Modulating
Close Properly: ☐ Yes ☐ No
Open Properly: ☐ Yes ☐ No
Sealed All Sides: ☐ Yes ☐ No

Return Air

Return Air Damper:
Position:
☐ Minimum
☐ Full open
☐ Modulating
Close Properly: ☐ Yes ☐ No
Open Properly: ☐ Yes ☐ No
Sealed All Sides: ☐ Yes ☐ No

Exhaust/Relief Air

Exhaust Air Damper:
Position:
☐ Minimum
☐ Full open
☐ Modulating
Close Properly: ☐ Yes ☐ No
Open Properly: ☐ Yes ☐ No
Sealed All Sides: ☐ Yes ☐ No

Filters:

Type:
Fiber:
☐ Viscous
☐ Dry
☐ HEPA
☐ Bag
☐ Continuous Roll

Renewable:
☐ Viscous
☐ Dry
☐ Electronic

Filter Condition:
☐ Dirty
☐ Clean
☐ Sealed All Sides: ☐ Yes ☐ No

Filter Pressure:
Entering Air   (D)_____   (M) _____
Leaving Air   (D) _____   (M) _____
Drop   (D) _____   (M) _____

Coils

Cooling Coil Type:
☐ Refrigeration DX   Refrigerant _____
☐ Chilled Water

Coil Size:
Height _____
Width _____
Depth _____
Rows _____
Fins Per Inch _____

Piping, Chilled Water:
☐ Counter flow
☐ Parallel flow
☐ Supply: ☐ Top ☐ Bottom
☐ Return: ☐ Top ☐ Bottom

Coil Condition:
☐ Dirty
☐ Clean
☐ Combed
☐ Sealed All Sides: ☐ Yes ☐ No

Coil Pressure:
Entering Air   (D)_____   (M) _____
Leaving Air   (D) _____   (M) _____
Drop   (D) _____   (M) _____

Coil Bypass Factor:
_____

Coil Face Velocity:
Feet Per Minute   (D) _____   (M) _____

Entering Air Temperature:
Dry Bulb   (D) _____   (M) _____
Wet Bulb   (D) _____   (M) _____

Leaving Air Temperature:
Dry Bulb   (D) _____   (M) _____
Wet Bulb   (D) _____   (M) _____

Water Temperature:
Entering Water   (D)_____   (M) _____

Leaving Water   (D) _____   (M) _____
Rise   (D) _____   (M) _____

Water Flow:
GPM   (D) _____   (M) _____

Heating Coil type:
☐ Steam
☐ Hot water
☐ Other

Coil Size:
Height _____
Width _____
Depth _____
Rows _____
Fins Per Inch _____

Piping, Heating Water:
☐ Counter flow
☐ Parallel flow
☐ Supply: ☐ Top   ☐ Bottom
☐ Return: ☐ Top   ☐ Bottom

Piping, Steam:
☐ Counter flow
☐ Parallel flow
☐ Supply: ☐ Top   ☐ Bottom
☐ Condensate: ☐ Top   ☐ Bottom

Coil Condition:
☐ Dirty
☐ Clean
☐ Combed

Coil Pressure:
Entering Air   (D)_____   (M) _____
Leaving Air   (D) _____   (M) _____
Drop   (D) _____   (M) _____

Coil Face Velocity:
Feet Per Minute   (D) _____   (M) _____

Entering Air Temperature:
Dry Bulb   (D) _____   (M) _____
Wet Bulb   (D) _____   (M) _____

Leaving Air Temperature:
Dry Bulb   (D) _____   (M) _____
Wet Bulb   (D) _____   (M) _____

Water Temperature:
Entering Water   (D)_____   (M) _____
Leaving Water   (D) _____   (M) _____

Drop   (D) _____   (M) _____

Fluid Flow:
Water GPM   (D)_____   (M) _____
Steam Lbs/Hour   (D) _____ (M) _____

Fan Motor

☐ Single phase
☐ Three phase
Nameplate HorsePower (D)_____   (M) _____
Nameplate Amperage   (D) _____ (M) _____
Nameplate Voltage   (D) _____ (M) _____

Belt

Condition:
☐ Good
☐ Worn
☐ Tight
☐ Loose

Belt Position in Drive Sheave:
☐ High
☐ Low
☐ Center

Sheaves

Sheave Type:
Motor:
☐ Adjustable
☐ Fixed
Fan:
☐ Adjustable
☐ Fixed

Sheave Manufacturer/Size:
☐ Motor _____
☐ Fan _____

Chiller

Manufacturer/Model:

_____

Water Pressure:
Entering Water   (D)_____   (M) _____
Leaving Water   (D) _____ (M) _____
Pressure Drop   (D) _____ (M) _____

Water Temperature:
Entering Water   (D)_____   (M) _____
Leaving Water   (D) _____ (M) _____
Drop   (D) _____ (M) _____

Water Flow:
GPM   (D) _____   (M) _____

Compressor

Motor:
☐ Single phase
☐ Three phase

Nameplate HorsePower	(D)_____	(M) _____
Nameplate Amperage	(D) _____	(M) _____
Nameplate Voltage	(D) _____	(M) _____

Refrigerant: _____

Charge: _____

Type of metering device _____

Pressure:

Oil	(D) _____	(M) _____
Suction	(D) _____	(M) _____
Discharge	(D) _____	(M) _____

High Pressure Control:

Cut-in	(D) _____	(M) _____
Cut-out	(D) _____	(M) _____

Low Pressure Control:

Cut-in	(D) _____	(M) _____
Cut-out	(D) _____	(M) _____

Condenser

Manufacturer/Model:

_____

Water Pressure:

Entering Water	(D)_____	(M) _____
Leaving Water	(D) _____	(M) _____
Pressure Drop	(D) _____	(M) _____

Water Temperature:

Entering Water	(D)_____	(M) _____
Leaving Water	(D) _____	(M) _____
Drop	(D) _____	(M) _____

Water Flow:
GPM   (D) _____   (M) _____

Cooling Tower

Manufacturer/Model:

_____

Water Temperature:

Entering Water	(D)_____	(M) _____
Leaving Water	(D) _____	(M) _____
Range	(D) _____	(M) _____

Water Flow:
GPM   (D) _____   (M) _____

Air Temperature:
Entering Air Dry Bulb   (D)_____   (M) _____
Entering Air Wet Bulb   (D) _____   (M) _____
Leaving Air Dry Bulb   (D) _____   (M) _____
Approach   (D) _____   (M) _____
Cooling Tower Fan:
Number of Fans   (D) _____   (M) _____
RPM   (D) _____   (M) _____

Cooling Tower Motor:
☐ Single Phase
☐ Three phase
Nameplate HorsePower   (D)_____   (M) _____
Nameplate Amperage   (D) _____   (M) _____
Nameplate Voltage   (D) _____   (M) _____

Boiler

Manufacturer/Model:

_____

Water Pressure:
Entering Water   (D)_____   (M) _____
Leaving Water   (D) _____   (M) _____
Pressure Drop   (D) _____   (M) _____

Water Temperature:
Entering Water   (D)_____   (M) _____
Leaving Water   (D) _____   (M) _____
Rise   (D) _____   (M) _____

Water Flow:
GPM   (D) _____   (M) _____
Temperature Controls:
Cut-in   (D) _____   (A) _____
Cut-out   (D) _____   (A) _____

Safety Controls:
Low-Water Cut-out Condition and Operation _____
Pressure Relief Valve PSI   (D) _____   (A) _____
Pressure Relief Valve Condition and Operation_____

Combustion Analysis:
Percent Oxygen (M)_____
Percent Carbon Dioxide (M)_____
Percent Excess Oxygen (M)_____
Carbon Spot Test_____

Flue Gas Temperature (M)_____
Room Temperature (M)_____
Boiler Efficiency (M)_____

Boiler Fan Motor:
☐ Single Phase
☐ Three phase
Nameplate HorsePower .   (D)_____   (A) _____
Nameplate Amperage   (D) _____   (M) _____
Nameplate Voltage   (D) _____   (M) _____

Water Pump

Type:
☐ Single suction
☐ Double suction

Manufacturer/Model:
_____

Service:
_____

Pump Speed:
RPM (D) _____ (M) _____

Pump Static Head:
Suction   (D) _____ (M) _____
Discharge (D) _____ (M) _____
TDH   (D) _____ (M) _____

Motor:
☐ Single Phase
☐ Three phase
Nameplate HorsePower (D)_____ (S) _____
Nameplate Amperage   (D) _____ (M) _____
Nameplate Voltage   (D) _____ (M) _____

Air Distribution

Main Duct Pressure Classification
☐ High _____
☐ Medium _____
☐ Low _____

Medium or High Pressure Duct Condition:
Leak Tested   ☐ Yes ☐ No
Leakage Class _____
Leakage Rate _____
Sealed: ☐ Yes ☐ No
☐ Extensive air leakage
☐ Nominal air leakage
☐ Negligible air leakage

Low Pressure Duct Condition:
Sealed   ☐ Yes ☐ No
☐ Extensive air leakage
☐ Nominal air leakage
☐ Negligible air leakage

Insulation:
☐ Wrapped
☐ Lined

Do supply outlets have balancing dampers?   ☐ Yes ☐ No
Do return inlets have balancing dampers?   ☐ Yes ☐ No
Do exhaust inlets have balancing dampers?   ☐ Yes ☐ No
If not, how many are needed?   Supply _____
                                            Return _____
                                            Exhaust _____

Are there balancing dampers at the zones?   ☐ Yes ☐ No
If not, how many are needed?   _____

General Construction and Condition:
☐ Good
☐ Fair
☐ Poor
☐ Aspect ratios
☐ Use of Fittings

Building or Space Pressurization

☐ Positive _____ ″WG
☐ Negative _____ ″WG

AIR DISTRIBUTION—TEMPERATURE VERIFICATION

Procedure

As applicable, gather the following information and record on the appropriate reporting form.

- Place thermostat for each zone on full cooling.
- Record air temperature in cold deck.
- Record air temperature in hot deck.
- Record air temperature in main supply duct.
- Record air temperature entering and leaving reheat or recool coil.
- Record air temperature leaving supply outlet.
- Record air temperature entering return inlet.
- Record air temperature in main return duct.
- Place the thermostat on full heating and repeat measurements.

HEATING AND COOLING COIL PERFORMANCE VERIFICATION

Procedure

As applicable, gather the following information and record on the appropriate reporting form. The system has been balanced for both air and water flow.

- Place thermostat(s) for normal operation.
- Record air dry-bulb and wet-bulb temperatures entering and leaving the coil.
- Record water temperatures entering and leaving the coil.
- Calculate and compare air heat transfer (Btuh) to water heat transfer (Btuh).

APPENDIX F *Optimization Opportunities*

The following is a general list of areas to investigate to find energy management, occupancy comfort, and process function optimization opportunities as they relate to HVAC systems.

ENERGIZED SYSTEMS

Air balancing	Cooking practices	Heat pumps	Relief air
Air distribution	Cooling towers	Heat recovery	Solar systems
Air handling units	Current leakage	Heat storage	Space heaters
Air infiltration	Dampers	Humidification	Steam distribution
Air purging	Dehumidification	Hydronic systems	Steam traps
Air volume	Demand limiting	Infiltration	Swimming pools
Ballasts	Diffusers	Instrumentation	System air leakage
Boilers	Dishwashing	Kitchen	System interaction
Boiler auxiliaries	Domestic water heating	Lamps	System redesign
Chillers	Duct resistance	Laundry	Thermostats
Chiller auxiliaries	Elevators	Lavatory fixtures	Time-of-day rates
Chiller heat recovery	Equipment relocation	Outside air control	Transformers
City water	Exhaust system	Pipe insulation	Utility rates
Cogeneration	Fan-coil units	Pipe resistance	Ventilation layout
Coils	Filters	Power distribution	Vending machines
Computers	Fixtures	Power factor	Water balancing
Condensate systems	Fuel acquisition	Pumps	Water coolers
Condensing units	Fuel systems	Radiators	Water treatment
Controls	Fume hoods	Refrigeration	Weatherstripping

Human Systems

Documentation	Occupant indoctrination
Financial practices	Occupant training
Maintenance personnel training	Staff training
Management structure	

Nonenergized Systems

Building design	Exterior color	Shading	Vapor barrier
Building geometry	Glazing	Space planning	Vegetation
Building insulation	Interior color	Space insulation	Vestibules
Building materials	Interior design	Space segregation	Walls
Building orientation	Roof color	Surface color	Windows
Building location	Roof insulation	Thermal shutters	Window treatment
Caulking			

Index

0